Subject To Recall After 2 Weeks

WITHDRAWN

# INDUSTRIAL SOLID WASTES

# INDUSTRIAL SOLID WASTES

A Textbook by

NELSON LEONARD NEMEROW

BALLINGER PUBLISHING COMPANY
Cambridge, Massachusetts
A Subsidiary of Harper & Row, Publishers, Inc.

Copyright © 1984 by Ballinger Publishing Company. All rights reserved. No part of this publication may be reproduced, stored in a retrieval system, or transmitted in any form or by any means, electronic, mechanical, photocopy, recording or otherwise, without the prior written consent of the publisher.

International Standard Book Number: 0-88410-876-7

Library of Congress Catalog Card Number: 82-13866

Printed in the United States of America

**Library of Congress Cataloging in Publication Data**

Nemerow, Nelson Leonard.
  Industrial solid wastes.

  Includes index.
  1. Factory and trade waste. 2. Recycling (Waste, etc.) I. Title.
TD897.N378   1983        363.7'28        82-13866
ISBN 0-88410-876-7

# DEDICATION

To our son, Dr. Glen Robert Nemerow, cancer research scientist and immunologist at the Scripps Research Institute, La Jolla, California, and to all the scientists throughout the world whose patience, dedication, and hours of research lead to discoveries for the improvement of the health and welfare of mankind and our total environment.

# CONTENTS

List of Figures     xiii

List of Tables     xvii

Preface     xxiii

Acknowledgment     xxv

**Chapter 1**
**Basic Data on Solid Wastes**     1

Introduction     1
U.S. Public Health Service Data     2
Basic Information     4

**Chapter 2**
**Volume and Size Reduction**     7

Introduction     7
Methods of Reduction     7
References     12

## Chapter 3
## Collection of Solid Wastes 13

Introduction 13
Hoist-Truck Systems 15
Tilt-Frame Systems 15
Trash-Trailer Systems 16
Collection Operations 18
Collection Routes 19
References 19

## Chapter 4
## Resource Recovery 23

Introduction 23
Paper and Paper Products 24
Recovery of Products After Treatment 25
Ferrous Metals 27
Aluminum 27
Other Nonferrous Metals and Ash 28
Glass 30
Rubber 30
Plastics 31
Automobiles 31
Process Flow Sheets 34

## Chapter 5
## Solid-Waste Disposal Microbiology 43

Introduction 43
Bacteria 44
Fungi 46
Algae 46
Protozoa 46
Viruses 46
Carbon Compounds in Soil Reactions 47
Nitrogen Changes in Soil 47
Sulfur Changes in Soil 51
Biological Action Within Sanitary Landfills 51
Biological Changes During Composting 52
References 53

## Chapter 6
## Composting 55

| | |
|---|---:|
| Theories Involved | 55 |
| Window Composting | 57 |
| Continuous Mechanical Composting | 59 |
| References | 60 |

## Chapter 7
## Sanitary Landfills 61

| | |
|---|---:|
| Basic Definitions | 61 |
| Public Attitude | 63 |
| Engineering Planning | 63 |
| Administration and Control | 63 |
| Operation of a Sanitary Landfill | 63 |
| Operating Conditions | 64 |
| References | 76 |

## Chapter 8
## Incineration 77

| | |
|---|---:|
| Introduction | 77 |
| Advantages | 77 |
| Disadvantages | 78 |
| Costs | 78 |
| Design and Operating Features | 80 |
| Air-Pollution Control | 84 |
| Fly Ash | 85 |
| Final Cost Relative to Other Treatments | 85 |
| Reference | 85 |

## Chapter 9
## Pyrolysis 87

| | |
|---|---:|
| Definition and Products | 87 |
| Pyrolysis Pilot Plant and Test Procedures | 87 |
| Yields of Products from Pyrolysis of Refuse | 89 |

## Chapter 10
## Hazardous Wastes 93

| | |
|---|---|
| Introduction | 93 |
| Fire and Explosions | 95 |
| Spontaneous Combustion | 95 |
| Explosions | 98 |
| Gas Production | 98 |
| Leachate Production | 98 |
| Wastebaskets | 99 |
| Refuse Chutes | 99 |
| Compactors | 100 |
| Incinerators | 100 |
| Shredders | 101 |
| Reclamation Plants | 101 |
| Landfills | 101 |
| Hazardous Industrial Wastes | 101 |
| Industrial Insurance Requirements | 119 |
| References | 120 |

## Chapter 11
## Legal Aspects of Solid Wastes 121

| | |
|---|---|
| Introduction | 121 |
| Resource Recovery Act of 1970 | 121 |
| Resource Conservation and Recovery Act of 1976 | 123 |
| References | 133 |

## Chapter 12
## Health Hazards in Connection with Solid Wastes 135

| | |
|---|---|
| Introduction | 135 |
| Asbestos Dust | 137 |
| Health and Protection of Workers | 137 |
| Public Health Hazards | 139 |
| References | 141 |

## Chapter 13
### Agricultural Waste 143

Introduction 143
Sewage Contaminated Sludge Agricultural Waste 143
Contaminants and Treatment 146
References 156

## Chapter 14
### Hospital Solid Wastes 157

Introduction 157
Characteristics 158
Treatment of Hospital Wastes 164
References 165

## Chapter 15
### Industrial Wastes 173

Introduction 173
Meat Industry in General 194
Metal-Plating Solid Wastes 195
Steel Mills 197
Food Solid Wastes 201
Poultry 250
Plastic Plants 253
Textile Plants 257
Cement, Construction, and Demolition 282
Power Plants 285
Leather Tanning and Finishing 293
Pulp and Paper Mills 297
Phosphate Fertilizers 300
Rubber Plants 303
Mining Industries 303
Chemical Industries 312
Nonferrous Metals 323
References 327

**Chapter 16
Energy-Process Systems
from Solid Wastes** 331

References 351

Index 353

# LIST OF FIGURES

| | | |
|---|---|---|
| 2-1 | Schematic of a Hammermill | 10 |
| 2-2 | Schematic of a Rasp Mill | 11 |
| 2-3 | Schematic of Crushing Apparatus | 11 |
| 3-1 | Letter to Customers | 17 |
| 3-2 | Article on Illegal Use of Transfer Stations | 20 |
| 3-3 | Letter Regarding Typical Fees of Urban Collection and Disposal | 22 |
| 4-1 | Pictorial Flowsheet for Materials Recovery Systems | 35 |
| 4-2 | Flowsheet for Materials Recovery System (Central Contra Costa Sanitary District and Brown and Caldwell Consulting Engineers) | 36 |
| 4-3 | Schematic of Process for Treatment of Uniform Solid Waste Composition and Utility Alternatives | 40 |
| 5-1 | The Carbon Cycle | 48 |
| 5-2 | The Nitrogen Cycle | 50 |
| 5-3 | Sulfur Changes in Soil | 52 |
| 7-1 | Sanitary Landfill Operating Costs | 70 |
| 9-1 | Flow Diagram of Pilot Plant Used to Pyrolyze Municipal and Industrial Refuse | 88 |
| 9-2 | Pyrolysis System | 96 |

xiv  LIST OF FIGURES

| | | |
|---|---|---|
| 12-1 | Health Hazards of Solid Waste—A Question-and-Answer Decision | 136 |
| 13-1 | Agricultural Waste Sources | 144 |
| 13-2 | Odor Offensiveness of Chicken Manure as a Function of Moisture Content | 151 |
| 14-1 | Operations of Hospitals That Produce Solid Wastes | 162 |
| 14-2 | Economic Evaluation of Hospitals | 170 |
| 15-1 | Distribution of Waste for Disposal: Employee Ratios in Twenty-one SIC Code Groups | 192 |
| 15-2 | Distribution of Waste for Disposal Among Twenty-one SIC Code Groups | 193 |
| 15-3 | Flow Diagram of Standard Manufacturing Process for Bottled and Canned Soft Drinks | 209 |
| 15-4 | Process Diagram for Malt Liquor Production | 211 |
| 15-5 | Brewery Input–Output Characteristics | 213 |
| 15-6 | Typical Sugar Factory with Cane Wash | 217 |
| 15-7 | Water Usage in a Typical Sugar Cane Factory | 218 |
| 15-8 | Typical Carbon Refinery | 220 |
| 15-9 | Wastewater Flow Diagram for a Crystalline Refinery | 221 |
| 15-10 | Materials Flow in Beet Sugar Processing Plant with Typical Water Utilization and Waste Disposal Pattern | 222 |
| 15-11 | The Corn Wet Milling Process | 224 |
| 15-12 | The Corn Dry Milling Process | 225 |
| 15-13 | The Bulgur Process | 226 |
| 15-14 | The Wheat Milling Process | 227 |
| 15-15 | The Rice Milling Process | 228 |
| 15-16 | The Parboiled Rice Process | 229 |
| 15-17 | Animal Feed Manufacturing | 230 |
| 15-18 | Flaked or Crisp Cereal Production | 231 |
| 15-19 | Wheat Starch and Gluten Manufacturing | 232 |
| 15-20 | White Wine Production Diagram | 234 |
| 15-21 | Red Wine Production Diagram | 235 |
| 15-22 | Brandy and Wine Spirits Production Diagram | 239 |
| 15-23 | Beef Cattle Industry Structure and Mass Flow Diagram | 242 |
| 15-24 | Swine Industry Structure and Mass Flow Diagram | 243 |
| 15-25 | Broiler Industry Structure and Mass Flow Diagram | 244 |
| 15-26 | Sheep Industry Structure and Mass Flow Diagram | 245 |
| 15-27 | Turkey Industry Structure and Mass Flow Diagram | 246 |

## LIST OF FIGURES xv

| | | |
|---|---|---|
| 15-28 | Duck Industry Structure and Mass Flow Diagram | 247 |
| 15-29 | Horse Industry Structure and Mass Flow Diagram | 250 |
| 15-30 | Flow Chart of Poultry-Processing Plant | 252 |
| 15-31 | Estimated Quantities of Total Waste to Land Disposal, 1974 (Dry/Wet Weight) | 260 |
| 15-32 | Category A—Typical Wool Scouring Process | 262 |
| 15-33 | Category B—Typical Wool or Wool Blend Fabric Dyeing and Finishing Process | 266 |
| 15-34 | Category C—Typical Greige Goods Process | 268 |
| 15-35 | Category D—Typical Woven Fabric Dyeing and Finishing Process | 270 |
| 15-36 | Category E—Typical Knit Fabric Dyeing and Finishing Process | 274 |
| 15-37 | Category F—Typical Tufted Carpet Dyeing and Finishing Process | 276 |
| 15-38 | Category G—Typical Yarn and Stock Dyeing and Finishing Process | 278 |
| 15-39 | Gypsum Pond Water Seepage Control | 301 |
| 15-40 | Managing Chemical Wastes | 304 |
| 16-1 | Typical Waterwall Furnace for Unprocessed Solid Waste | 333 |
| 16-2 | Simplified Flow Diagram Showing How the Dry Processing Approach Is Used To Produce Fluff, Densified, or Dust RDF | 334 |
| 16-3 | Wet Process Energy Recovery System | 336 |
| 16-4 | The Monsanto Landgard System | 337 |
| 16-5 | Torrax Slagging Pyrolysis System | 339 |
| 16-6 | Union Carbide Purox System | 340 |
| 16-7 | Production of "Oil" from Solid Wastes Using the Occidental Process | 341 |
| 16-8 | Production of Electricity from Landfill Gas | 342 |
| 16-9 | Simplified Plant Configuration in Pompano Beach, Florida | 344 |
| 16-10 | RefCOM Process | 345 |
| 16-11 | Gas Turbine Generating System Using Refuse as a Fuel (CPU-400) | 346 |
| 16-12 | Wet Process Fiber Recovery System | 347 |
| 16-13 | Dry Process Paper and Materials Recovery | 349 |
| 16-14 | Magnetic Separator Configuration | 350 |

# LIST OF TABLES

| | | |
|---|---|---|
| 1-1 | Average Per Capita Quantities of Solid Wastes Collected from Urban Sources in the United States, 1968 | 2 |
| 1-2 | Costs of Solid-Waste Collection and Disposal in the United States, 1971 to 1985 | 3 |
| 1-3 | Components of Municipal Solid Wastes Generated in the United States, 1971 | 5 |
| 1-4 | Constituents of Urban Waste in Spain in 1977 | 5 |
| 3-1 | Typical Container Capacities Available for Use with Various Collection Systems | 14 |
| 4-1 | Average Municipal Solid-Waste Composition | 24 |
| 4-2 | Yield from Pyrolysis of a Ton of Typical Refuse | 25 |
| 4-3 | Market Metal Prices in September 1979 | 26 |
| 4-4 | Market Prices for Various Commodities Commonly Found in Urban Wastes (Excerpts from The Wall Street Journal) | 28 |
| 4-5 | Energy Saved by Recycling | 29 |
| 4-6 | Typical Materials Specifications That Affect Selection and Design of Processing Operations | 33 |
| 4-7 | Estimated Recoverable Quantities for Various Components in Solid Wastes Using Mechanical Equipment | 34 |

xviii    LIST OF TABLES

| | | |
|---|---|---|
| 4-8 | Light and Heavy Fractions of Solid Waste Components After Shredding | 38 |
| 4-9 | Chemical Processes Used for Conversion of Solid Wastes | 39 |
| 5-1 | Soil Population (Number of Organisms Per Gram in a Fertile Agricultural Soil) | 44 |
| 6-1 | Some Characteristics of Compost Products | 60 |
| 7-1 | Environmental Rate and Conversion Products of Domestic Wastes in Sanitary Landfill | 74 |
| 7-2 | Costs for Solid-Waste Disposal | 76 |
| 8-1 | Some Approximate Heat Value of Components of Refuse | 82 |
| 9-1 | Pyrolysis Pilot Study Data | 91 |
| 10-1 | Characteristics of Hazardous Waste | 94 |
| 10-2 | U.S. Federal Laws and Agencies Affecting Toxic Substances Control | 96 |
| 10-3 | Water Quality Limits for Toxic Pollutants for Three Uses | 104 |
| 10-4 | Representative Hazardous Substances Within Industrial Waste Stream | 113 |
| 10-5 | Waste Materials Classified as Toxic | 114 |
| 12-1 | Troublesome Metals and Organic Solids | 138 |
| 12-2 | Specific Diseases Caused by Flies, Mosquitoes, and Rodents | 140 |
| 13-1 | Diseases Associated with Human Fecal Waste | 145 |
| 13-2 | Principal Agricultural Waste Components | 147 |
| 13-3 | Agricultural Wastes | 148 |
| 13-4 | Average Amount and Type of Trash in Seed Cotton Harvested by Various Methods | 148 |
| 13-5 | Production of Wastes by Livestock in the United States, 1965 | 149 |
| 13-6 | Population Equivalents of Animal Wastes (BOD Basis) | 149 |
| 13-7 | Per Capita Animal Contribution of Indicator Microorganisms | 150 |
| 13-8 | Animal Waste Defecation (Per 1,000 Pounds of Live Weight) | 152 |

## LIST OF TABLES xix

| | | |
|---|---|---|
| 13-9 | Characteristics of Animal Manures | 154 |
| 13-10 | Nutrients in Animal Wastes | 155 |
| 14-1 | Classification Schemes for Hospital Solid Wastes as Summarized from Various Sources | 159 |
| 14-2 | General Classification System | 160 |
| 14-3 | Problems Associated with Waste Type | 161 |
| 14-4 | Hospital Solid Wastes Classification by Origin | 162 |
| 14-5 | Advantages and Disadvantages of Hospital Internal Transportation Systems for Solid Waste | 163 |
| 14-6 | Equipment Cost for Wet Scrubbers | 166 |
| 14-7 | Cost of Incinerators | 167 |
| 14-8 | Advantages and Disadvantages of Waste-Handling Systems | 168 |
| 14-9 | Service Life of Solid-Waste Processing Equipment | 171 |
| 15-1 | Solid Industrial Waste Quantities | 176 |
| 15-2 | Principal Industrial Waste Components | 178 |
| 15-3 | Relative Quantities of Solid Waste Per Year Generated from Durable and Nondurable Industries for 1967 | 188 |
| 15-4 | Large Firm Multipliers | 189 |
| 15-5 | Small Firm Multipliers | 190 |
| 15-6 | Estimated Production of Solid Wastes in the Processing of Each 100 Kilograms of Various Foods | 205 |
| 15-7 | Percentage of Solid Wastes from the Canning and Frozen Foods Industry Handled by Various Disposal Methods in California | 205 |
| 15-8 | Waste Analysis of the Effluents from Soft Drink Bottling Plants | 209 |
| 15-8A | Waste Loads Discharged from Soft Drink Bottling Plants | 210 |
| 15-9 | Brewery Effluent Characteristics Obtained from a Survey of 75 Plants | 213 |
| 15-9A | Sources of Pollutants from a Brewery | 213 |
| 15-10 | Typical Concentrations of Wastes Discharged from Specific Brewery Operations | 214 |
| 15-11 | Overall Plant Raw Waste Characteristics | 215 |
| 15-12 | Characteristics of Wastewaters from the Grain Milling and Processing Industry | 233 |
| 15-13 | Characteristics of Winery Wastewater Sources | 236 |
| 15-14 | Wastewater Characteristics for Nondistilling California Wineries | 237 |

## LIST OF TABLES

| | | |
|---|---|---|
| 15-15 | Wine Stillage Characteristics | 238 |
| 15-16 | Physical Characteristics of Livestock Manure | 248 |
| 15-17 | Characteristics of Runoff from Cattle Feedlots | 251 |
| 15-18 | Coefficients Used in Estimating Byproducts, Water Use, and Waste Loads of Poultry-Slaughtering Plants | 254 |
| 15-19 | Densities and Calorific Values of Typical Plastics | 255 |
| 15-20 | The Quantity of Plastic Wastes Generated in the United States | 256 |
| 15-21 | Potentially Hazardous Waste (in 1000 kilograms per year) | 258 |
| 15-22 | Summary of Land-Destined Total and Potentially Hazardous Wastes from the Textiles Industry (1000 kilograms per year) | 259 |
| 15-23 | Category A—Wool Scouring Sludge Analysis (milligrams per kilogram of dry sludge) | 264 |
| 15-24 | Category B—Wool Fabric Dyeing and Finishing Sludge Analyses (milligrams per kilogram of dry sludge) | 265 |
| 15-25 | Category D—Woven Fabric Dyeing and Finishing (milligrams per kilogram of dry sludge) | 272 |
| 15-26 | Category E—Knit Fabric Dyeing and Finishing Sludge Analysis (milligrams per kilogram of dry sludge) | 273 |
| 15-27 | Category F—Tufted Carpet Dyeing and Finishing Sludge Analyses (milligrams per kilogram of dry sludge) | 277 |
| 15-28 | Category G—Yarn and Stock Dyeing and Finishing Sludge Analysis (milligrams per kilogram of dry sludge) | 280 |
| 15-29 | Estimation of Industrial Wastes Generation Rates | 284 |
| 15-30 | Generation by Fuel Type | 286 |
| 15-31 | Range of Ash Content in Coal | 287 |
| 15-32 | Average Amounts of Trace Elements in Coal Ashes (ppm) | 289 |
| 15-33 | Trace Element Analyses of Lime and Limestone Samples (ppm) | 291 |
| 15-34 | Typical Residual Oil Ash Analysis | 292 |
| 15-35 | Ash Utilization | 293 |
| 15-36 | Origins and Concentrations of the Six Major Processing Areas | 295 |
| 15-37 | Total Process and Potentially Hazardous Solid Wastes | 296 |
| 15-38 | Tannery Solid Waste Disposal | 298 |

## LIST OF TABLES    xxi

| | | |
|---|---|---|
| 15-39 | Disposal of Solid Wastes from Some Tanneries in India | 299 |
| 15-40 | Typical Waste Automotive Oil Composition | 302 |
| 15-41 | Generation, Destination, and Disposal of Waste Oils, 1972 (Millions of Gallons) | 306 |
| 15-42 | Solid Waste Generated by Tire and Tire Products Manufacturing | 307 |
| 15-43 | Porjected Annual Ore Production By Selected Industries | 307 |
| 15-44 | Projected Annual Generation of Mining Solid Wastes by Selected Industries | 308 |
| 15-45 | Disposal Methods for Mining Wastes | 310 |
| 15-46 | Data on Pennsylvania Antracite Refuse Banks | 311 |
| 15-47 | Gross Chemical Composition of Most Slates | 311 |
| 15-48 | Classification of Chemicals by Composition | 318 |
| 15-49 | The Most Significant Organic Chemicals | 320 |
| 15-50 | Government Estimates of Amounts of Wastes Per Year | 322 |
| 15-51 | Aluminum Solid Wastes Generated in the United States and Europe | 325 |
| 15-52 | Residual Generation Factors for Nonferrous Metal Smelting and Refining | 325 |
| 15-53 | Hazardous Waste Generated by the Nonferrous Metals Industries | 326 |

# PREFACE

Although this book was written to provide information and assistance for the solutions of industrial solid waste problems, much data are also given on urban solid wastes. The first three chapters summarize material generally useful on municipal solid waste, its quality and quantity, methods of reducing its volume, and means of collection.

In Chapter 4 I emphasize the recovery of various materials from all solid wastes. In the long run, recovery will provide the most cost effective method of ultimate disposal of solid residues. Since biological activity plays such a vital role in the utilization of solid waste as compost and in its decomposition in landfills, Chapter 5 is entirely devoted to microbiology. This is followed by a discussion of composting in Chapter 6 and sanitary landfills in Chapter 7.

The two heat-intensive methods of treating solid wastes, incineration and pyrolysis, are explored in Chapters 8 and 9. Since recent evidence points to the increasing hazards of solid wastes, these hazards are delineated in Chapter 10. Legal edicts are described in Chapter 11, which includes the latest federal legislation. Specific health hazards connected with handling and disposing of solid wastes are described in Chapter 12. Agricultural and hospital wastes, two major sources of solid residues, are described in Chapters 13 and 14.

The major emphasis of this textbook is industrial solid wastes, explored in Chapter 15. Seventeen broadly classified industries are covered by giving as much of their waste origin, characteristics, and treatment as possible. The literature is still very sparse in relation to solid wastes from industry. It is expected, however, that because of federal requirements more data as well as treatment will be made public in the near future. Finally, more complete energy process systems for disposal and reuse of solid wastes are given in Chapter 16.

I am grateful to all previous authors dealing with certain aspects of refuse for permitting me to use some of their data pertinent to this subject. Special mention should be made of George Tchobonaglou's book, *Solid Waste*, published by McGraw-Hill in 1977, and E. Joe Middlebrook's book, *Industrial Pollution Control*, published by Wiley-Interscience in 1979.

Once again, I must express my appreciation to my graduate students. They assisted me in researching literature and checking references, and most of all they participated actively in classes as the manuscript developed, using the unfinished notes of this book.

Without the help of my wife, Joan, who helped with editing, typing, and the cover art designs, this text would not have been possible.

**Nelson Leonard Nemerow, Ph.D.**
May 20, 1983

# ACKNOWLEDGMENT

Without the patience and assistance of my wife, Joan—teacher, artist, secretary, and travel associate—and the many graduate students at the University of Miami, this textbook would not have been published. They helped me to research and develop the sparse existing literature on industrial solid wastes.

I also wish to thank my wife for the many months of assembling, transcribing, and typing my vast notes and research on solid wastes into an accurately written manuscript. Joan also edited, and designed the artistic cover for this textbook.

# 1 BASIC DATA ON SOLID WASTES

## INTRODUCTION

Littering of food and other solid wastes in medieval towns, the practice of throwing wastes into the unpaved streets, roadways, and vacant land, led to the breeding of rats, with their attendant fleas carrying the germs of disease, and to the outbreak of the plague. The relationship between public health and the improper storage, collection, and disposal of solid wastes is quite clear. Public health authorities have shown that rats, flies, and other disease vectors breed in open dumps, as well as in poorly constructed or poorly maintained housing, in food-storage facilities, and in many other places where food and harborage are available for rats and the insects associated wih them. Ecological impacts, such as water and air pollution, also have been attributed to improper management of solid wastes. (For instance, liquid from dumps and poorly engineered landfills has contaminated surface waters and groundwaters. In mining areas the liquid leached from waste dumps may contain toxic elements, such as copper, arsenic, and uranium, or may contaminate water supplies with unwanted salts of calcium and magnesium.) Everyone is familiar with solid wastes, especially those generated in municipalities, such as food wastes and rubbish, abandoned vehicles, demolition and construction wastes, street sweepings, and garden wastes. Far greater amounts, however, result from agricultural, industrial, and mineral sources. (Although the data are varied, recent estimates indicate that an average of 4.4 billion tons of solid wastes are generated each year in the United States alone. Of this total, municipal wastes represent approximately 230 million tons; industrial wastes, 140 million tons; and agricultural wastes, 640 million tons. By far the greatest amount of solid wastes comes from mines and minerals and from animal wastes, each with an average of 1.7 billion tons per year. The total amount generated from all sources by the year 2000 may approach 12 billion tons per year. Looking at just the urban and industrial wastes, the generation rate in the United States is approximately 3,600 pounds per capita per year.)

## U.S. PUBLIC HEALTH SERVICE DATA

In 1968 the U.S. Public Health Service published data obtained in its National Survey of Community Solid Waste Practices. The average generation rates for urban sources in the United States are shown in Table 1-1. When demolition and construction wastes are added to industrial wastes, a total of 2.62 pounds per person per day or about

Table 1-1. Average per Capita Quantities of Solid Wastes Collected from Urban Sources in the United States, 1968.

| Source | Lb/Capita/Day |
|---|---|
| Combined residential and commercial | 4.29 |
| Industrial | 1.90 |
| Institutional | 0.16 |
| Demolition and construction | 0.72 |
| Street and alley cleanings | 0.25 |
| Tree and landscaping | 0.18 |
| Park and beach | 0.15 |
| Catch basin | 0.04 |
| Sewage treatment plant solids | 0.50 |
| Total | 8.19[a] |

a. The corresponding total per capita quantities for all areas (7.92 lb/capita/day) are somewhat lower than those from urban areas.

Note: lb/capita/day × 0.4536 = kg/capita/day.

Source: Adapted from Publication no. 1867, U.S. Department of Health, Education and Welfare, by Tchobanoglous (1977).

**Table 1-2.** Costs of Solid-Waste Collection and Disposal in the United States, 1971 to 1985.

| Item | 1971 (Estimated) | 1980 (Projected) | | | 1985 (Projected) | | |
|---|---|---|---|---|---|---|---|
| | | Low | Medium | High | Low | Medium | High |
| Collected wastes (tons, millions)[a] | 120 | 150 | 160 | 175 | 165 | 190 | 220 |
| Unit costs ($/ton) | | | | | | | |
| Collection | 18 | 18 | 18 | 18 | 18 | 18 | 18 |
| Disposal | 4 | 5 | 5 | 5 | 5 | 5 | 5 |
| Total | 22 | 23 | 23 | 23 | 23 | 23 | 23 |
| Total national costs, millions of dollars (1971) | | | | | | | |
| Collection | 2,160 | 2,700 | 2,880 | 3,150 | 3,150 | 3,420 | 3,960 |
| Disposal | 480 | 750 | 800 | 875 | 875 | 950 | 1,100 |
| Total | 2,640 | 3,450 | 3,680 | 4,025 | 4,025 | 4,370 | 5,060 |

a. It is assumed that 95 percent of the projected waste generation will be collected.

Note: tons × 907.2 = kg
$/ton × 0.0011 = $/kg.

Source: Adapted from EPA's *Resource Recovery and Source Reduction*, Publication no. sw-122, by Tchobanoglous (1977: 13).

32 percent of total urban wastes can be classed as industrial. The character and quantity of this urban fraction influences the manner which municipal refuse is collected and disposed. It is increasing at about 4.5 percent per year. A more reasonable figure, based on the greater awareness of the public with respect to resource recovery and recycling, might be 3.5 percent. If major efforts at resource recovery and recycling were to be effective, the rate of increase might be as low as 2.5 percent. Unfortunately, the standard of living in the United States is inevitably tied to the generation of solid wastes—the squandering of natural resources from this country and abroad, and one-time use of material of many types, and the philosophy of wastefulness and rapid obsolescence of products. It is reasonable to presume that a departure from this philosophy of wastefulness must reduce the tonnage of wastes to be managed. This concept inevitably leads to the need for resource and recovery and the recycling of recovered materials to the mainstream of industry. Another alternative is to continue the wasteful practices of modern industrial society and pay the penalty. As indicated in Table 1-2, the difference between the high and low cost figures for 1985 could be $1 billion. It could be even greater if conservation (recovery and re-use) were to become a way of life for citizens, commerce, and industry.

## BASIC INFORMATION

The material components of municipal solid wastes as found in 1971 are given by Tchobanoglous in Table 1-3. He presents quantities generated during that year of paper, glass, metal, plastic, rubber, textile, wood, and food wastes. Although the origin of these contaminants is industrial, they are wasted by urban customers in municipal refuse. Paper represents almost 40 percent, while glass and metal each contribute about 12 percent. It is interesting to note the component fractions found existing in the urban solid wastes in Spain in 1977. These figures are given in Table 1-4. Compared to the United States data shown in the previous table, we see the lower paper (18 versus 31.3 percent), the lower glass (3 versus 9.7 percent), and the lower metals (4 versus 9.5 percent). Since these materials are potentially recoverable, it is significant to note that United States apparently wastes two to three times more of these constituents than Spain does.

Table 1-3. Components of Municipal Solid Wastes Generated in the United States, 1971.

| Component | Total Generated | | Total Disposed | |
|---|---|---|---|---|
| | Tons, Millions | Percent | Tons, Millions | Percent |
| Paper | 39.1 | 31.3 | 47.3 | 37.8 |
| Glass | 12.1 | 9.7 | 12.5 | 10.0 |
| Metal | 11.9 | 9.5 | 12.6 | 10.1 |
|   Ferrous | 10.6 | 8.5 | — | — |
|   Aluminum | 0.8 | 0.6 | — | — |
|   Other nonferrous | 0.5 | 0.4 | — | — |
| Plastic | 4.2 | 3.4 | 4.7 | 3.8 |
| Rubber and leather | 3.3 | 2.6 | 3.4 | 2.7 |
| Textiles | 1.8 | 1.4 | 2.0 | 1.6 |
| Wood | 4.6 | 3.7 | 4.6 | 3.7 |
| Food | 22.0 | 17.6 | 17.7 | 14.2 |
|   Subtotal | 99.0 | 79.2 | 104.8 | 83.9 |
| Yard wastes | 24.1 | 19.3 | 18.2 | 14.6 |
| Miscellaneous inorganics | 1.9 | 1.5 | 2.0 | 1.5 |
|   Total | 125.0 | 100.0 | 125.0 | 100.0 |

Note: tons × 907.2 = kg.

Source: Adapted from EPA's *Resource Recovery and Source Reduction*, Publication no. sw-122, by Tchobanoglous (1977: 10).

Table 1-4. Constituents of Urban Waste in Spain in 1977.

| Constituents Composition | Percent |
|---|---|
| Metals | 4 |
| Paper-cardboard | 18 |
| Glass | 3 |
| Plastics | 4 |
| Organic matter | 50 |
| Other materials | 21 |

Source: Subsecretarial Office of Planning, Presidency of the Government, 1977. "The Spanish Environments."

# 2 VOLUME AND SIZE REDUCTION

## INTRODUCTION

Reduction is important because any operation which reduces the volume collected or transported will reduce the total cost as well. Densification by compaction prior to collection or transportation will reduce collection time and result in fewer transportations to the final disposal site. Volume reduction also means less handling and hence significant savings in operating costs.

Compaction of refuse also aides in removal of disease vectors such as mosquitoes, flies, rats, and roaches. Fire hazards are reduced also because less material has to be stored for shorter times. Compaction and the accompanying volume reduction leads to cleaner and neater storage.

## METHODS OF REDUCTION

These methods depend upon the size and ownership of the refuse.

- In older times burning in backyards was practiced, but in most locations it is no longer allowed.
- Composting can be used with some proper concern for health problems.

- Bottomless cans are set into the ground and tightly covered in which food wastes decompose about 75 percent. The cans are cleaned out every six months. Objections include odors, difficulty of handling, and can rusting.
- Garbage grinders reduce volume about 10 percent, but will not handle many types of organic matter such as bones and fruit skins (especially bananas and oranges), poultry, fat, and so on. They transfer the load to municipal sewers and sewage treatment plants. As cost of the latter skyrocket, grinders for the home are becoming objectionable.
- Compaction units offer promise for homeowners, but they place quite an expense on them without sufficient personal rewards.

*Groups of residences*, such as apartment complexes or development projects or townhouses, can afford more expensive means of volume reduction which may not be possible for individual residences. The usual means of volume reduction for residential complexes is incineration. Two types are commonly being used:

1. One type uses a burner to heat plates around which refuse is placed, moisture evaporated, and refuse ignited.
2. Another type stores the refuse until it is ignited by some preset timer which activates a burner supplied with an oil or gas fuel. In some apartment complexes the chutes for feeding refuse also serve as the flue for carrying away the combustion gases. This aides in drying the incoming refuse before incineration. However, these incinerators usually operate inefficiently because of limited combustion chamber, poor air supply, low burning temperature, and poor dispersion characteristics which all cause incomplete combustion and resultant air pollution. Because of all these problems with incinerators they are not considered entirely acceptable as a means of refuse volume reduction.

*Very large groups of people*, such as large housing projects, institutions, and hospitals, may be in a position to use newer volume-reducing techniques such as wet pulverization and pulping. Capital expense of these systems may be high, and good maintenance and operation are necessary. Volume reduction in pulping of refuse may be as high as 80 percent. Pulping can include salvaging as an integral part of the system. Pulp, paper, magnetic and nonmagnetic metals, and glass are salvagable.

In large systems compaction may also be used in conjunction with pulpers. Large pneumatic or hydraulic compactors have been in use for large complexes. The 75 percent compaction attainable reduces the volume which must be collected and transported. Automated equipment used in these compactors reduces labor costs and results in neater, quieter, and easier controlled systems despite high maintenance costs involved when clogging results. Therefore, for large institutions, the combination of compaction at the collection point and pulping at the receiving station can reduce the volumes at both points and hence reduce both transportation and treatment costs.

*Volume reduction at central site* at some distance from the source of the refuse is desirable under certain circumstances. When central sites are used, refuse is collected and transported to the focal point where it is ground, baled, or compacted for volume reduction. The longer the haul is from the disposal area to the central site, the more feasible this type of volume reduction is.

### Prior to Ultimate Disposal Systems

*Milling, Grinding, or Shredding.* This type of treatment is used prior to ultimate disposal systems such as in landfilling, incineration, and composting. The Heil-Gonard System which developed in France uses a unique hammermill grinding action to reject objects that cannot be disintegrated by the hammer rotation. The reject material is delivered to a recovery chute, collected, and is available for additional sorting for potential reuse or discharge.

The milling allows greater subsequent compaction of two to three times that of raw municipal refuse. The ground refuse becomes odor- and rodent-free for landfilling or storing prior to incineration or composting. It evidently destroys the insect eggs and larvae contained normally in municipal refuse. Rats are discouraged from feeding primarily because a greater energy effort is needed to search for ample concentration of organic matter. Burroughing animals also have difficulty finding shelter in shredded refuse since the structure of the refuse is no longer stable and falls in on an animal.

10   INDUSTRIAL SOLID WASTES

Other types of shredders include the Hydrasposal (Black Clawson Company) which has been adapted from the pulp and paper industry, and various kinds of vibrating or rotating knife-like hammers which reduce the size of amenable objects such as brush, paperboard, cartons by about 80 percent.

*Composting* requires extensive accessories of size reduction of six different types: (1) hammer mills, (2) flail mills, (3) vertical-axis rasps, (4) drum rasps, (5) roll crushers, and (6) pulpers.

Hammermills (Figure 2-1) are most common and usually use fixed or pivited hammers or cathers generally with a simple horizontal or vertical rotor. Rasp types (Figure 2-2) have a higher initial cost and are somewhat larger than hammermills, but they use lower power and have lower maintenance costs. The Pulverator (Figure 2-3) is a drum rasp used in the United States.

Some of the major crushers and grinders are shown in the schematic views of Figures 2-1, 2-2, and 2-3.

*Landfilling*, accomplished mainly by hammermills, is enhanced by size reduction and is mainly practiced in Europe and lately in the United States. The hammermills serve a dual purpose of grinding and rejecting material. The nongrindable matter is forced through a recovery chute to be collected. Because the density is increased by about three times by the increased compatability, the disposal by landfill is improved. The resulting refuse does not provide as good a

Figure 2-1.   Schematic of a Hammermill.

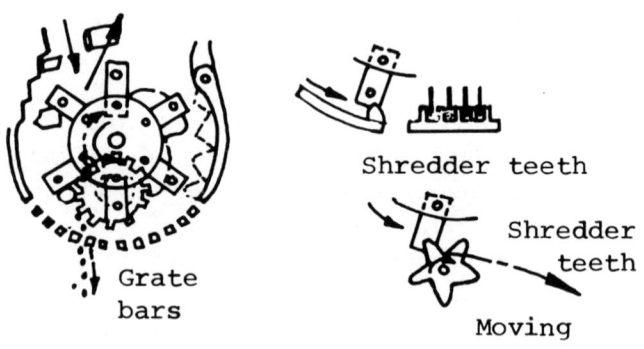

Source: *Recovery and Utilization of Municipal Solid Waste.*

VOLUME AND SIZE REDUCTION 11

Figure 2-2. Schematic of a Rasp Mill.

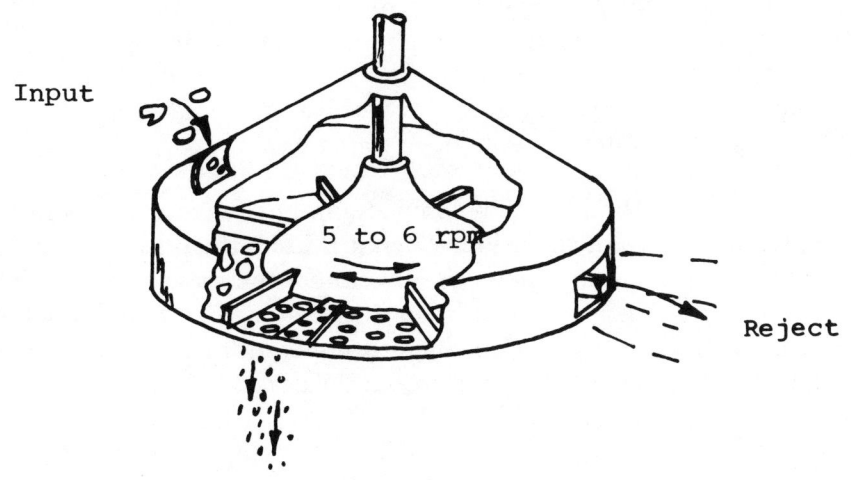

Source: *Recovery and Utilization of Municipal Solid Waste.*

Figure 2-3. Schematic of Crushing Apparatus.

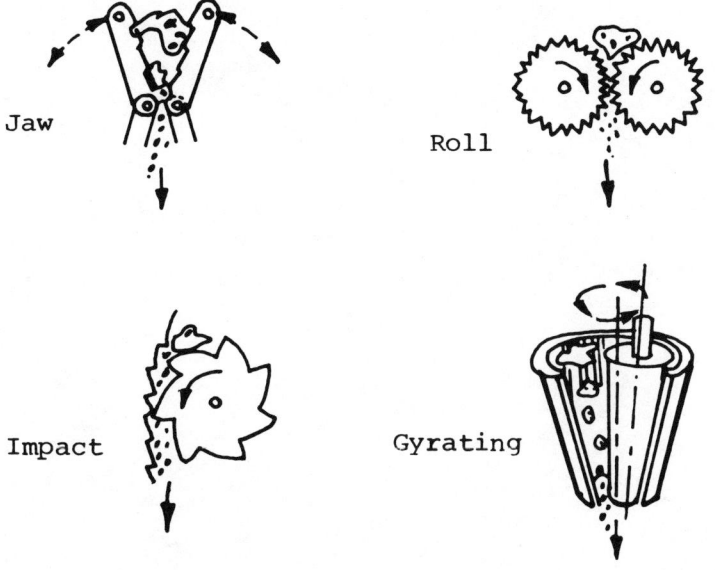

Source: *Recovery and Utilization of Municipal Solid Waste.*

habitat for disease-carrying vectors. Grinding destroys much of the larvae and insect eggs which make the resulting landfilled refuse relatively free from the breeding of flies or other insects. Void spaces are minimized; hence there is also an almost complete absence of mice and rats.

Paperboard box shredders can be used at the landfill site or the site of origin to reduce the volume of paper waste.

*Incineration.* Grinding and compaction or even milling alone is rarely used prior to incineration. When it is used, the purpose is more for uniformity than anything else. It reduces the bulky items which are difficult to feed and burn in an incinerator. Size reduction and uniformity reduces cost of incinerator operation. It also prevents explosions of cans containing hazardous volatile chemicals.

## REFERENCES

*Recovery and Utilization of Municipal Solid Waste.* n.d. USPHS Publication #1908.

# 3 COLLECTION OF SOLID WASTES

## INTRODUCTION

Urban solid-waste collection is challenging because the generation of both residential and commercial–industrial solid wastes takes place in all homes, all apartment buildings, and all commercial and industrial facilities as well as in all streets, parks, and even the vacant areas of every community. The industrial and commercial development of suburbs all over the country makes collection even more difficult.

Collection has now become more critical because of the high operational cost of fuel and labor. Of the total amount of money spent for the collection, transportation, and disposal of solid wastes in 1975, approximately 60 to 80 percent was spent on the collection phase. This fact is important because any small percentage improvement in the collection operation will result in significant savings in the overall daily cost. Collection systems can be classified according to their mode of operation into the two categories of hauled-container systems and stationary container systems:

1. Hauled-container systems (HCS) are collection systems in which the containers are used for storage of wastes, then hauled to the disposal site, emptied, and returned to either their original location or some other location.

Table 3-1. Typical Container Capacities Available for Use With Various Collection Systems.

| Vehicle | Collection Container Type | Typical Range of Container Capacities, yd$^3$ |
|---|---|---|
| Hauled-container systems | | |
| Hoist truck | Used with stationary compactor | 6–12 |
| Tilt-frame | Open top, also called debris boxes | 12–50 |
| | Used with stationary compactor | 15–40 |
| | Equipped with self-contained compaction mechanism | 20–40 |
| Truck-tractor | Open-top trash-trailers | 15–40 |
| | Enclosed trailer-mounted containers equipped with self-contained compaction mechanism | 20–40 |
| Stationary container systems | | |
| Compactor, mechanically loaded | Open top and enclosed top and side-loading | 1–8 |
| Compactor, manually loaded | Small plastic or galvanized metal containers, disposable paper and plastic bags | 20–55 (gal) |

Note: yd$^3$ × 0.7646 = m$^3$; gal × 0.003785 = m$^3$.
Source: Tchobanoglous (1977).

2. Stationary container systems (SCS) are collection systems in which the containers used for the storage of wastes remain fixed at the point of generation, except for occasional short trips to the collection vehicle.

Hauled-container systems are recommended for the collection of wastes from various sources producing large quantities because they are relatively large containers (see Table 3-1). By using large containers we can eliminate manpower and the environmentally objectionable conditions which usually exist when using numerous smaller containers. Hauled-container systems are flexible containers, come in many different sizes and shapes, and are available for the collection of all types of wastes. Only one truck is required to make a collection cycle. However, a round trip to the site of disposal must be made so that each container can be picked up.

Obviously, then, container size and utilization are of great economic importance. Furthermore, when highly compressible wastes are to be collected and hauled over considerable distances, the economic advantages of compaction are obvious. Tchobanoglous (1977) describes three types of HCS systems: (1) hoist truck, (2) tilt-frame container, and (3) trash-trailer.

## HOIST TRUCK SYSTEMS

These systems collect wastes from small operations with only a few pickup where significant amounts of wastes are stored. These are also used to collect bulky items such as heavy metal and lawn trimmings as well as industrial rubbish which does not lend itself to collection with vehicles with compaction capabilities. Such a system is shown in Tchobanoglous (1977, p. 116).

## TILT-FRAME SYSTEMS

"Drop boxes" are ideally suited for the collection of all types of solid waste and rubbish from locations where the very high generation rate calls for the use of larger storage bins. These large containers used in conjunction with stationary compactors are common at multistoried apartment buildings, shopping centers, small commer-

# INDUSTRIAL SOLID WASTES

cial industries, and transfer stations. The tilt-frame hauled container system (Tchobanoglous (1977, p. 117)) has become popular for private collectors servicing commercial accounts. These customers generally produce large volumes of solid wastes.

## TRASH-TRAILER SYSTEMS

Trash-trailers are used for the collection of excessively heavy rubbish such as sand, timber, and metal scrap. They are similar to the tilt-frame trucks and often are used for the collection of demolition wastes at construction sites.

To aid in proper storage and collection, some urban agencies suggest practices for their customers to optimize the collection services. One example is given in a letter from Dade County, Florida, to its customers (see Figure 3-1).

Containers vary from relatively small sizes (1 yd$^3$ to 50 yd$^3$) to sizes such as those handled with a hoist truck. Because truck bodies are difficult to maintain and because of the weight involved, these systems are not ideally suited for the collection of heavy industrial wastes such as that produced at construction and demolition sites. When high volumes of rubbish are produced, they are difficult to service because of the space requirements for the large number of containers. The major application of manual transfer and loading methods is in the collection of residential wastes and litter. Manual loading is more suitable than mechanical loading in residential collection because most individual pickup points are inaccessible to the collection vehicle. Transfer operations are those in which the wastes, containers, or collection trucks holding the wastes are transferred to a transfer or haul vehicle for economic considerations. These transfer operations may prove economical when (1) relatively small, manually loaded collection vehicles are used for the collection of residential wastes and requiring long-haul distances for disposal; (2) extremely large quantities of wastes must be hauled over long distances, and (3) a single transfer station can serve a number of collection vehicles. The unit time required to perform each task must be determined to ascertain vehicle and labor requirements for the various collection systems and methods. The steps involved in the collection of solid

Figure 3-1. Letter to Customers.

# METROPOLITAN DADE COUNTY · FLORIDA

ADMINISTRATIVE DIVISION  
140 WEST FLAGLER STREET  
MIAMI, FLORIDA 33130  
PHONE: 579—5176

**DEPARTMENT OF SOLID WASTE COLLECTION**

Dear Householder:

You can help us keep your community clean, eliminate physical environmental pollution, enhance good air and water quality, by utilizing the following

<u>HELPFUL HINTS</u>

<u>GARBAGE</u>

* Please have your garbage cans out the night before your scheduled pickup days, or by 7:00 A.M.

* Please make sure that the weight of each can or bag does not exceed fifty (50) lbs. Heavy cans caused 279 back injuries this year alone. If your cans are overloaded, please use plastic bags for the remainder of your garbage.

* A spoonful of ammonia or vinegar placed in the garbage can or bag will be helpful in detracting dogs and cats.

* Small quantities of trash, containerized or tied in a bundle, will be picked up along with the garbage.

<u>TRASH</u>

* Please place only old furniture, appliances and garden trash in the neighborhood trash transfer stations near your home. Some have dumped garbage in these stations, attracting rats and causing a health problem in the area. Remember these stations are there for your convenience and to improve the appearance of your homesite (i.e., no big trash piles in front of your home). Please use them properly.

* A major uplifting of all transfer stations is taking place, as well as an increase in security by the Public Safety Department.

Thanks for your cooperation in this effort.

Sincerely,

A. D. Moore  
Acting Director

wastes are classed into four separate operations—pickup, haul, disposal, and associated by Tchobanoglous (1977, p. 122).

## COLLECTION OPERATIONS

### Pickup

Pickup is the time spent driving to the next container after an empty container has been deposited, the time spent picking up the loaded container, and the time required to redeposit the container after its contents have been emptied. This may also be the time spent loading the first container into the collection vehicle until emptying the final container into the truck.

### Haul

This is the time required to reach the disposal site, plus the time after leaving the disposal site until the truck arrives at the location where the empty container is to be redeposited or at the location of the first container to be emptied on the next collection route. It does not include any time spent at the disposal site.

### Disposal

This is the time spent at the disposal site and includes the time spent waiting to unload as well as the time spent unloading.

### Associated

This includes all time spent on activities that are nonproductive from the point of view of the overall collection operation. Therefore, the time spent on offf-route activities may be subdivided into necessary and miscellaneous. Necessary off-route time is time spent checking in and out in the morning and at the end of the day, time spent driving to the first pickup point or from the approximate location of the last pickup point because of unavoidable congestion, and time spent on equipment, maintenance, repairs, and so on. Miscellaneous off-route time includes time spent for food breaks in excess of the stated lunch period, time spent on taking unauthorized coffee breaks, and general time taken to engage in "small talk" with any persons along the way.

## COLLECTION ROUTES

Routes for collection must be established so that both work force and equipment are used effectively. In general, collection routes are determined empirically. Firm rules for collection routes do not exist. Some general policies for municipal pickups that require investigation include: (1) existing regulations for refuse such as the point of collection and frequency of collection; (2) coordination of the crew size and vehicle types; (3) formulation of routes to begin and end near arterial streets, using topographical and physical barriers as route boundries; (4) formulation of routes to start at the top of the grade and proceed downhill as the vehicle becomes loaded; (5) collection of the last container on the route to be located nearest the disposal site; (6) collection of wastes generated at traffic-congested locations to be timed before business hours (which varies from one area to another but usually is early in the morning); (7) service of extremely large quantity wastes to be done at the beginning of the collection route; (8) collection of extremely small quantities of solid wastes at widely separated locations to be planned for one trip of the vehicle (See Tchobanoglous (1977, p. 140).

Use and maintenance of transfer stations are not without their problems. In Dade County, Florida, for example, illegal dumping has been a serious problem. Convenience for commercial establishments and a lower cost of disposal have been the major reasons for the illegal use of this and other transfer stations. Figure 3-2 gives a recent and more complete discussion of this problem. Typical fees for urban collection and disposal are also shown for Dade County, Florida, in Figure 3-3.

## REFERENCES

Tchobanoglous, G. 1977. *Solid Wastes, A Text.* New York: McGraw-Hill.
Tchobanoglous, George, and G. Klein. 1962 *An Engineering Evaluation of Refuse Collection Systems Applicable to the Shore Establishment of the U.S. Navy.* Sanitary Engineering Research Laboratory, University of California, Berkeley.

Figure 3-2. Article on Illegal Use of Transfer Stations.

## Illegal dumping contributes to higher rates from Metro
### Businesses blamed for illegal dumping

*By KENNETH CAMPBELL* — Herald Staff Writer

Illegal trash dumping has become an uncontrolled after-dark business in unincorporated Dade County, according to sanitation workers.

About 40 per cent of the trash being dumped at the 26 neighborhood trash transfer stations throughout Dade County does not belong there, according to A. D. Moore, acting director of Metro Solid Waste Collection.

Small private businesses and trash collectors are the main culprits, sanitation workers say. The transfer stations were established for residents in unincorporated areas to carry garden debris and large household trash such as broken furniture.

Metro and commercial trash collectors are supposed to haul their trash to one of five county facilities—three landfills, a countywide transfer station and an incinerator.

Residents and businesses of incorporated communities are not allowed to use any of the county facilities.

The excess trash is partly responsible for a $5 fee increase per household, from $47 to $52 over the next six months, imposed on residents in unincorporated Dade.

"Basically, the only reason transfer station [rates] increased recently is the cost of disposal has increased so much in the last couple of years," said Ron Tepper, senior administrative officer in Metro Solid Waste Collection.

"The commercial establishments have found an advantage in dumping at the trash stations"—saving money.

At Metro's five disposal sites, all users pay the same rates: $4.50 per ton at the three landfills, $8.50 per ton at the countywide transfer station and $8 per ton at the incinerator.

The rates will go up in January to $16.50 per ton at the transfer station and $12 per ton at the landfills and incinerator.

One small trash collector, Kenneth Melvin of Fortune Contractors, 10211 SW 102 Terr., said money is not the only reason small collectors have been avoiding the landfills.

He said their trucks are often damaged and tires punctured there.

They like to dump in the neighborhood stations, along the roadsides and in vacant lots, Melvin said, because the disposal sites are far from where they collect trash.

He said the neighborhood trash transfer stations are usually closer and have shorter waiting lines.

The only disposal site in Southwest Dade, a landfill, is at SW 248th Street and 97th Avenue.

The illegal dumping has been a serious problem since the neighborhood transfer stations opened in 1972, Tepper said. When the security guards at the sites go off duty, it's open house. The dumps are not fenced in.

Figure 3-2. Article on Illegal Use of Transfer Stations (*continued*).

Manola Tilman, the guard at the transfer station at SW 80th Street and 107th Avenue, said she often finds appliances and other debris when she reports to work in the mornings.

"Chickens, chicken feathers, rotten eggs, food—that's garbage. It's not supposed to be dumped here," she said. "But once I'm gone, there is nobody to stop them."

She said regular users of the station often tell her that cars have dumped illegally, but fail to get tag numbers.

Felix Jones, who hauled trash from the stations to the landfills for years, agreed that most of the problem is created by private businesses and small dumpers.

"It's not the neighbors; it's the commercial establishments," he said at the neighborhood station at SW 101st Street and 97th Avenue.

He said the items businesses most often dump at neighborhood stations are roofing materials, tiles, rocks, bricks, air conditioners and large kitchen appliances.

The problem could be solved with stiffer penalties and power to issue on-the-spot citations to violators, according to Tepper.

The county manager's office is working on an ordinance that would delegate such authority, according to Dennis Carter, assistant county manager. It might be ready by the end of the year, he said.

Source: This article by Kenneth Campbell is retyped verbatim from the *Miami Herald*, July 13, 1980.

**Figure 3-3.** Letter Regarding Typical Fees of Urban Collection and Disposal.

# METROPOLITAN DADE COUNTY · FLORIDA

ADMINISTRATIVE DIVISION  
140 WEST FLAGLER STREET  
MIAMI, FLORIDA 33130  
PHONE: 579—5176

DEPARTMENT OF SOLID WASTE COLLECTION

January 18, 1980

Dear Homeowner:

In order to maintain the present level of service of solid waste collection and meet new Federal and State mandated rules and regulations on solid waste disposal, it has become necessary to increase the semi-annual waste fees you now pay.

There has not been an increase in waste fees since July 1, 1977. During this time we have all felt the effects of the high rate of inflation. This has caused an increase in the cost of equipment varying from 25% to 114% and our cost of fuel, like yours, has increased by more than 68% over the last two years.

In addition to the increased cost of operations, the new rules and regulations on solid waste disposal mandated by both the Federal and State governments have forced Dade County and other communities throughout the Nation to adopt new methods of waste disposal, such as recycling, which are environmentally acceptable. These new methods are far more costly than those methods used in the past, but will provide for a cleaner environment.

Last fiscal year it cost $17,162,000 to collect and dispose of the solid waste generated. It will cost $19,279,000 to do the same in this fiscal year, and for the next fiscal year, we project the cost to be $25,175,000.

The new semi-annual fees for twice-weekly containerized waste service and two (2) bulky waste collections per year of up to 25 cubic yards per collection, are as follows:

| EFFECTIVE DATE | AMOUNT | EFFECTIVE DATE | AMOUNT |
|---|---|---|---|
| February 1, 1980 | $48.00* | January 1, 1981 | $57.00 |
| July 1, 1980 | $52.00 | July 1, 1981 | $62.00 |

*Note: For the billing period January 1, 1980 through June 30, 1980, the fee will be Forty-Seven Dollars ($47.00) which is one month at the January 1, 1980 rate and five months at the February 1, 1980 rate.

Charges for bulky waste in excess of 25 cubic yards will be $5.00 per cubic yard effective February 1, 1980 and $6.25 per cubic yard effective January 1, 1981.

Payment for waste fees will become delinquent if not paid in full within sixty (60) days of the due date. Please feel free to contact this office (579-5176) if you need additional assistance or information.

Sincerely,

A. D. Moore  
Acting Director

# 4 RESOURCE RECOVERY

## INTRODUCTION

Systems involving recovery of resources from solid wastes depend primarily upon economics and local situations such as characteristics of refuse and land available for landfilling. Most systems utilize sanitary landfilling or incineration as tried-and-true ultimate disposal methods of the residual solids after recovery. At least eight potential overall systems are used: (1) incineration with recovery of materials from the ash residue; (2) incineration with heat recovery to produce industrial steam; (3) incineration with heat recovery and recovery of materials from ash residue; (4) incineration with heat recovery used to generate electric power for industry or within an existing municipal or industrial utility; (5) pyrolysis, with recovery of oil, char, and inorganic materials; (6) composting that produces humus and inorganic materials from the nonbiodegradable solids; (7) material recovery (paper, aluminum, ferrous metals, and glass) involving separation of mixed refuse into its marketable parts; (8) recovery of organics for use in public utility boiler furnaces as supplemental fuel and ferrous metal recovery.

Average composition of municipal refuse in the United States can be assumed to approximate the figures in Table 4-1.

The composition shown in Table 4-1 will vary regionally and seasonally and will change with the economic conditions and scientific

Table 4-1. Average Municipal Solid-Waste Composition.

| Waste Component | Percent by Weight |
|---|---|
| Paper | 33.0 |
| Glass | 8.0 |
| Ferrous metals | 7.6 |
| Nonferrous metals | 0.6 |
| Total metals | 8.2 |
| Plastics, leather, rubber, textiles, wool | 6.4 |
| Garbage and yard wastes | 15.6 |
| Miscellaneous (ash, dirt, and so on) | 1.8 |
| Total dry weight | 73.0 |
| Moisture | 27.0 |

and industrial progress. For example, the advent of lightweight aluminum cans has increased that component; the increase in home garbage grinders has decreased that portion; there is some indication that garbage grinders for new homes are decreasing because of extra cost for sewage treatment. Hence garbage, once again may increase in the near future. The pressure by conservationists for using returnable bottles has decreased the glass portion.

Some discussion about the recoverable refuse materials before disposal techniques is in order at this point in our text.

## PAPER AND PAPER PRODUCTS

It has been approximated that about 50 million tons of paper and paper products have been used by Americans annually. About 80 percent received one-time use and were then discarded. Recently, (late 1970s), paper products constitute 40 to 50 percent of all solid wastes (even higher than shown in Table 4-1). Therefore, of the 40 million tons of paper waste, 25 percent or 10 million tons were recycled. If recycled paper prices were more competitive, about 80 percent of all our paper raw materials could be met with waste paper and hence could relieve the pressure on our forest resources. Paper products are among the easiest to recycle. Separation systems are available, but their use must be encouraged by higher prices for re-used paper. Their re-use is not without production problems, however; the uniformity of composition and quality as well as dependability of quantity are a couple of them.

## RECOVERY OF PRODUCTS AFTER TREATMENT

### Pyrolysis for Reuse

Table 4-2 shows what *pyrolysis* of a ton of typical refuse has been shown to yield.

Despite all these potential recoveries, recycling and reclamation cannot be expected to handle a major portion of municipal solid wastes for at least a few more years due to both processing problems and lack of markets for the output. A major impetus can be given to reuse if and when a monetary value is given to the environmental consequences of nonrecovery.

### Composting for Re-use

Composting is an admirable solution for solid-waste treatment because it converts municipal refuse into a re-usable product, a useful soil conditioner. Sorting and salvage precedes composting. Sorting removes most inorganics, noncombustibles, bulky material, and salvagable items. Generally, handsorting is used to salvage re-usable

Table 4-2. Yield from Pyrolysis of a Ton of Typical Refuse.

| Urban Refuse | Industrial Refuse of Paper, Rags, and Cardboard |
|---|---|
| 154-425 pounds of solid residue | 613-838 pounds residue |
| 0.5-6.0 gallons of tar | |
| 1-4 gallons of light oils (source of benzene and toluene) | 1.5-3 gal. of light oil |
| 97-133 gallons of liquor | 68-75 gals. of liquor |
| 16-32 pounds of $(NH_4)_2SO_4$ | 12-23 lbs. $(NH_4)_2SO_4$ |
| 7,380-18,058 cubic feet of gas | 9,270-14,065 cu. ft. gas |

Note: The energy from the gas is more than enough to provide heat for pyrolysis.

products, but sometimes inertial or magnetic separation is used. The private market for composting is small and not fully developed. Compost from municipal refuse is not ideally suited by itself as a fertilizer since it is relatively low in nitrogen, phosphorous, and potash to be used for heavy agriculture. Home gardening, landscaping, and park utilization are major re-uses for compost. Recycling both before and after composting may not be feasible now (1979), but with the development of better markets and the increase in value of both the products and the value of a clean environment, it is only a question of time before it becomes practical.

Some typical and recent (1979) market prices are given in Table 4-3 from Tampa, Florida, *Tribune*. In Table 4-4 various commodities such as food, grains, fats, textiles, metals, rubber, hides, fuel oil, and so on are given for September 18, 1979. These prices change

Table 4-3. Market Metal Prices.

---

*Tampa Tribune, Tuesday, September 18, 1979*

*Metal Prices*

---

New York (UPI)—Latest metal market prices as quoted Monday by the American Metal Market, authoritative metals publication:

*Aluminum, primary, 99.5 percent plus pure 50 lb. ingots 58.50-63.00c/lb.*
Antimony, domestic, refined in alloy, 200.00-200.02c/lb.
*Copper, electrolytic, delivered U.S. 92.625-93.625c/lb.*
*Lead, common, U.S. primary producers 58.00c/lb.; U.S. nonprimary (secondary) producers 58.00c/lb.*
*Magnesium, 99.8 percent, ingot 105.50c/lb.*
Manganese, 99.9 percent boxed regular 58.00c/lb.
Mercury, $310-320 76 lb. flask.
*Nickel, electrolytic cathodes, f.o.b. Port Carborne, Ont., $3.00-3.05 lb.*
Palladiaum, N.Y. Am. Met. Mkt. dealer $131.50-133.60 per troy ounce.
Platinum, soft, 99.5 fine, producer $380; dealer-approx., $422-425 per troy ounce.
*Steel, No. 1 heavy scrap, Pittsburgh $90 per ton (consumer buying price); Am. Met. Mkt. composite Scrap Price $87.67 per ton.*
*Tin, N.Y. Am. Met. Mkt. ex-dock price $701.50c/lb.*
*Tin, N.Y. Am. Met. Mkt. alloyer price $750.75c/lb.*
Tungsten powder (H-Red), 98.8 percent minimum pure $13.90 per lb.
*Zinc, prime western, U.S. 35.50-37.00c/lb.*

hourly for each trading day. In Table 4-4 for example, aluminum ingots sell for sixty and one-half to sixty-three cents, whereas one year before they sold for fifty-five cents. Aluminum prices quoted for September 17, 1979, were fifty-eight and a half to sixty-three cents per pound for 99 percent plus pure fifty-pound ingots. This provides the reader with two different ways for quoting prices of the same commodity on similar days. It is interesting to see from Table 4-4 the change which has occurred recently in the price for newspaper (a recyclable commodity) from $35 to $50 per ton from September 1978 to July 1980 and April 1981.

Aluminum also increased to seventy-two cents per pound in July 1980. Steel scrap, on the other hand, decreased from $103 to $72 per ton. Fluctuations in market values for recoverable solid wastes make the economics of recovery and re-use difficult to predict. But, in the long run, it represents a challenge similar to that encountered in any industrial production venture.

When considering the economics of any recovery of material from refuse, the specifications of the recovered material will be of utmost importance. If the recovered material does not meet product specifications with a minimum of renovation, the re-use process may not be economically feasible.

### FERROUS METALS

Ferrous metals mainly come from tin cans in the refuse. The principal market for tin cans is the copper-smelting industry (where tin contamination is not a hindrance). However, recently the costs associated with separating, recovering, and processing the scrap have in most cases surpassed the value of the scrap. Once again economics prevail, and the cost of environmental degradation by tin cans is not included in the price of recovery and reuse.

### ALUMINUM

The use of aluminum cans has increased and so have the value of recovered and re-used cans. The production of raw aluminum from bauxite ore is expensive and uses large amounts of energy as shown later in Table 4-5. A method of separating aluminum from other

Table 4-4. Market Prices for Various Commodities Commonly Found in Urban Wastes (*Excerpts from The Wall Street Journal*).

|  | Cash Prices | | |
| --- | --- | --- | --- |
|  | Tuesday September 18, 1979* | | |
|  | Tues. | Mon. | Yr. Ago |
| *Fats and Oils* | | | |
| Grease | .21$^1$/$_4$ | .21$^1$/$_4$ | .20 |
| *Metals* | | | |
| Aluminum | p.60$^1$/$_2$ | .60$^1$/$_2$ − .63 | .55 |
| Copper Scrap | k.68$^1$/$_2$ | .70 | .53 |
| Nickel | p3.05 | 3.05 | z |
| Steel Scrap | 90.00 | 90.00 | 79.00 |
| Tin composite price lb | s7.5071 | 7.4764 | 6.8715 |
| Zinc Prime Western lb | p.35$^1$/$_2$ − .36 | .35$^1$/$_2$ − .36 | .32 |
| *Miscellaneous* | | | |
| Hides | .74 | .73 | .61 |
| Newspapers | z | z | 35.00 |
| Rubber | .66 | .65 | .55$^1$/$_2$ |

*Quotations as of 4 p.m. Eastern time.    k- Represents dealer selling price in lots of 40,000 pounds or more, p-Producer price, s-Source: Metals Week, z-Not quoted.

nonferrous metals must be developed. Collection locations have been established for the return of aluminum cans, with industry (Reynolds Aluminum Co.) assisting. In Miami, Florida, Reynolds is offering $.27 per pound for aluminum discarded cans (see *Miami Herald*, Neighbors Section, January 22, 1981).

## OTHER NONFERROUS METALS AND ASH

Lead, copper, zinc, and tin all have high salvage values. However, they are generally wasted from industries and municipalities in small quantities and, as impure alloys, they make recovery difficult.

Table 4-4. continued

| | Cash Prices | | | | | |
|---|---|---|---|---|---|---|
| | Tuesday, July 8, 1980* | | | Thursday, April 16, 1981* | | |
| Tues. | Mon. | Yr. Ago | Thurs. | Wed. | Yr. Ago |
|---|---|---|---|---|---|
| b.13 $3/4$ –.14 | .14 | .22 $1/2$ | .18 $1/2$ – $5/8$ | .18 $1/2$ – $5/8$ | .17 $1/4$ |
| p.72 | .72 | .60 $3/4$ | p.76 | .76 | .68–.72 |
| k.73 | .73 | .66 | k.67 | .66 $1/2$ | .68 |
| p3.52 | 3.52 | 3.05 | p3.50 | 3.50 | 3.50 |
| 72.00 | 72.00 | 103.00 | 110.00 | 110.00 | 94.00 |
| 8.5336 | 8.3660 | 7.6426 | 6.7244 | 6.7905 | 8.8720 |
| p.35 $1/2$ | .35 $1/2$ | .40 $1/2$ | p.43 $3/4$ | .43 $3/4$ | .37 $1/2$ –.39 |
| .45 | .45 | .76 | n.55 | .55 | .50 |
| 50.00 | 50.00 | z | 40.00 | 45.00 | 65.00–70.00 |
| n.68 $3/8$ | .68 $5/8$ | .66 $3/4$ | n.59 $3/8$ | .59 $5/8$ | .72 $3/4$ |

b-Bid, k-Dealer selling price in lots of 40,000 pounds or more, n-Nominal, p-Producer price, z-Not quoted.

n-Nominal, p-Producer price.

Table 4-5. Energy Saved by Recycling.

| Original Production | |
|---|---|
| Energy Required (Kwh/ton) | Recycling |
| Steel 4.3 | 1.7 |
| Copper 13.5 | 1.7 |
| Aluminum 51.4 | 2.0 |
| Paper 5.0 | 1.5 |

One method being researched and piloted for separating these valuable metals from ferrous metals and nonmetallic solids is that of cryogenics. Low temperatures make some material brittle while other material remains malleable. Crushing the frozen material to shatter the embrittled matter and subsequently subjecting it to magnets, screening, or gravity separation has been used to recover specific components. Dry ice (solid $CO_2$) can cause low enough temperatures to effect acceptable separation of insulated wire and mixed nonferrous refuse.

Fly ash and bottom ash from power plants are beginning to be re-used. For example, recovered fly ash from stack gases can be used as a cement additive. Bottom ash from power plants can be used as an additive in the manufacture of concrete blocks. See Chapter 15, page 287 for some examples.

## GLASS

Glass has a low salvage value, and industrial processors have been resistant to re-using it unless it is one color and grade and is free from impurities such as metals. Economic limitations (low glass value) seems to control its recovery potential. Sand is not very expensive as a raw material for glass manufacture, but fuel and energy to fuse it into the product may be vital and costly enough to justify recovery of glass.

One use of glass from incinerator residue is for the manufacture of glass wool. This is used for house attic and ceiling insulation and blanket insulation in building walls. It also can be used as a flux for common clays. When added to clays before curing it, it can lower the maturing temperature significantly to save considerable kiln fuel— usually natural gas.

Glass has also been used successfully as an aggregate additive for highways. It has been reported to increase the life and structural characteristics of the roads. It also sparkles, thus aiding the driver's vision.

## RUBBER

Rubber, usually in the form of old tires, is easily separated from refuse. However, because of the great variety and grade of tires no large-scale recycling is taking place. The increased use of a variety of steel-belted radial tires has made the recovery of rubber all the more difficult. They do burn with very high heat, however, and after shredding can be used as fuel for boilers to recover heat or power.

## PLASTICS

Plastics are comprising a greater percentage of refuse because of their utility in modern, affluent societies. Most are nonbiodegradable, and hence landfilling and composting of them are impractical. Recoveries of almost all plastic can be accomplished by electrodynamic separation techniques. Plastics when incinerated may yield HCl and hence contaminate the air surrounding such an operation.

## AUTOMOBILES

The salvage of old automobiles is now profitable. The major problem with recovering the valuable components from autos (steel, iron, rubber, and so on) is the efficient separation of the various components. Crushers have been designed and constructed for recovery of scrap metal in dense form. Magnetic separators aid in separating ferrous from nonferrous metals. Higher prices of scrap iron for the steel industry have made salvage economically feasible. Re-use potential depends upon the health of the steel industry. However, recovering of other components of autos has not yet been practical or economical.

Recovery is practiced by both auto dismantlers and scrap processors. The former is a retailer who buys vehicles to obtain usable parts for resale. The dismantler must be concerned with the year, make, model, and condition of the autos purchased. After the saleable parts have been removed, the remainder is sold by weight to the scrap processor who is purchasing unprepared scrap for processing and final recycling.

A new $2.5 million plant has been designed by using a 2,400 horsepower hammermill constructed to grind up 6000 cars a day into ninety-pound chunks of steel and smaller quantities of aluminum, brass, and copper according to the *Miami Herald*, November 10, 1980, p. 6). The Scrap Metal Company pays scavengers of junk cars about $2.50 per 100 pounds. The company expects the average

3,000-pound old car minus the fuel tank to give approximately 2,250 pounds of ferrous metals, 90 pounds of nonferrous metals, and 660 pounds of nonmetallic waste. The ferrous scrap is extracted from the shredded debris by giant electromagnets, while nonferrous metals are segregated in a series of flotation tanks. Copper, aluminum, and zinc are sold to metal brokers. The company also expects to install equipment designed to reclaim platinum from catalytic converters.

Table 4-6 gives some typical material specifications for those items usually recovered.

Many consultants and urban agencies are concluding that incineration and composting are losing favor as treatment methods because of technical, economic, and environmental reasons. More specifically these are (1) the increase in capital and operating costs of these treatments, at least partially due to environmental controls being demanded; and (2) the decrease in potential value of constituents present in solid wastes, such as BTU value for incineration and organic matter for composting. The reason for this may be because of our increased interest in conservation of resources such as paper and plastics. Instead our objectives are turning to: (1) developing new technology aimed at recovering raw materials or the energy-producing portion of urban solid wastes; (2) reducing the quantity of contaminants causing impacts in both water and air from other methods of disposal; and (3) obtaining methods for using the solid waste constituents for resale, including energy production.

Recycling signifies the return to the consumption cycle of materials (partially or completely renovated and finished) recovered from solid wastes. This can be accomplished either by direct or indirect means. The former implies re-use of materials without altering their physical, chemical, and biological character. For example, the re-use of broken glass directly by the glass industry or paper to make pulp by this industry represent direct recycling. Indirect recycling involves transforming the recovered material for re-use for a different purpose than originally used: for example, re-using glass after crushing as a filling material for construction items such as roads or building blocks or re-using organic matter as fertilizer after composting. Recycling materials is also known to save energy as shown in Table 4-5. In reusing aluminum, a major energy consumer, about 97 percent energy is saved during production.

Table 4-6. Typical Materials Specifications that Affect Selection and Design of Processing Operations.

| Re-use Category and Materials Component | Typical Specification Items |
|---|---|
| *Raw Material* | |
| Paper and cardboard | Source; grade, no magazines; no adhesives; quantity, storage, and delivery point |
| Rubber | Recapping standards; specifications for other uses not well defined |
| Plastics | Type (for example, ABS, PVC); degree of cleanliness |
| Textiles | Type of material; degree of cleanliness |
| Glass | Amount of cullet material; color, no labels or metal; degree of cleanliness; freedom from metallic contamination; quantity, storage, and delivery point |
| Ferrous metals | Source (domestic, industrial, etc.); density; degree of cleanliness; degree of contamination with tin, aluminum, and lead; quantity, shipment means, and delivery point |
| Aluminum | Particle size; degree of cleanliness; density; quantity, shipment means, and delivery point |
| Nonferrous metals | Vary with local needs and markets |
| *Fuel Source* | |
| Combustible organics | Composition, Btu content; moisture content; storage limits; firm quantities; sale and distribution of energy or byproducts |
| Wastepaper | Vary with local needs and markets |
| *Land Reclamation* | |
| Organics | Local and state regulations; method of application; control of methane gas migration; leachate control; final land-use designation |
| Inorganics | Local and state regulations; final land-use designation |

Table 4-7. Estimated Recoverable Quantities for Various Components in Solid Wastes using Mechanical Equipment.

| Fraction or Component | Recoverable Portion of Original Components, Percent | | Comments |
|---|---|---|---|
| | Range | Typical | |
| Light fraction | 80-95 | 90[a] | Recoverable portion will vary with the composition of the solid wastes and the characteristics of the wastes after shredding. |
| Heavy fraction | 90-98 | 96[b] | |
| Ferrous metal | 65-95 | 85 | Varying amounts of light and heavy fraction material will also be removed with these components depending on the specific process and equipment used. |
| Glass | 50-90 | 80 | |
| Aluminum | 55-90 | 70 | |

a. Varying amounts of the light fraction will be retained with the heavy fraction.
b. Varying amounts of the heavy fraction will be carried over with the light fraction.

## PROCESS FLOWSHEETS

Both in Figures 4-1 and 4-2 a dry-process flowsheet has been adopted. Dry processing is less costly than wet processing since a hydropulper is used in the wet process. Standard equipment used in the mineral-processing industries can be adapted for use in the dry process. In both flowsheets air classification follows primary shredding, and cyclone separators remove the contaminated air from the light fraction. From a review of Figures 4-1 and 4-2 it is evident that many variable flowsheets can be prepared. Manual separations of specific waste components are also commonly used. Important factors that must be considered in the design and layout of such systems include: (1) process performance efficiency; (2) reliability and flexibility; (3) ease and economy of operation; (4) aesthetics; and (5) environmental controls. The degree of separation achieved for the various components determines the efficiency of the system.

RESOURCE RECOVERY 35

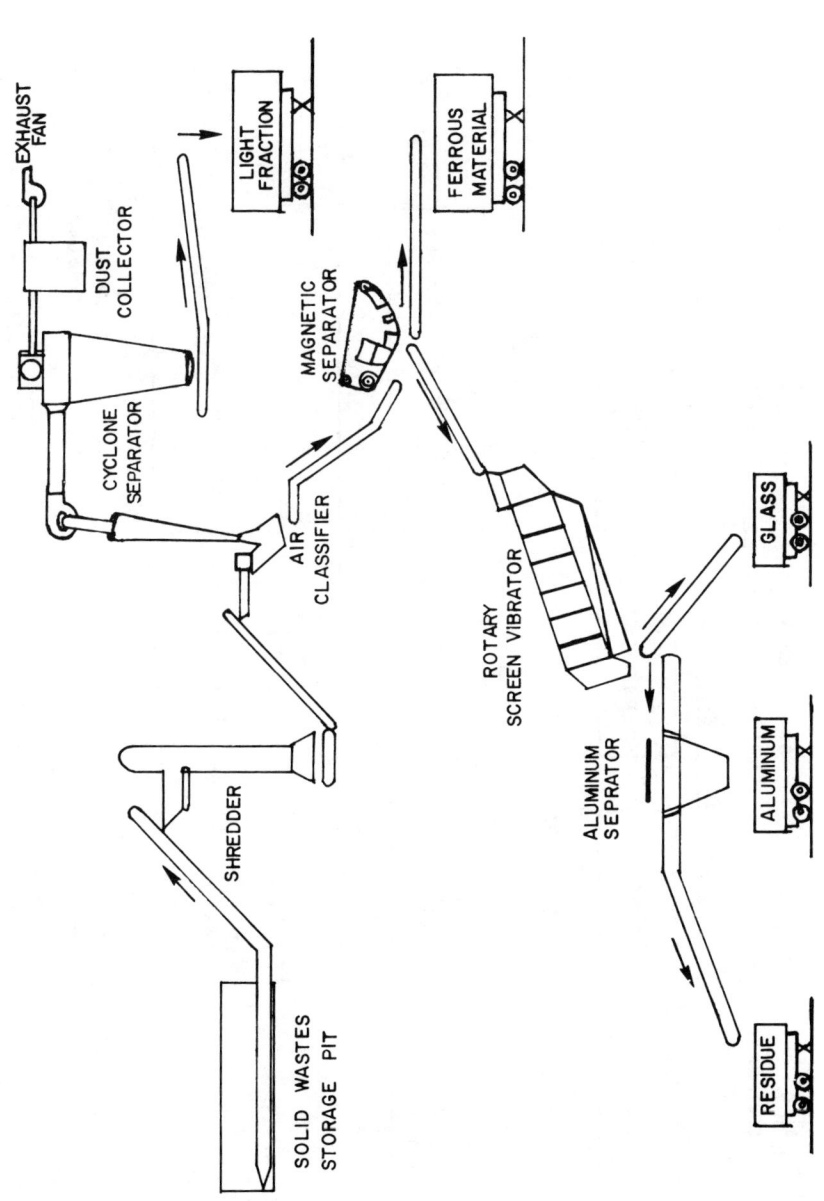

Figure 4-1. Pictorial Flowsheet for Materials Recovery Systems.

36  INDUSTRIAL SOLID WASTES

Figure 4-2. Flowsheet for Materials Recovery System.

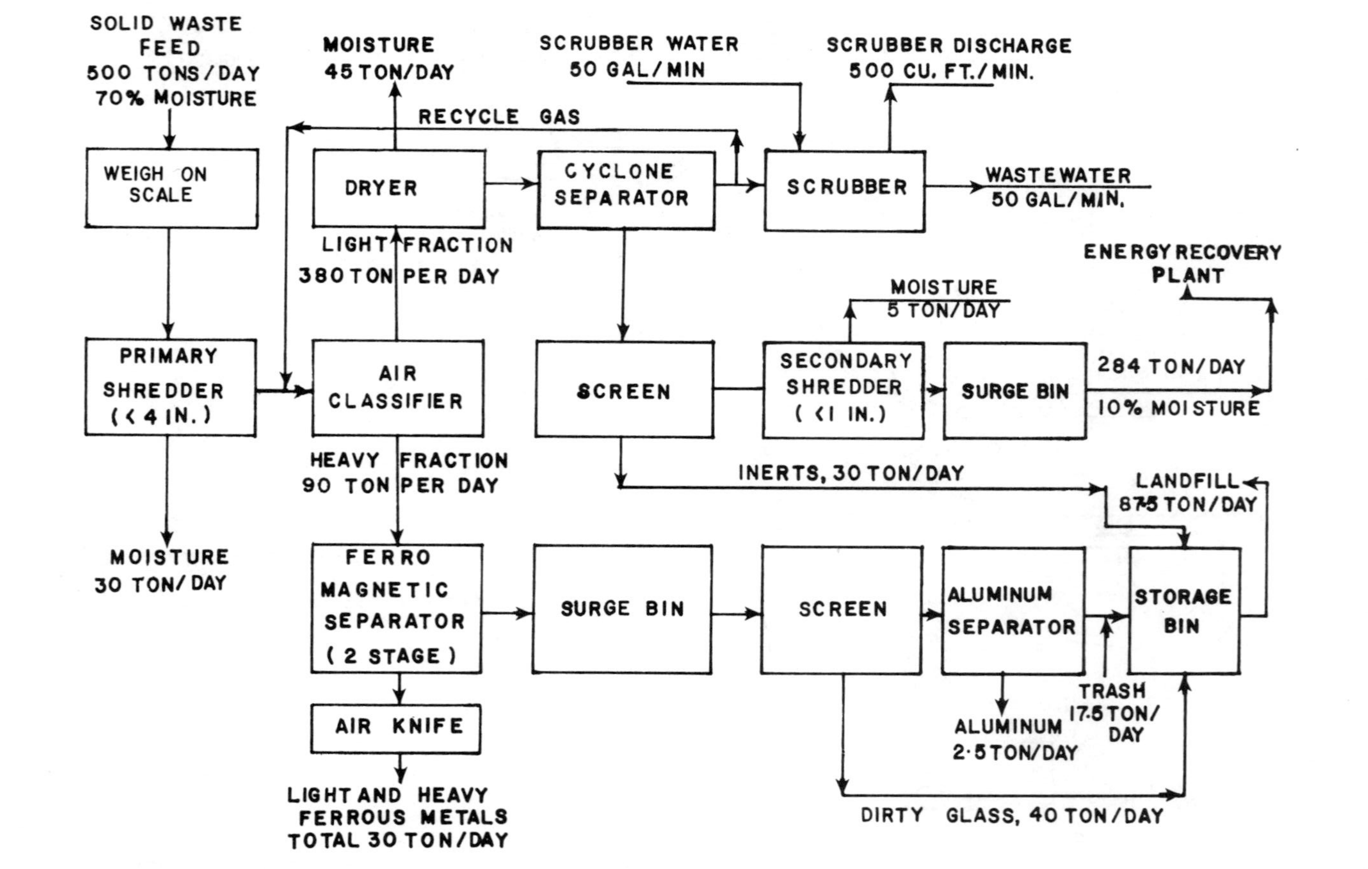

Source: Central Contra Costa Sanitary District and Brown and Caldwell Consulting Engineers.

The quantity of material to be handled needs to be determined as a first step in the design of a processing facility. In situations when processed wastes are to serve as a source of fuel, the design quantities will usually depend on the amount of continuous power that must be developed. Units are sized according to the loading rates which are determined on the basis of the characteristics of the solid wastes and the separation process to be used. Estimation of the quantities of materials that can be recovered and of the appropriate design loading rates represents an important part of any recovery system. Data and information that can be used to estimate the required quantities are presented in Table 4-7. The components that normally make up the light and heavy fractions after shredding and air classification are identified in Table 4-8.

Belt and chain belt conveyors of unprocessed solid wastes have proved especially troublesome. Conveyors can be easily and frequently damaged by heavy, bulky and cumbersome solid wastes dropped onto them. Wire and rope in the wastes become snagged on the equipment, causing waste spillage and overflows. This is comparable to a clogged or obstructed sewer which backs up on streets and house basements. Binding and wedging of conveyor systems have also been a problem. Because of the abrasive nature of many of the components such as metal parts—some broken and jagged—found in solid wastes, the processing equipment wears out rapidly. Duplicate units are now recommended to be operated independently, especially where continuous power is produced from the heat generated. Equipment should be selected for easy repair with standard parts and components. Sprockets, gears, pins, belts are especially troublesome. Oils, gases, organic compounds, and heat can be produced by chemical conversion of solid wastes. Incineration and pyrolytic processes are the few full-scale installations which have been producing these by-products. With pyrolysis, most of the full-scale experience exclusive of pilot plants is in the petroleum and wood-processing industries. Some of the chemical processes which can be used for the conversion of solid wastes are given in Table 4-9.

ENADISMA (Spain) proposes a new resource recovery process (1980) of pneumatic separation accompanied by the basic operations of grinding and magnetic and complementary mechanical separating procedures. The schematic drawing of the process is shown in Figure 4-3.

Table 4-8. Light and Heavy Fractions of Solid Waste Components After Shredding.

| Component | Percent by Weight[a] | Fraction by Weight, Percent | | Comment |
|---|---|---|---|---|
| | | Light | Heavy | |
| Food wastes | 15 | 15 | — | Components assumed to make up the light fraction after shredding. After air classification the light fraction will contain from 2 to 8 percent of the components from the heavy fraction by weight. |
| Paper | 40 | 40 | — | |
| Cardboard | 4 | 4 | — | |
| Plastics | 3 | 3 | — | |
| Textiles | 2 | 2 | — | |
| Rubber | 0.5 | 0.5 | — | |
| Leather | 0.5 | 0.5 | — | |
| Garden trimmings | 12 | 12 | — | |
| Wood | 2 | 2 | — | |
| Glass | 8 | — | 8 | Components assumed to make up the heavy fraction after shredding. After air classification the heavy fraction will contain from 5 to 20 percent of the light fraction components by weight. |
| Tin cans | 6 | — | 6 | |
| Nonferrous metals | 1 | — | 1 | |
| Ferrous metals | 2 | — | 2 | |
| Dirt, ashes, brick, and so on | 4 | — | 4 | |
| Total | 100 | 79 | 21 | |

a. Moisture loss during shredding not considered.

Table 4-9. Chemical Processes Used for Conversion of Solid Wastes.

| Process | Conversion Product | Preprocessing Required | Comment |
| --- | --- | --- | --- |
| Incineration with heat recovery | Energy in the form of steam | None | Markets for steam must be available; proved in numerous full-scale applications; air-quality regulations may prohibit use. |
| Supplementary fuel firing | Energy in the form of steam | Shredding, air separation, and magnetic separation | If least capital investment desired, existing boiler must be capable of modification; air-quality regulations may prohibit use. |
| Fluidized bed incineration | Energy in the form of steam | Shredding, air separation, and magnetic separation | Fluidized bed incinerator can also be used for industrial sludges. |
| Pyrolysis | Energy in the form of gas or oil | Shredding, magnetic separation | Technology proved only in pilot applications; even though pollution is minimized, air-quality regulations may prohibit use. |
| Hydrolysis | Glucose, furfural | Shredding, air separation | Technology on laboratory scale only. |
| Chemical conversion | Oil, gas, cellulose acetate | Shredding, air separation | Technology on laboratory scale only. |

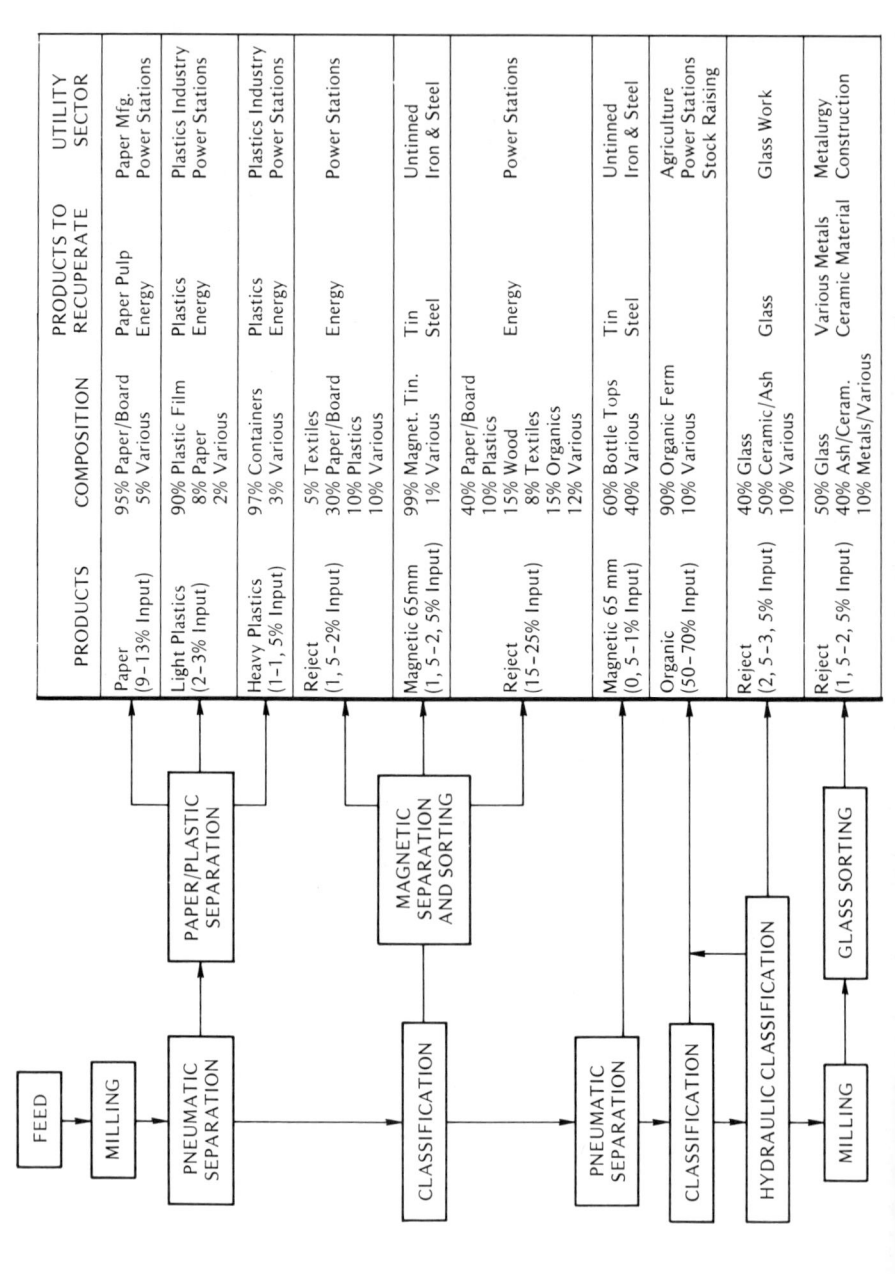

Figure 4-3. Schematic of Process for Treatment of Uniform Solid Waste (Enadimsa) Composition and Utility Alternatives.

The recovery and re-use of solid waste is not always without its own problems—and its wastes. Many times recovery techniques result in further pollution of the environment, and often by insidious discharge of hazardous wastes. Such a situation was illustrated by the Sapp Battery Plant of Alford, Florida (*Miami Herald*, July 26, 1981), which recovered spent auto batteries. The recovery plant removed the tops of as many as 50,000 old batteries per week, also removed the metal plates in each battery, and dumped out sludge bottom deposits. The heavy plates and sludge contained lead that Sapp sold to smelting companies for about $1 million per month. The guts of the batteries awash in water from spraying pits were heaped outside the plant. The water drained into a pond on Sapp's property, then found its way into a culvert that ran under a dirt road, trickled into a swamp. Gradually, the trees began to take on a steely look, turning from green to brown and finally to cold, gray stumps. Sapp's 40 acres has become a "sponge" soaked with sulphuric acid, lead, and heavy metal residues. When it rains, the contaminants seep out into the creek and flow fifty miles downstream into another river and a series of lakes. The contaminants have killed the trees and fish in the creek, and unusually high levels of lead and heavy metals have been found in the tissues of clams in the lakes. The Florida Department of Environmental Regulation assessed a $11,159,940 judgment on Sapp—$4.5 million in punitive damages and $6.4 million in actual damages to the environment. Whether the lead can ever be reclaimed and returned to an uncontaminated state is unknown. Even plants recovering and re-using contaminants must be aware of their subsequent pollution potential.

# 5 SOLID-WASTE DISPOSAL MICROBIOLOGY

## INTRODUCTION

Biological waste treatment is an old system which involves the decomposition of solid waste in soil by micro-organisms. The bulk of organic matter of these wastes disappears by conversion to stabilized soil material commonly referred to as humus. Soil micro-organisms break down long-chain organic compounds into simple compounds. These are now largely minerals and serve as nutrients for plants. As we will explore later in this text, microbiological systems used today for solid waste disposal are sanitary landfilling and composting.

In order to understand the fundamentals of the former we must study the soil environment. The fertile soil environment provides an ideal habitat containing the most varied media of micro-organisms. The major groups of micro-organisms and numbers are given by Burges (1958) in Table 5-1.

Soil micro-organisms are microscopic and include bacteria, protozoa, rickettsia, viruses, algae, and fungi. Bacteria are single-celled vegetative organisms; protozoa are single-celled animals; rickettsia are similar to both bacteria and protozoa but are classified under a single genus; viruses are ultramicroscopic, and generally considered to be proteins capable of multiplication; algae are microscopic plants containing chlorophyll; and fungi are microscopic plants devoid of chlorophyll. No single analytical method will yield the absolute total

Table 5-1. Soil Population (*Number of Organisms per Gram in a Fertile Agricultural Soil*).

| Organism | Number |
|---|---|
| Bacteria | 2,500,000,000 |
| Actinomycetes | 700,000 |
| Fungi | 400,000 |
| Algae | 50,000 |
| Protozoa | 30,000 |

Source: Burges (1958).

microbial population because of the variety of micro-organisms living in the soil. The microbial flora existing in a specific soil will depend mainly on the chemical and physical character of the earth. The major constituents which make up all soil systems may be broadly categorized in terms of solids, liquids, and gases. The solids in soil contains a mixture of large numbers of micro-organisms, animals such as rodents, worms, insects, and so on, and the root systems of higher plants. All the remains of these plants and animals along with refuse deposited in the soil are converted to organic substances, better known as humus, as a final decomposition product. The mineral matter of soil varies from fine particles to large rocks and chemically represents the composition of the original stone.

Water is the solvent in soil for all plant and animal nutrients supplying them with the necessary food for life. Gases such as nitrogen, oxygen, carbon dioxide, and even methane and hydrogen sulfide are also present in the interstices of soil solids. They are utilized, transformed, and given off by plants and animals living in the soil. If we consider each group of micro-organism present in soil, it becomes obvious that certain varieties and numbers prevail.

## BACTERIA

The bacterial population of soil is the largest of all the micro-organism groups in both number and variety. Direct microscopic counts as high as several billion per gram are common in soil containing the proper food supply, moisture, temperature, and physical conditions. These counts exceed even those encountered in one cubic centimeter (about one gram) of raw sewage. Since no single cultural medium will provide an environment nutritionally adequate and physically satisfactory for the growth of all physiological types present, plate counts will be much less than the real total number of micro-organisms present. Bacteria degrade complex organic substances in solid waste to simple compounds such as carbon dioxide, nitrogen, hydrogen, water, methane, and ammonia. Bacteria operate as a team; some initiate the decomposition process and others complete it. Some types of organic matter such as plastic bottles are nonbiodegradable and resist attack by bacteria.

Major bacteria in soil include aerobes (which require oxygen for growth), anaerobes (which grow in an oxygen-devoid environment), facultative (which exist either aerobically or anaerobically), psychrophiles (which are cold-loving bacteria, existing in temperatures as low as 0°C), mesophiles (which survive in a temperature range as high as 40°C), thermophiles (which survive in a temperature range of 40°C to 60°C), autotrophs (those micro-organisms that utilize only carbon dioxide as a carbon source and inorganic nutrients for energy), heterotrophs (those micro-organisms that utilize only organic compounds found in solid wastes for both a carbon source and energy). There are also special groups of bacteria which perform specific functions. Some of these bacteria are cellulose digesters, sulfur oxidizers, protein digesters, and nitrogen fixers, and certain reducing bacteria (which operate in an anaerobic environment) such as sulfate or nitrate-reducing bacteria.

Much of the decomposition of solid waste in soil is carried on by facultative organisms. These are able to degrade complex organic wastes to simple organic acids under anaerobic conditions. Because they are facultative they are able to convert the resulting water-soluble organic acids either aerobically to carbon dioxide and water, or anaerobically to ammonia and methane. Solid-waste composition will influence the time required for degradation. Starches, sugars, and proteins will decompose more easily than cellulose and lignin which are very resistant to biological attack.

Actinomycetes, a typical soil bacteria, are abundant in soil. They give warm soil the musty or familiar earthy odor of a freshly plowed field. These bacteria are carbohydrate lovers which are prevalent in typical solid waste material. Soil in the presence of solid waste has a relatively low nitrogen-to-carbon ratio which is so conducive to actinomycetes.

## FUNGI

These are many different species of mold or fungi in the soil. They are aerobic and exist predominantly near the surface. Mold mycelium imparts a physical soil structure of considerable agricultural importance. Fungi assist in decomposing cellulose and lignin of plants. Because of this, fungi play an important role in the stabilization of solid wastes in both landfills and composting processes.

## ALGAE

Algae are present in soil usually in smaller numbers than either bacteria or fungi. The major types consist of the green algae, the blue-green algae, and the diatoms. They concentrate near the soil surface because of their photosynthetic ability. Algae have the ability to extract and concentrate inorganic soil nutrients such as phosphorous and nitrogen.

## PROTOZOA

The one-celled animals (protozoa) can occur in millions per gram of soil. They function to keep the bacteria population at a maximum growth level by feeding on them. Most soil protozoa are amoeba or flagellates. Protozoa are important agents in maintaining the equilibrium of microbial flora in solid-waste disposal systems. The end product of this biological activity is humus, a dark-colored, amorphous substance composed of residual stabilized, nonbiodegradable organic material.

## VIRUSES

Some bacteriaphage are found in soil. When soil dries and turns to dust, it can be blown about by winds; in this manner it is possible to transmit disease to animals and humans. The function of virus in soil is not clearly understood.

The final solid product of all organic matter degradation, humus, is vital to the ability of soil to support growth of vegetation. Humus contributes to soil systems by improving soil texture, buffering the soil acidity, serving as a source of plant nutrients for plants and micro-organisms, and improving the water-holding capacity of the soil.

## CARBON COMPOUNDS IN SOIL REACTIONS

Solid wastes contain (at some times) certain amounts of organic carbon compounds such as cellulose, starch, sugar, lignin, and pectin. Glycogen from dead animals, as well as the lipids and proteins of all animal cells, also contain carbon in complex organic organization. Aerobic bacteria are likely to produce carbon dioxide, nitrates, and cell substance; whereas anaerobes produce ammonia, methane gas, a variety of fatty acids, alcohols, and other neutral products. Examples of carbon dioxide transformation or incorporation into organic compounds by bacteria include:

1. Utilization of carbon dioxide by autotrophic bacteria:

$$CO_2 + 2\ H_2 \longrightarrow (CH_2O)_x + H_2O$$

Photosynthesis can accomplish similar reactions by growing algae or plants:

$$CO_2 + \text{sunlight} \xrightarrow{H_2O} (CH_2O)_x$$

Burges (1958) gives the diagram shown in Figure 5-1 as the transformations of carbon in nature.

2. Bacteria which live on the organic matter in soil (heterotrophs) can incorporate carbon dioxide into an organic compound already present:

$$\underset{\text{(acetic acid)}}{CH_3COOH} + CO_2 \xrightarrow{\text{bacteria}} \underset{\text{(pyruvic acid)}}{CH_3COCOOH}$$

## NITROGEN CHANGES IN SOIL

Since solid wastes are simply a microcosm of all our foods and materials, nitrogenous matter is present in most components. In proteins, for example, nitrogen averages 16 percent. This protein nitrogen must be released first by proteolytic bacterial enzymes known as

48  INDUSTRIAL SOLID WASTES

Figure 5-1. The Carbon Cycle.

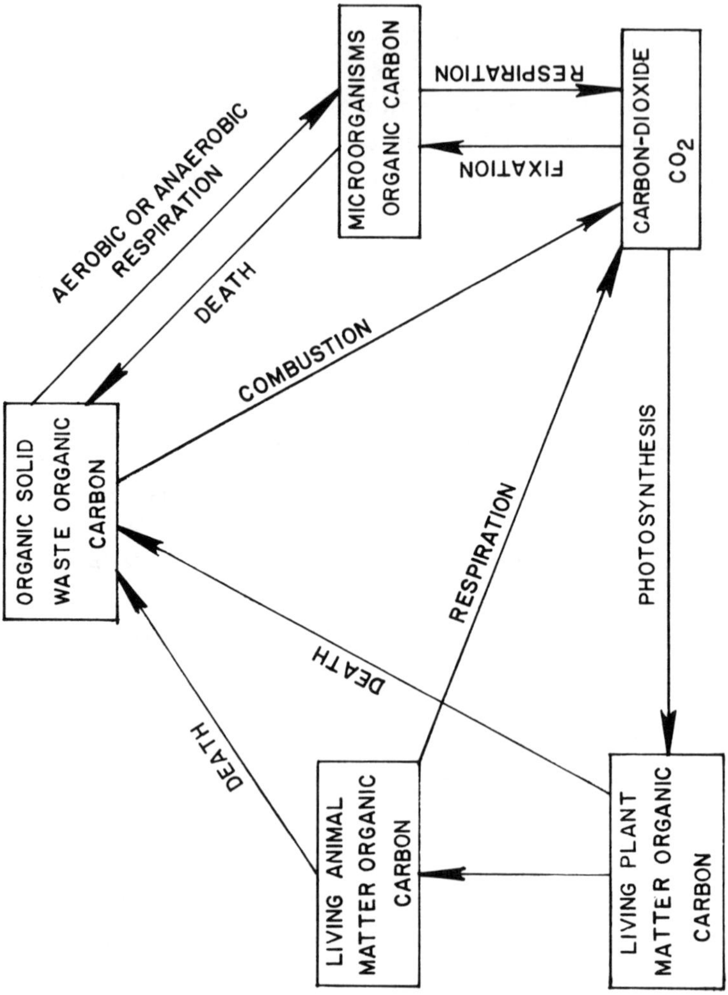

Source: Burges (1958).

proteinases. They degrade the protein to peptides where peptidases convert them further to their individual amino acid. Some bacteria and many fungi and soil actinomycetes possess proteolytic enzymes and hence are important in the degradation of solid wastes. Amino acids are further broken down (deaminated) to release ammonia ($NH_3$) and carbon dioxide ($CO_2$). These can serve as basic food components for both plants and bacteria. Plants are able to utilize the $CO_2$ along with sunlight and water to photosynthesize vegetation. Bacteria can use the $NH_3$ as a source of nitrogen necessary for vital energy of life for all growth. One such group of bacterial microorganisms is *Nitrosomonas* which can oxidize the $NH_3$ to form nitrite ($NO_2$) as follows:

$$2NH_3 + 3O_2 \xrightarrow{\text{Nitrosomonas}} 2 HNO_2 + 2H_2O$$

At this point another oxidative group of nitrogen bacteria (*Nitrobacter*) can use the nitrite to convert it to nitrate as follows:

$$HNO_2 + 0.5O_2 \xrightarrow{\text{Nitrobacter}} HNO_3$$

Both of these micro-organisms are known as strict autotrophs and aerobes using only mineral salts and ammonia or nitrite. When soils become anaerobic, another group of bacteria takes over to denitrify the nitrates to nitrogen gas. These organisms, however, are heterotrophic and hence will use organic matter for energy. Another group of heterotrophs will only reduce the nitrates to ammonia or nitrites as follows:

$$HNO_3 + 4 H_2 \xrightarrow{\text{organic matter}} NH_3 + 3H_2O$$

Still another important group of nitrogen bacteria are able to utilize nitrogen from air to manufacture plant-life. These are called nitrogen fixers. Since air contains about 80 percent nitrogen, the presence of these fixers in the absence of water-soluble nitrogen play an important part in producing plant life on land and waters (Burges 1958). The common version of the complete nitrogen cycle is show shown in Figure 5-2.

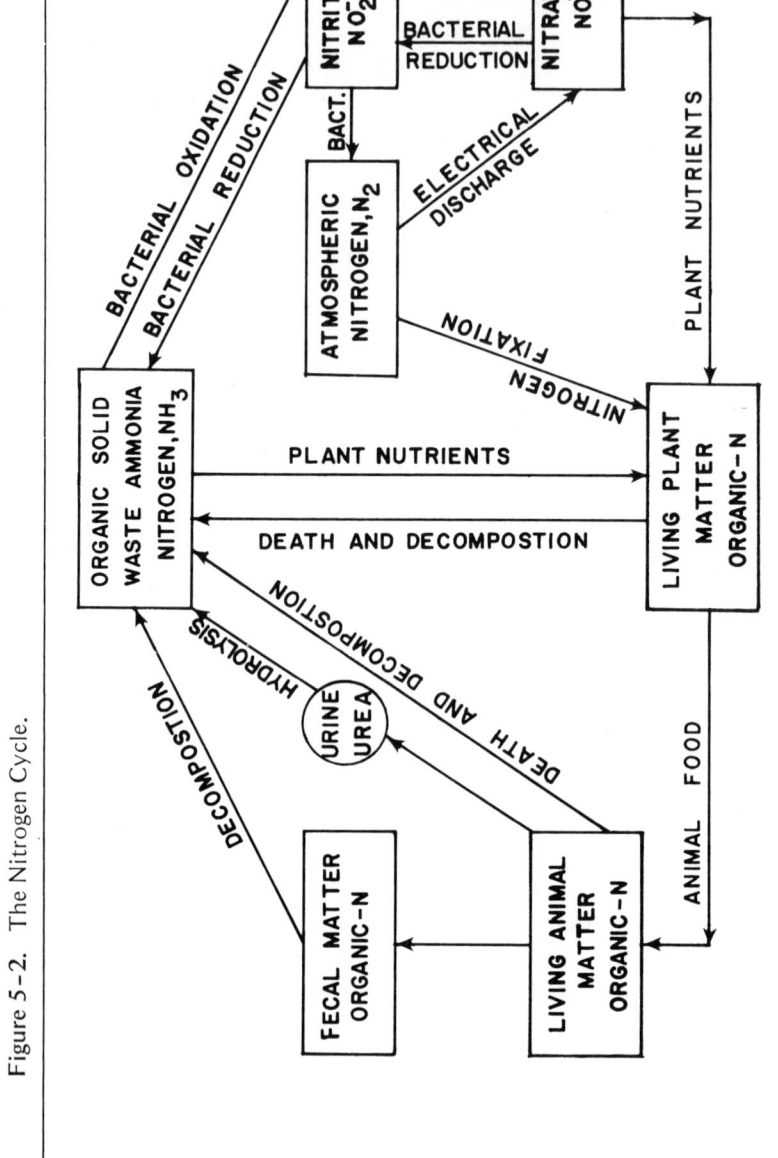

Figure 5-2. The Nitrogen Cycle.

Source: Burges (1958).

## SULFUR CHANGES IN SOIL

Sulfur compounds also undergo changes of oxidation and reduction in soils. The proportion of oxidative products to reduction compounds is of importance to the solid-waste engineer. Generally, reduction products are more volatile and malodorous, while oxidation products are more soluble, stable, and pleasant-smelling. The type of compounds present also dictate the type of microorganisms present as well as vice versa.

Some sulfur bacteria are able to utilize sulfur in its elemental form to convert it to an oxidized form which can be used directly by plants.

$$2S + 2H_2O + 3O_2 \longrightarrow 2H_2SO_4$$

When the sulfates are assimilated by plants, they can also be broken down to simpler reduction products such as various sulfur-bearing amino acids. Sulfur is released from these acids by heterotrophic bacteria. Elemental sulfur can also be formed from the oxidation of reduced sulfur such as $H_2S$. The hydrogen sulfide is formed from reduction of the sulfates by soil micro-organisms.

$$4H_2 + CaSO_4 \longrightarrow H_2S + Ca(OH)_2 + 2H_2O$$

The complicated cycle of sulfur transformations in soil by the various micro-organisms is shown in Figure 5-3.

## BIOLOGICAL ACTION WITHIN SANITARY LANDFILLS

A sanitary landfill, broadly described, is an engineered burial of solid wastes which are subsequently and ultimately degraded by soil micro-organisms. In fact, these biological forms slowly degrade the organic portion of solid wastes to stable compounds. The long-chained organic materials are degraded primarily by aerobic or facultative bacteria and fungi. Facultative organisms are of importance in a landfill ecosystem since they can function in both oxygen and oxygen-devoid atmospheres. Since in well-compacted soils oxygen may not be available much below the first few inches of a landfill, facultative types are vital. In anaerobic fills, anaerobic-loving micro-organisms convert organic matter to soluble organic acids which tend to lower the pH of the fill and solublize more organic matter chemically. The organic acids are further utilized by other bacteria which are able to oxidize them completely to water and carbon dioxide. Carbon dioxide, when dissolved in water as carbonic acid, tends to lower the pH of groundwater even further. Inorganic minerals such as limestone will be dissolved to some extent by this low pH water. When the fill environment is primarily anaerobic, thus favoring the anaerobic group of bacteria, ammonia and methane ($CH_4$) gas are produced. Although the $NH_3$ remains dissolved in the groundwater, methane gas permeates through the fill to the surface. Too much methane concentrated in certain porous areas of the fill can result in

**Figure 5-3.** Sulfur Changes in Soil.

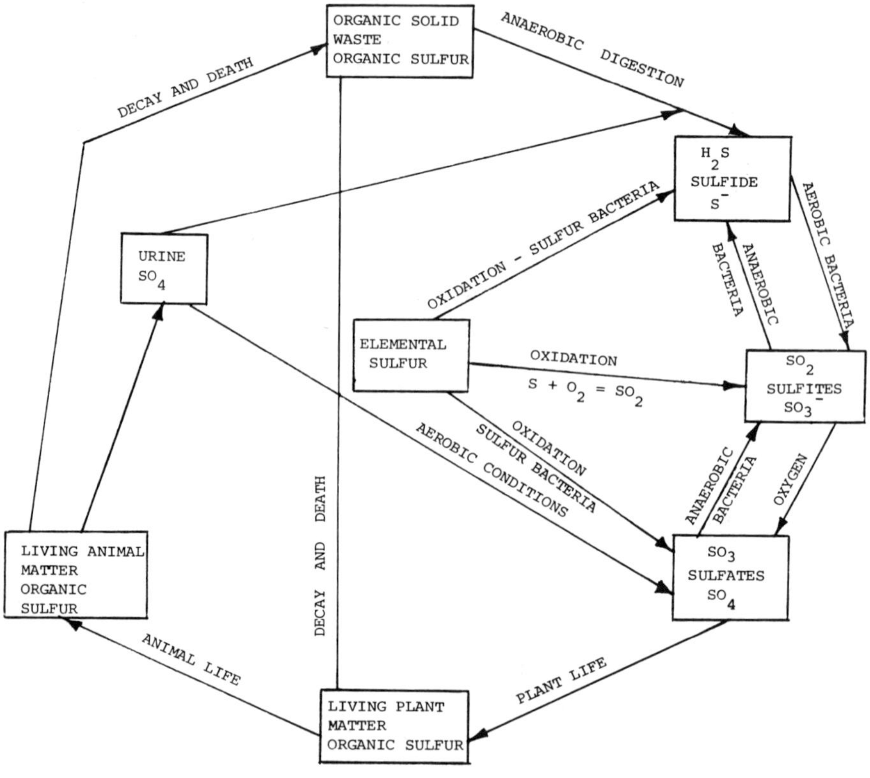

Source: Burges (1958).

dangerous fires and explosions. (More consideration of this problem is given in Chapter 7.) Water present in solid wastes enhances biological degradation, but excessive water fills interstices and promotes anaerobiosis. As the decomposition proceeds, fill densities change, and sometimes non-uniformly. This process makes for unequal consolidation of the fill with resulting settlement of the newly created soil. Any building on landfills of this type may be jeopardized by this differential settlement.

## BIOLOGICAL CHANGES DURING COMPOSTING

Although this treatment is described in detail in the next chapter, its biological nature warrents mentioning here. Composting consists of controlling and enhancing biological decomposition to produce a stabilized final material faster than in landfills. In addition, the compost is generally utilized for some beneficial purpose often at a site far removed from the actual composting operation. Mixtures of common wastewater bacteria, fungi, and actinomycetes combine to produce the compost from organic solid residues and water. The type of compost and time of production will depend primarily upon temperature, moisture, pH, oxygen, and size of solid refuse particles.

### REFERENCE

Burges, A. 1958. *Microorganisms in the Soil.* London: Hutchinson and Company, Ltd.

# 6 COMPOSTING

## THEORIES INVOLVED

In considering treatment of solid wastes, we deal first with composting, a process of natural fermentation that occurs in moist, mainly cellulosic materials and produces a granulated, dark-brown material resembling coarse coffee grounds. The finished product, after drying, contains organically bound nitrogen, humus, micro-organisms, and other important micro-nutrients. Compost after drying is usually 10 to 15 percent water.

The composting process has two primary objectives. The first is the inactivation of pathogenic micro-organisms, and the second is the production of a final material which is at least 60 percent dry, odorless, and easily handled. Composting is an aerobic process. Turning of piled refuse material incorporates air, allows carbon dioxide to escape, and, when continued, heats all parts of the pile to a high temperature. In composting it is essential to adjust carbon-to-nitrogen ratios, dry solid contents, and particle structures. The last can be accomplished by blending wet, compact, and nitrogen-rich sewage sludge cake with carbon sources and dry structuring bulking agents from industrial or municipal refuse.

Municipal refuse which contains a relatively high carbon-to-nitrogen (C/N) ratio can be composted successfully. However, more microbiological action is required to break down the carbonaceous material to $CO_2$ so that the C/N ratio will approximate twenty to one. In addition, it is sometimes desirable to blend compost with sewage sludge to increase the nitrogen content and to enhance the final product's usefulness as a fertilizer.

In general, compost is valued as a fertilizer from several points of view:

1. *Organic nitrogen* (generally 5 to 7 percent) is slowly converted by micro-organisms (since it is insoluble) to the active state.
2. *Humus* improves the water-holding capacity of the soil.
3. *Micro-organisms* convert insoluble minerals (nitrogen, carbon, sulfur, and so on) to soluble (active) forms which can be utilized readily by plants.
4. *Micro-nutrients* (trace elements) act as catalysts to insure plant vitality, especially boron and manganese.

In the United States composting is usually an aerobic process. Just as in aerobic liquid-waste treatment, living organisms use oxygen and feed on organic matter and produce cell growth from N, P, K, and C, and other micro-nutrients. About 65 percent of the carbon is converted to $CO_2$, and the rest combines with nitrogen in protoplasm. If too much carbon is present in the waste for the amount of nitrogen present, biological activity is less than optimum until the carbon is eventually oxidized to $CO_2$. When carbon is reduced sufficiently low, then nitrogen is reduced to $NH_3$, and under favorable (aerobic) conditions the $NH_3$ is further oxidized to $NO_3$.

This composting process is similar to that which occurs naturally on a forest ground where leaves and wood compost to humus soil. Optimum composting depends primarily upon the following six environmental factors:

1. Moisture. A desirable range is from 40 to 70 percent to enhance the growth of micro-organisms.
2. Temperature. Higher temperatures enhance biological degradation of organic matter. Thermophilic groups will operate at an optimum of 130 to 150°F.
3. pH. This is not too critical as long as the range remains between between 4.5 and 9.5. A neutral pH is the most desirable.

4. Air. Aerobic treatment is vital for oxidation of the organic matter at an accelerated rate. Usually twenty-five cubic feet of air are required per pound of organic matter composted per day.
5. Minerals. An optimum carbon-to-nitrogen-to-phosphorous ratio for composting is 100 to 40 to 1, respectively. This ratio will ensure active microbiological activity.
6. Size of refuse material. Smaller particles yield greater surface areas. Practical sizes of one-fourth to one inch are generally satisfactory for microbial degradation.

Heat is released as organic matter is degraded to $CO_2$. It is possible to raise the temperature of a compost pile to 70°C. if little or no heat loss takes place. Some heat will be dissipated by providing oxygen to the heap to keep it aerobic.

Aerobic conditions in a compost heap prevent the formation of $CH_4$ and $H_2S$ (odorous products). Aerobic decomposition can be carried out in silo digesters, pits, bins, stacks, or piles. Turning the piles at given times or adding jets of compressed air will maintain aerobic conditions.

Agripost Company proposes to take garbage from Dade County, Florida, and process it into compost that can be sold as soil or fertilizer additive ("Agripost Composts Garbage . . . " 1980). Dade County would pay Agripost $8.50 per ton to take the garbage, substantially under the then $12-per-ton cost to Dade County for disposing of its garbage.

Composting can be done in open windrow fermentation of in enclosed mechanical digester fermentation.

## WINDROW COMPOSTING

This is done in the open air and uses natural ventilation with periodic turning to keep the windrows aerobic. Grinding proceeds the piling in windrows, and turning can be done periodically with a front-end loader. This process is continued until the compost is ready. Some difficulties have been experienced in selling the product.

In ventilated cell composting a several-storied building is used with a vertical arrangement of progressive cells. Mixing and aeration occur during the lowering of the refuse from a higher cell to a lower one.

Variations of this process are used in the Naturizer Process, the Riker Process, the Frazier-Ericson Process, the Fairfield-Hardy Process, and the Scarab Process.

The *Naturizer Process* was used first in Norman, Oklahoma, in 1959. The newest plant of this type is located in St. Petersburg, Florida (100 tons refuse per day at $3.24 per ton). This composter is a five-storied building with conveyors running the length of each floor (155 feet). The conveyors take twelve hours to travel the length of each floor and then are stationary for twelve hours. After digestion and final screening, the compost is left in the yard for two weeks. The compost is marketed as "Cura" and is used for citrus groves, golf courses, commercial nurseries, and industrial landscape architects. The plant costs $1.5 million or —

$$\frac{1,500,000}{100} = \$15,000/\text{ton/day}$$

The *Riker Process* (from Williamston, Michigan) is similar to all open windrow processes except that it treats a mixture of garbage, vacuum-filtered raw sewage sludge, and corn cobs. Composting takes twenty-one days in two four-compartment vertical composters.

The *Frazier-Ericson Process* (Springfield, Massachusetts) is for refuse with high percentages of garbage which must be air-dried for a few days before composting.

The *Fairfield-Hardy Process* (Altoona, Pennsylvania) is a system that grinds refuse in a wet pulper prior to dewatering it in presses before digesting for five days. Stirring the heaps is done by augers suspended from a rotating bridge in a circular tank. Air is provided by means of a blower and air pipes embedded in the floor of the tank. Normal operating temperatures are 140 to 160°F. Cost to the city was $4.63 per ton at eighteen to twenty-three tons per day. Almost half the tons of refuse treated is recovered and sold as compost (at five dollars a ton). A 100-ton-a-day plant of this type would cost $900,000 including land (or $9,000 per ton per day). In Largo, Florida, there is a similar plant operating except that it turns the material by a rail-mounted bucket elevator moving in either direction. (The product sells for sixteen dollars a ton.) Houston, Texas, also has a plant of this type selling "Metrorganic 100" for rice and citrus growers and others engaged in agriculture in the Rio Grande Valley.

In the *Scarab System*, sludge cake from a municipal sewage treatment plant (about 20 percent solids) is combined with a bulking agent and formed into windrows. The bulking agent can be composted sludge (about 65 percent solids and 37 percent organic). In the Carson, California, study (*Report* 1979), windrows were about 12 feet wide, 4 feet high, 435 feet long, and usually on 26-foot centers. In this process, the windrows were mixed with a scarab machine which is self-propelled, and straddles a windrow. The scarab has a rotating drum equipped with flails (curved teeth), and literally picks up the material and deposits it behind the machine while moving through a windrow. Windrows were turned twice a day on the first day and usually once a day thereafter. This machine which allows complete mixing can turn more than a metric ton of organic matter per hour.

## CONTINUOUS MECHANICAL COMPOSTING

Continuous mechanical composting uses continuous mixing with gradual particle-size reduction and positive aeration. Although there are several systems of this type, only one, the DANO method, is used in the United States.

The *Dano Biostabilizer System* is a large drum, 9–12 feet in diameter and 60–100 feet long. It mixes continuously, aerates, grinds and decomposes all in one unit. Refuse is first sorted and separated magnetically before charging into one end of the slowly revolving unit (one-fourth to three-fourths rpm). Water and sludge are also added at the inlet. Slow grinding is done by the tumbling shearing action (similar to a Drum Dryer). The decomposed compost is discharged through outlet perforations of about four inches in diameter. The compost goes through a second magnetic separator, vibrating screen, and a ballistic or gravity separator before the partially decomposed refuse is put into windrows for final curing for a few days only. In the United States, although the process worked satisfactorily (Sacramento, California, and Phoenix, Arizona), it proved more costly than landfilling and has been discontinued in these cities. When the economic value of compost increases, this process will have great merit.

Table 6-1. Some Characteristics of Compost Products.

| Product | Characteristic |
| --- | --- |
| Moisture — 32% | Cu — 600 ppm |
| pH — 7.3 | Zn — 930 ppm |
| Organic Matter — 48.5% | Cr — 350 ppm |
| Nitrogen — 1% of total solids | Pb — 250 ppm |
| $PO_4$ — 0.5% of total solids | Cd — 11 ppm |
| K — 0.3% of total solids | Mn — 630 ppm |
| Boron — 30 ppm | |

The reader is directed to a compendium of papers on composting (*Staff of Compost Science/Land Utilization* 1981) containing material published on the subject over the last twenty years.

The *Ruther MSA* is a system developed in 1975 (Moos 1980). This composting plant in Salzburg, Austria, was begun in September 1977, processing 300 tons of refuse and about 100 tons of sewage per day. The plant costing 110,000,000 Austrian shillings is operated as a two-stage process: twenty-four hours of fermentation in a drum followed by four weeks of fermentation on an artificially aerated fermentation platform. Valuable compost is produced for a total cost of 244 Austrian shillings per ton of refuse (1 shilling = about $.07 U.S.). Some significant characteristics of the compost product are shown in Table 6-1.

## REFERENCES

"Agripost Composts Garbage to Make a Bundle." 1980. *Miami Herald* (August 11): 19.

Mooss, H. 1980. "Costs of and Innovative Solutions for Industrial Waste Treatment," In *Treatment and Disposal of Liquid and Solid Industrial Wastes*, Proceedings of The Third Turkish-German Environmental Engineering Symposium, Instanbul, Turkey, July 1979, edited by Kriton Curi. New York: Pergamon.

*Report on Municipal Sewage Sludge Composting Project (Tyroc Waste Management), Carson, California, Sept.-Nov., 1978.* 1979. Report prepared for the U.S. Department of Commerce.

Staff of Compost Science/Land Utilization, eds. 1981. *Composting—Theory and Practice for City, Industry, and Farm.* Emmaus, PA: The J.G. Press.

# 7 SANITARY LANDFILLS

### BASIC DEFINITIONS

The American Society of Civil Engineers (1959: 61) defines sanitary landfilling as "disposing of refuse on land without creating nuisances to public health or safety by utilizing the principles of engineering to confine refuse to the smallest practical area, to reduce it to the smallest practical volume, and to cover it with a layer of earth at the conclusion of each day's operation, or at more frequent intervals if necessary."

The American Public Works Association (1966: 528) adds further qualifications to sanitary landfilling:

1. It minimizes vector breeding or sustenance by eliminating all possible harborage and food supply for *rats, flies,* and other *vermin.*
2. It deals adequately with the possibility of direct disease transmission.
3. It controls effectively air pollution such as smoke and odor.
4. It minimizes fire hazard.
5. It minimizes the possibility of polluting surface and groundwaters.
6. It effectively controls nuisances (aesthetics and noise).

Factors that most affect the economic feasibility of a landfill project are: (1) availability of a suitable site at reasonable cost for purchasing or leasing; (2) distance that refuse must be hauled; (3) availability of cover material, either on-site or for purchase at a reasonable haul-distance and price; (4) local existing pay scale; (5) cost of landfill equipment; (6) prefill and postfill steps which must be taken to prepare and protect the surrounding environment and to enhance the final usefulness of the site.

Importance in ascertaining these factors will depend upon the volume and density and composition of refuse to be handled now and at some time in the future.

According to the American Public Works Association, a municipal landfill space requirement of 2 cubic yards per capita per year (= 1.25 acre-feet per 1,000 people per year) is a reasonable conservative estimate based on 5.5 pounds per capita per day at 1,000 pounds per cubic yard compaction density in the landfill. Industrial refuse equivalents have not yet been determined.

Urban waste generation rates vary widely, depending upon many factors such as habits of population, climate, season, and so on. A national average of 5.5 pounds per capita per day is just that—an average. The density of compacted solid waste in the fill will vary depending upon the degree of compaction and upon the moisture content of the solid waste. A solid waste density of 1,000 pounds per cubic yard is common for reasonably well-compacted shallow landfills (less than twenty feet deep). Soil cover requires volume as well as compacted refuse of four parts usually to at least one part of soil cover—or 20 percent of fill volume will be soil cover. However, in actual practice it has been found that soil cover will fill interstices of refuse so that only about 5 percent of fill volume is soil. When milling and baling refuse, refuse volumes are reduced considerably. A recommended ten years of landfill use should be planned for a given site or for a portion of the land for use by a municipality.

Site requirement considerations must include hauling costs—the greater the distance, the greater the cost. Also, road conditions and maintenance requirements must be carefully evaluated. A haul distance of ten to fifteen miles from the farthest point is generally economically feasible without use of a transfer station.

As far as cover material is concerned, the best type of soil for landfilling is a sandy or silty land containing rocks no larger than six inches in diameter. High clay content generally makes poor cover material.

Other considerations are: (1) zoning and political location of the sites; (2) depth to bedrock and degree of fracturing of bedrock which could transmit leachates; (3) depth to groundwater and direction of its flow and use; (4) types of soils which are traversed by leachates in reaching and moving with groundwaters; (5) potential for explosive $CH_4$ moving through fractured rock or porous soil strata toward nearby structures and people.

## PUBLIC ATTITUDE

Support of community for this "low" use of land is vital and difficult. Generally, a long-term public relations program designed to educate the public is the most effective way of obtaining general public support.

## ENGINEERING PLANNING

A sanitary landfill is an engineering project and should be planned like one. Therefore, proper surveys of properties, construction of roads for all-weather conditions, depth of fill on the site, location, and amount of cover material at the site, required grades and culverts to permit proper drainage, water supply for fire and dust control, and the type of sanitary landfill best suited for a certain location all need consideration.

## ADMINISTRATION AND CONTROL

Nearly every type of administrative and operational arrangement has been used. No one system is preferable. Records and accounting systems are important.

## OPERATION OF A SANITARY LANDFILL

There are three general methods of landfilling—area, trench, and ramp methods.

The area method is usually employed on sloping land, in ravines, canyons, marshes, quarries, and other natural or man-made depressions. Refuse is dumped in or near the fill site, then spread and compacted by a bulldozer. Cover dirt is usually obtained from a nearby high point and moved to the fill by a scraper, front-end loader, or bulldozer. At least six inches of cover material compacted is recommended on the top and exposed sides of the compacted refuse. Usually the cover material is placed at the end of a working day; the resulting covered area becomes a "cell." A general principle is to keep the working face area as small as practically possible. Generally, the deeper the cells, the greater the degree of refuse compaction which can be achieved. Often trucks can be passed over filled areas to get about 30 percent greater compaction; thus settling of the fill site is reduced. The final cover for an area landfill should consist of a minimum of twenty-four inches of compacted earth.

The trench method is usually used on flat or slightly sloping land. It involves the excavation and filling parallel trenches separated by a three- to four-foot dirt wall. Usually dirt from the first trench is used to construct berms for windbreakers and remains as stockpile to cover the final trench. As dirt is needed to cover the first trench, the second is opened. Trenches vary greatly in length and should always be dug at least twice as wide as the tractors which must work in them. The depth varies with soil type and groundwater level. Normal depth is eight to ten feet although some reach depths of fifty feet.

Cover material rules are the same as for the area method.

The ramp method is a variation of the area method suited best to a sloping land. Solid refuse are deposited on the side of the slope, spread, and compacted by a bulldozer. However, the difference lies in the cover material which is generally excavated from below and in front of the working face of the fill as it is needed. The advantage of this method is that one piece of equipment—usually a bulldozer—can be used to operate all the functions. Therefore, this method is used especially for small operations.

## OPERATING CONDITIONS

Many of the following factors are described in detail in (National Center 1974, and American Public 1966).

## 1. Compaction

Normally refuse is compacted to about 1,000 pounds per cubic yard in all three methods of landfilling. Thin layers of a maximum of two feet deep are passed over by several passes with heavy landfill equipment. Refuse is approximately three to four times as dense in landfills as in residential trash cans and nearly twice as dense as in compacter-collection vehicles. Separation of items such as tree stumps in the refuse before filling yields higher compaction densities. Good compaction prevents excessive later settling and the creation of gas pockets and cracks. Cracks allow surface water to flow directly into the refuse; gas pockets give rise to fires, odors, and so on.

## 2. Wind-blown Litter

A fixed fence should circumvent the landfill area to prevent blowing paper as well as limit access and visibility from outside. Sometimes temporary fences are used downwind of filling operations to control windblown paper.

## 3. Wet Weather Conditions

Although earth-moving can be dispensed with during rain, collecting, transporting, and disposing of refuse must continue. Special piles of cover should be prepared and covered for rain periods when digging is not possible.

Landfills should be designed and constructed to maintain proper drainage during construction and operation for the life of the fill. This can be done with the aid of dikes, culverts, ditches, drainage pipes, and even pumps when necessary.

One percent slope is the minimum required of the finished landfill site. However, excessive slopes are avoided because of erosion.

## 4. Dust and Fire Control

Dust at the site should be avoided for health and aesthetic reasons. Water should be available for dampening the landfill. Light oil-spreading, grass-seeding, as well as dampening with water can be used to keep dust down.

Fires can best be prevented by good landfill practices but can be extinguished by smothering with dirt fill or cutting off the oxygen supply in any convenient way or by water with high pressure.

Dirt fill in each cell helps to contain fires and prevent them from spreading.

Fires create political as well as environmental risks. Open burning is banned by most states and violate the term "sanitary" landfill.

### 5. Salvage

Salvaging at the landfill site is not recommended since it interferes with the proper operation of a sanitary landfill.

### 6. "Afterfill" Maintenance

Settlement and erosion and potential gas explosions should be inspected and protected after the land is filled. Twenty-five years is a reasonable period for this observation and protection.

### 7. Equipment

The type of equipment needed at a landfill depends primarily upon amount of refuse handled, its composition, the geological characteristics of the site, and the fill method used. The most common piece of equipment is a bulldozer. For small operations it may be the only piece of equipment. For large operations it can be equipped or attached with and to a dozer blade, front-end loader, ball clam, bucket, scraper, or special trash blade.

A wheeled, rubber-tired, tractor can also be rigged with these attachments to excavate and haul cover material and spread and compact refuse.

A bulldozer is a track-type and therefore is more durable, has better traction, makes more rapid and even compaction, and yields a predictable trade-in value.

A wheeled tractor possesses more speed and maneuverability, especially from one location to another without being hauled.

For 15,000 people the requirement would be one tractor equipped with a ballclam (multipurpose blade) or front-end loader of one yd$^3$ capacity. For 15-30,000 people, the requirement would be a front-end loader of 2 yd$^3$ capacity. For 30-75,000 people, two tractors of 3 yd$^3$ capacity would be required.

Draglines and self-propelled scrapers are sometimes needed for large operations when digging is difficult and haul distance is long.

Compactors apply very high pressures by a combination of heavy weight and breaker bars or teeth mounted on the wheels to crush and compact the solid wastes. Units weigh from 30-100,000 pounds, are useful with bulky and industrial refuse, and are quite maneuverable.

### 8. Milling the Refuse

This method is chosen for the following reasons: (1) better acceptance by neighbors; (2) longer landfill site life (10-75 percent); (3) diminished cover needs; (4) improved operations; and (5) better adaptability for re-use and recovery and further processing.

Density after milling increases by about 15 percent. Practice dictates that 90 percent of the milled refuse (by weight) should pass a three-inch sieve to achieve the best landfill characteristics.

Magnetic separation after milling is used to recover ferrous metals for re-use in the steel industry. Air classification to further separate light from heavy fractions is generally practiced.

Incineration, composting, and pyrolysis can also be used following milling.

Milling processes are also located at transfer stations. There are three reasons for not milling: cost; administrative problems; and lack of research and practice in the United States. If landfill site volume is critical and cost can be afforded, milling is a potential advantageous process to use. It requires a feeding system, mill, and milled refuse transport system. The most common type of mill is the hammermill in which rapidly rotating hammers strikes refuse until it can pass a given grate size opening (usually three inches). (See Figure 2-1).

## 9. Baled Refuse Landfilling

This process involves compression of solid wastes into bales by application of high pressure prior to landfill disposal. Since it is used primarily when long-distance hauling is called for to remote landfill sites, it saves cost on railway cars. Cambridge, Massachusetts, San Diego, California, and St. Paul, Minnesota all use this system. Reasons for baling include the following: (1) refuse is compressed for more economical transportation; (2) refuse is rendered more acceptable to the public during transportation and at the landfill; (3) landfill operations are improved; and (4) landfill life is increased.

Baling is just one of several methods to reach the allowable weight limits on trucks. On rail the volume of refuse is critical so the increased levels of compression accomplished by baling are especially desirable. Baling causes the disappearance of original objectionable qualities of refuse, and public acceptance of it is therefore enhanced. Operational problems in landfilling have been reduced by baling. Blowing debris is eliminated driving over fill in wet periods is improved; odors are reduced; fires and vectors are decreased. Bale densities of 1600-1800 pounds per cubic yard (wet weight) increases the filled site life which normally receives refuse at 1000-1250 pounds per cubic yard. Baling has the same objections as milling does.

There are several types of balers used. The single stroke, continuous-operation baler forms one long continuous bale which is cut into smaller sections as desired for handling. Two or three stroke units are also used which compress the refuse from other perpendicular directions as well. It is not generally agreed upon as to whether tying or wrapping of the compressed bales is necessary. Bales are usually about two tons each and are loaded by rollers onto flat-bed trucks also loaded with rollers for easier discharge with the aid of overhead cranes at the landfill site.

## 10. Economics

Each landfill project is independent and varies in cost as well, ranging from fifty cents per ton to four dollars per ton. Collection and hauling range from twelve to twenty dollars, and incineration from eight to fifteen dollars. Variation in costs are also due to labor costs, equipment used, and land costs and must be updated from 1970.

Land costs may average from one cent per ton of refuse disposed (remote county) to ten cents per ton (Los Angeles). Therefore, land costs usually range from 2-10 percent of total landfill costs, a relatively small amount. Also, land sometimes increases in value after filling, and its sale price following land filling must be considered. Leasing of the land is a common method used to help both the user and the landowner.

A rough rule of equipment capital costs is one dollar per pound of refuse handled.

An approximate capital cost of an enclosed refuse-milling facility is about $5,000 per ton of rated capacity per shift. This cost would include total costs of building and foundation, dumping floor refuse feeding system, mill, refuse transport system to nearby landfill, and operating facilities. Landfill capital costs are not included.

Baling costs are difficult to assess, but a plant processing 900 tons per sixteen-hour day of high pressure baling (September 1971) cost $1 million excluding landfill costs. Labor or wages comprise about half the total operating costs or about a dollar per ton. Equipment costs make up about 40 percent of total operating costs or about fifty cents per ton. Cover material costs seventy-five cents to three and one-half dollars per yd.$^3$ Water content of refuse can affect cost of landfilling. Cost-accounting methods differ, and items included are different and hence not comparable from one system to another.

National average costs for the late '60s show two and one-half dollars per ton for small operations and one dollar a ton for large operations (Figure 7-1).

## 11. Public Health Aspects

The health of the public is the main reason for sanitary landfilling. State inspections of landfills are becoming general practice. Three main areas of public health matter in landfills are: (1) potential spread of disease from contaminated refuse; (2) potential contamination of ground- or surface water; and (3) production of gases which may be hazardous to surrounding environment. These are considered in the following sections.

**Figure 7-1.** Sanitary Landfill Operating Costs.

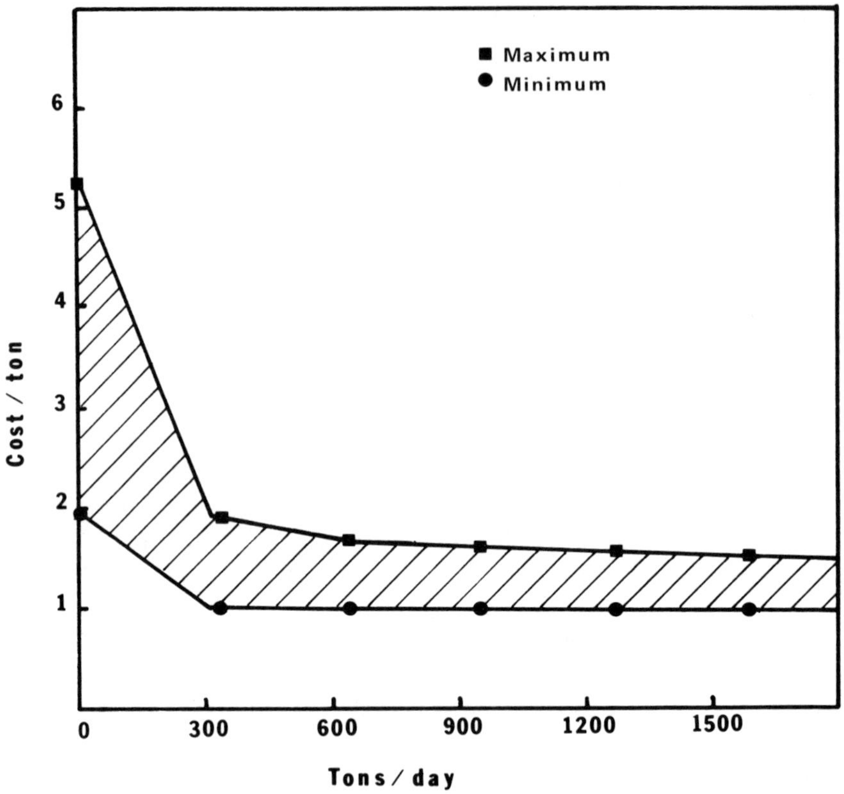

Source: Adapted from "Sanitary Landfill Facts" U.S. Department of Health, Education, and Welfare Solid Waste Report No. S.W. 4ts (1978) p. 23.

## 12. Vector Control

Cover placed over compacted refuse reduces the presence of carriers such as birds, insects, and vermin. A minimum of six inches of cover also prevents flies and fly eggs in the refuse from reaching the surface.

Vermin extermination should be practiced on all dumps converted to sanitary landfills. Care should be used to make certain that toxic chemicals used in exterminating do not run off in water supplies during rains. Therefore, flooding the landfills with chemicals should be done during dry periods and refuse compacted immediately after filling. Fills around airports should be avoided unless done in a sanitary manner to eliminate birds which have caused problems with jet-aircraft.

## 13. Water Pollution

It is inevitable that some part of rainwater will seep through the soil cover and into the refuse. This water plus the water absorbed already in the refuse will slowly migrate to surrounding soil and eventually into the ground- or surface water in the area. This is known as *leachate.*

Little direct evidence exists of harm done to water supplies by leachates. Distance between the base of a landfill and any aquifer has been limited by some agencies: for example, Illinois requires thirty feet, California requires three to ten feet of soil between the base of the landfill and the groundwater table. Some parameters which indicate groundwater contamination by leachates are pH, specific conductance, total hardness, calcium, alkalinity, chloride, and nitrate. Iron, Chemical Oxygen Demand (COD), and total bacteria can also be helpful in pinpointing pollution from refuse fills. Pathogens buried with refuse (for example, from hospital wastes) can be transported out of the fill into groundwater. However, because of soil absorption and high temperatures in fills, pathogen survival is rare.

## 14. Leachate Control

Placing refuse in high water table areas or overporous rock should be avoided. If these areas cannot be avoided, diversion of flowing waters away from the refuse is required. This minimizes the production of leachates. Using a liner or sealing the ground prior to filling will prevent leachate from getting out of the fill. Leachate production can also be minimized by preventing infiltration by using a tight, smooth, and sloped soil cover; by preventing percolation by using a tight, impermeable material such as clay under the surface; by enhancing transpiration by planting cover crops over the fill. Also, sometimes it is necessary to use culverts or pipes to channel underground or surface waters around or under the refuse fill area.

Clay liners or berms have been used to regulate or control the leachate that is formed in the fill as a natural degradation phenomenon aided by even the small amount of rainwater which falls directly on the fill.

Liners to prevent leachate must be used simultaneously with controlled release of ponded water to treatment facilities. Some sites can be designed to collect leachates as a natural consequence for treatment before discharge or reuse.

## 15. Gas Hazards

Gas from decomposing refuse in landfills has caused much property damage and has led to both injuries and deaths. Generally, biological decomposition takes place quite readily until the oxygen is used up and then proceeds more slowly under anaerobic conditions. The temperature rise is greatest during the first few weeks of decomposition and begins to decrease slowly when the oxygen is gone. $CO_2$ results primarily during the aerobic stage, while $CH_4$, $NH_3$, and $H_2S$ are prevalent later under anaerobic conditions. The length of time of gas generation can be expected to be from five to twenty years depending on refuse composition, depth of fill, and local climate.

Table 7-1 shows what happens to the various components of refuse during a sanitary landfill.

The $CH_4$ and other gases tend to rise through porous cover material or through cracks in soil material and then escape into the air. The main concern is to limit the $CH_4$ emission into the air since it is explosive at concentrations of 5-15 percent in air. The purpose of proper design is to insure that the 30 percent $CH_4$ in the fills does not mix with 0 percent $CH_4$ in normal ambient air to make a 5-15 percent explosive mixture. Methods used range from the two extremes of preventing its rise to the surface to venting it in controlled discharges.

Improper venting is especially dangerous and is made even more difficult by impervious cover material. If soils are permeable, leachate increases and gas rise also increases. If soils are wet, gas evolution is hindered by a factor of about twenty or more. Total gas flow is largely a function of type of soil. Therefore, if surrounding soils are impermeable, it is necessary to provide venting on the landfill site itself. The lateral movement of the gas to adjoining soils must be cut off to prevent damage. If landfill is located in impermeable soils, gases will not travel laterally, and gas pressure will build up and finally burst through a break in soil.

The most appropriate gas control method is to use a permeable cover material and to avoid paving or building on the surface for many years. A network of perforated pipes inserted in the landfill to penetrate the refuse cells should be used for venting, but they should reach high enough out of the fill to be well out of the reach of the passer-by. It is preferable to burn off the vented gas (TIKI burner). This extends ten feet into the air equipped with a windshield and manual or automatic ignited pilot flame. A gravel liner under and around the site is another method of providing free venting. A series of gravel filled holes may be used rather than the trench. Barriers to gas flow are used to prevent gas going in certain directions. Gunite is sometimes used.

One type of industrial refuse being landfilled is that from the building construction industry. It is common practice to bury construction debris, including tree limbs, damaged plywood, and wood pellets. However, Dade County, Florida, disallows burial of any materials which can decompose underground. Further, the county requires a public hearing to excavate any hole deeper than eighteen inches. If the excavation is approved, regulations specify the type of materials that can be buried. County officials are concerned with settling of homes, breeding of termites, and groundwater pollution.

Rail-haul and sanitary landfills have been shown to be most practical and economical in nearly all situations. Large coastal cities such as New York City and San Francisco face difficult political problems in finding acceptable landfill sites. Some comparative 1970s costs for landfilling, incineration, and composting are given in Table 7-2.

Land cultivation as defined by SCS engineers (EPA 1978) as "a process whereby waste is spread and incorporated into the soil surface" can be incorporated with crop production or used solely as a disposal practice. The report states that 3 percent of all industrial waste can be disposed of by land cultivation. They found that solid-waste loadings were limited by soil texture, drainage, and permeability, and the waste pH, bulk density, soluble salt and metal contents, and nitrogen and phosphorous contents. Costs range from two to eighteen dollars per cubic meter of industrial waste, exclusive of transportation.

**Table 7-1.** Environmental Rate and Conversion Products of Domestic Wastes in Sanitary Landfill.

*Metals (Approximately 7% of total)*

1. Remain in landfill as inert or relatively inert compounds unless fill eroded by surface runoff: Uncombined metals; $Al_2O$; $Al(OH)_3$; $FeO$; $FeO_2$; $Fe(OH)_2$; $Fe(OH)_3$; $CrO_2$; $Cr_2O_3$; $Cr(OH)_2$; $Cu_2O$; $Cu(OH)_2$; $SnO$; $ZnO$; $Zn(OH)_2$; $V_2O$; $Be$; $BeO$; $NiO$; $Ni_2O_3$; $Ni(OH)_2$; $CdO$; $Cd_2O$; $Cd(OH)_2$; $PbO$; $Pb_2O_3$; $Pb_3O_4$; $Pb_2O$; $Se$; $SeO_2$; $Hg$; $HgO$; $TiO_2$; $MgO$; $Mg(OH)_2$; $Ca(OH)_2$; $CaCO_3$.

2. May leach to groundwater: Ca and Mg sulfates; Fe, Ca, and Mg bicarbonates; $CO_2$; also oxides of Sn, Zn, Cu in acid water.

3. Escape to atmosphere: essentially none.

*Cloth-Natural and Synthetic (4% of total)*

1. Remain in landfill as inert or relatively inert materials:
   a. Materials fabricated of synthetic fabrics.
   b. Oxidized and reduced minerals in "natural" fibers.

2. Remain in landfill by incorporation into microbial protoplasm; $NH_4^+$; reduced sulfur compounds: C; P; K.

3. May leach out to groundwater: $CO_2$; aldehydes; ketones; organic acids; sulfates; phosphates; $NH_4^+$; $NO_2^-$; $NO_3^-$.

*Rubber-Natural and Synthetic (1.1% of total)*

1. Synthetic rubber is essentially inert.

2. Natural rubber breaks down extremely slowly.

*Glass (8% of total)*

Inert in landfill.

*Wood (2% of total)*

1. May leach to groundwater: $CO_2$; aldehydes; ketones; organic acids; phenol; $NH_4^+$; $NO_2^-$; $NO_3^-$.

2. Remain in landfill through incorporation into microbial protoplasm: $NH_4^+$; C; P; K.

3. May escape into atmosphere: $CO_2$; $CH_4$; volatile short-chain fatty acids; $N_2$; $NH_3$.

*Garbage (15.5% of total)*

1. Possible leachates to groundwater; See *WOOD* (above); plus sulfates, phosphates, and carbonates.

2. See Item 2, *WOOD* (above).

3. May escape into atmosphere: $CO_2$; $CH_4$; volatile short-chain fatty acids; $H_2S$; mercaptans; $N_2$; $NH_3$.

# SANITARY LANDFILLS

4. Escape to atmosphere: $CO_2$; $CH_4$; volatile short-chain fatty acids; $N_2$; $NH_3$; $H_2S$; mercaptans.

*Plastics (1.9% of total)*

Essentially inert in landfill.

*Leather (1% of total)*

1. May leach out to groundwater: $CO_2$; aldehydes; ketones; organic acids; sulfates; phosphates; $NH_4^+$; $NO_2^-$; $NO_3^-$.

2. Remain in landfill by incorporation into microbial protoplasm' $NH_4^+$; reduced sulfur compounds; C; P; K.

3. May escape into atmosphere: $CO_2$; $CH_4$; volatile short-chain fatty acids; $N_2$; $NH_3$; $H_2S$; mercaptans.

*Paper (51.5% of total)*

1. May leach to groundwater: $CO_2$; aldehydes; organic acids; phenol; $NH_4^+$; $NO_2^-$; $NO_3^-$.

2. Remain in landfill through incorporation into microbial protoplasm: $NH_4^+$; C; P; S.

3. May escape into atmosphere: $CO_2$; volatile short-chain fatty acids; $H_2S$; mercaptans; $N_2$; $NH_3$.

*Unclassified (5% of total)*

1. Relatively inert.

2. Ashes in fill may leach soluble minerals to groundwater reducing its chemical quality.

Source: Sanitary Engineering Research Laboratory (1969).

Table 7-2. Costs for Solid-Waste Disposal.

|  | Capital Cost (Land Excluded) Dollars per Ton per Day | Operating Cost Dollars per Ton Disposed |
|---|---|---|
| Sanitary landfill | 1000-2000 | $1.25-2.25 |
| Central incineration | 3500-7000 | 3.50-5.00 |
| Composting | 1500-10,000 | 2.00-7.00 |

Note: Maintenance and operation of open dumps is estimated at five to twenty-five cents per ton.

# REFERENCES

American Public Works Association. 1966. *Municipal Refuse Disposal*, 2nd ed. Danville, Ill.: APWA.

Committee on Sanitary Landfill Practices of the Sanitary Engineering Division. 1959. *Manual of Practices, No. 39.* New York: American Society of Civil Engineers.

Sanitary Engineering Research Laboratory. 1969. *Comprehensive Studies of Solid Waste Management, Second Annual Report.* Berkeley: University of California Press.

National Center for Resource Recovery, Inc. 1974. *Sanitary Landfills.* Lexington, MA: Lexington Books, D.C. Heath and Co.

U.S. Environmental Protection Agency. 1978. *Land Cultivation of Industrial and Municipal Solid Wastes—A State of the Art Study*, EPA 600/2-78-1402. Washington, D.C.

# 8 INCINERATION

### INTRODUCTION

Incineration is essentially a process of reducing combustible wastes to inert residue by high-temperature burning. The process is over 100 years old. (In 1874, the first "crematory" in England was designed specifically for municipal use.) In the United States the first incineration plant was built in 1885 for the Army installation at Governors Island, New York.

The purpose of central incineration is to provide an economic, nuisance-free, sanitary method of disposing of refuse. Weighing the following advantages and disadvantages must precede a decision to select this process.

### ADVANTAGES

1. Much less land is required for disposal than when using a landfill.
2. A central location for an incineration plant is possible. A carefully operated plant in well-designed building with well-landscaped grounds which is accepted by neighbors can be used as contrasted to landfills.
3. An incinerator can produce ash residue that contains a negligible amount of organic matter and is nuisance-free. Residue is acceptable as fill material.

4. Modern incinerators can burn many kinds of refuse.
5. An incinerator is unaffected by climate.
6. An incinerator is flexible and can handle increasing or decreasing volumes of refuse as seasons and circumstances change.
7. Revenue can be obtained by recovering steam, metals, and glass.

## DISADVANTAGES

1. Incineration is costly—$1,500-$6,000 per ton of rated twenty-four-hour capacity in 1955. For example, a plant serving 100,000 people required a capital cost of about half a million dollars in 1955.
2. Operating costs are relatively high. Wages are high; maintenance and repair costs are high.
3. Site acceptance may be difficult because of potential nuisances from trucks, smokestacks, noise, and odor, and because of aesthetics.
4. Incineration is not an ultimate disposal method. Fly ash and residual ash must still be disposed of in other ways.

## COSTS

Incineration systems must include all facilities necessary such as (1) refuse handling and storage facilities, (2) furnaces and chambers where refuse is burned, (3) chimneys or stacks, (4) facilities for handling the gaseous and ash products of combustion, as well as the buildings and grounds.

Operational or service costs are usually paid for out of taxes, but construction costs are usually financed by a special bond issue. If capital costs are amortized over a twenty-year period, annual payments per $1,000 of total cost will be about $125.67 computed at a 1979 interest rate of about 11 percent.

Costs (capital) can be influenced up or down considerably by: (1) site preparation and subsoil conditions; (2) specific ordinances of the municipality for aesthetics and air pollution control; (3) peculiar labor and material costs of the area; (4) effect weather has on closure and structure insulation; (5) extra auxiliaries such as steam or power generation and the automation equipment; (6) percentage of twenty-four-hour rated capacity plant actually used; (7) provisions for expansions; and (8) amount of maintenance-free equipment (such as stainless steels) used in construction.

Operating costs will depend upon: refuse quality, degree of burning attained, degree of environmental protection used (air, water, noise, and land protection), type of incineration plant, wage scale and employee benefits provided, and, finally, the productivity of administrators.

Operating costs for New York City incinerators (1958) averaged about six and a half dollars per ton refuse destroyed.

Size and capacity are determined by: volume, weight, and character of refuse to be burned; hours a day to be operated and number of hours; and peak loads for seasonal and standby requirements.

Incinerator plant units included in usual systems are as follows:

1. Truck scales for weighing refuse trucks are either mechanic or manual.
2. Tipping floor is generally a paved (impervious) area on which trucks maneuver during dumping. The floor should extend far enough to allow several trucks to maneuver and unload simultaneously.
3. Storage pits are required if the collection and quantity of refuse reaching the incinerator building is too great to burn immediately. The storage may be required either because the refuse arriving at any one time is too great or the furnace used is too small. Pit storage volumes vary from capacities of twelve to thirty-six hours. Storage for much longer may lead to decomposition and accompanying odors.
4. Charging equipment consists of various methods including conveyors, rams, bulldozers, clamshells, and so on.
5. Charging hoppers keep an even flow of refuse going into the furnace and are generally located above the top of the furnace.
6. Monorail cranes (with capacities of 100 to 300 tons) are sometimes used to lift the refuse from the storage pit to the charging hoppers. In the larger furnaces (300 tons or more) a bridge crane is necessary. Cab controls for bridge cranes are preferred for optimum visibility and safety. Dust and wide variations in temperatures make air-conditioning necessary in the crane cab. The size and number of cranes depends on the magnitude of the plant.

Alternatives are recommended because breakdowns in crane operations can result in complete plant idleness. Accessories for cranes include grapples or buckets (one to four yd$^3$) of clamshell type with tines or teeth built to withstand considerable abrasion. An automatic lubrication system is also desirable for the crane.

7. Incinerator unit includes the furnace (drying hearth, grates, ignition chamber), combustion chamber, subsidence chamber, and breeding.

The furnace is an enclosed, refractory-lined structure equipped with grates and supplied with excess amounts of air in which the burning takes place. The combustion chamber is an enclosed, refractory-lined structure, sometimes combined with the furnace, in which the secondary, more complete burning of air-borne particles and gases take place. The subsidence chamber is a large separate, insulated chamber in which exhaust gases are expanded and slowed down so as to settle out the particulate matter before gases are discharged to the air environment.

## DESIGN AND OPERATING FEATURES

Further discussion of the design and operating features of each of the three major components of an incinerator system is necessary for a more complete understanding of this method of refuse disposal.

### Furnace

The ignition and primary burning of refuse takes place here. Also, preheating, drying, ignition, and most of the burning occurs. In the furnace we include the hearth, grates, the ash hopper, and gate. Stoking is almost always automatic (mechanical) in modern furnaces.

The continuous feed (most common) grate is of a traveling type, usually slightly inclined (although sometimes flat).

The furnaces are lined with refractory bricks and insulating bricks. Furnaces operate as low as 1200°F and as high as 2000°F, but an optimum depends on type of refuse and environmental standards and is usually about 1,750°F. At higher temperatures refractories tend to deteriorate, and gas velocity is increased so high that settling of fly ash is hindered. The two primary design criteria are the grate area and combustion volume. For optimum burning the loading on the grates is designed on the release of 300,000 BTUs of heat per hour per square foot of grate. On the other hand 200,000 BTUs of heat per hour per cubic foot of combustion volume are released from typical burning of urban refuse. However, these general values depend upon the type and composition of refuse being burned. The BTUs of heat released per pound of refuse should be determined for each refuse type prior to designing the furnace. In general, for the pounds per square feet per hour, values of 75 for municipal, 50 for rubbish only, and 65 for combustible refuse (garbage plus burnable rubbish), can be used for grate area computations when 300,000 BTUs per hour per square foot of heat are released or 25 cubic feet of combustion volume per 24 hours per ton of refuse for municipal refuse; 45 for rubbish and 35 for combustible refuse, which corresponds to a heat release of about 12,500 BTU per hour/ft$^3$ of volume.

### Combustion Chamber

This is a secondary chamber in which complete combustion of the gaseous products occurs. It can be located separate from or included with the primary furnace. An important function of the combustion chamber in addition to completing the burning of gaseous and incandescent particular matter, is to settle out the heavier large particles of fly ash. Gas velocities should be controlled within the range of ten to forty feet per second. Designing the chamber with baffles, bridge walls, and other obstacles to force the combustion gaseous products to collide thus affects the settling of fly ash.

### Subsidence Chamber

This is used to complete the fly-ash removal and sometimes to complete the combustion of gaseous products. If fly-ash removal is required to a high degree, wet collectors and scrubbers are sometimes used in place of the subsidence chamber. Here gas velocities are kept low at five to ten feet per second.

## Fly Ash Screens

Fly-Ash Screens are used to trap large particulate light fly ash which escaped the subsidence chambers and are placed in the breeching area or stack. The wire should have mesh opening of a maximum of one-quarter inch.

Air for complete combustion is required for the furnace. It is introduced under the grates, over the fire by blowers, or both. Air required for combustion varies from 5.4 to 12.1 pounds per pound of moisture and ash-free waste (MAF). In general, it requires 0.75 pounds of air per 1000 BTUs for complete combustion. Some approximate heat value (high) of components of refuse are shown in Table 8-1.

The "as-fired" BTU/# value is ontained by the following:

$$\text{BTU/\# as-fired} = \left[1 - \left(\frac{\% \text{ moist} + \% \text{ ash}}{100}\right)\right] 7300$$

for garbage with 35 percent moisture and 5 percent ash.

$$\begin{aligned}
\text{BTU/\# as fired} &= 7300 - (.35 + .05) \\
&= 7300 - .4\,(7300) \\
&= 7300 - 2920 \\
&= 4380
\end{aligned}$$

Table 8-1. Some Approximate Heat Value of Components of Refuse.

| Refuse | Heat Value (BTU/#MAF waste) |
|---|---|
| Paper | 7,900 |
| Wood | 8,400 |
| Leaves and grass | 8,600 |
| Rags, wool | 8,900 |
| Rags, cotton | 7,200 |
| Garbage | 7,300 |
| Rubber | 12,500 |
| Suet | 16,200 |

Usually in the complete combustion we use an excess of air of 200 percent of that required.

### Refractories

Refractories are made of fired clay and other materials such as brick which are highly resistant to heat. These are used to line furnaces, combustion chambers, subsidence chambers, breechings and chimneys. Design materials for refractories include cost, physical strength, resistance to heat, heat and physical shockability, spalling, slagging, and wear abrasion, low coefficient of expansion, and low-bonding properties. Ideal refractory will contain a minimum of fused glassy material, be a porous structure, and possess a high pyrometric cone equivalent. Bricks of high density, low porosity, and high pyrometric cone equivalent are most suitable for refractories. A 1959 cost of a standard refractory brick (nine by four and a half by two and a half inches) was two dollars each for the SiC (Silicon Carbide) optimum type.

### Chimneys

Chimney (Stacks) are the vertical flues which carry the gases to the atmosphere and, at the same time, provides natural draft to aid combustion. The gases during operation are hotter than surrounding air and hence are less dense and rise in the stack. The pressure at the bottom of the stack is less than that at the top in contact with outside air, and hence change in pressure causes a natural draft. The draft depends on the height of the stack and the change of temperature. The height and diameter of the stack are designed by computing the draft required, the gas velocity, and maximum quantity. If short chimneys are used for one reason or another, induced draft fans must also be used to compensate for reduced draft, and hence cost savings are eliminated. Stacks may be steel, masonry, or reinforced concrete.

Steel stacks are lower in cost, lighter in weight, require less space, allow less infiltration of outside air, are unsightly, and require guy wires, higher maintenance, and costly heat-resistant alloys, and need protective coatings.

Masonry stacks are used for high, natural draft stacks, for which special foundations are required. They use two surfaces (an outer shell of structural steel and an inner shell of brick) separated by a four-inch air space. The outer shell is thicker at the base of the stack than at the top. They do not require guying and have a longer useful life with lower maintenance costs.

Reinforced concrete stacks are strong and can withstand greater wind and ground stresses. They can be less expensive for more than 200 feet high stacks, and their surface may crack because of temperature and stress conditions.

## Accessories

Accessories include lightning arresters, warning lights, test openings, clean-outdoors, ladders, catwalks, and caps. Lightning rods are usually three-quarter-inch lead-covered copper rods extending about six feet above the chimney cap and spaced about eight feet apart.

Residues consist of 5 to 25 percent which remains as ash after combustion. They must be cooled or quenched: dust and steam are rapidly removed from the ash tunnel by either suction fans or blowers, and the exhaust is used for combustion air. In large plants (continuous feed) ash is discharged continuously in a water trough. An endless conveyor system drags the settled ash from the bottom up an incline to allow the quenching water to run off. The drained ash is put in trucks and hauled to a final disposal site. Recovery and re-use of unburned matter (cans, steel rods, and so on) in the residue can be carried out on the conveyor before depositing it in the trucks. Conveyors are subject to a great deal of wear, and moving parts coming in contact with abrasive ash should be made of durable alloys.

## AIR-POLLUTION CONTROL

Although this is a subject for another study, it must also be considered here. The variability of character of refuse and the air environment surrounding the plant make establishing standards and hence abatement equipment difficult. Most municipalities require a limit of particles of suspended matter per unit of gas volume. These are measured by a wide variety of procedures varying from general observation of smoke to the use of a smoke density chart (Stern 1962) or to the use of a Hi-Volume filter to entrap the particles in a given gas volume.

## FLY ASH

The amount of fly ash is dependent upon the character of refuse, the design of the incinerator, and the method of operating the plant. Forced draft also causes more fly ash than natural draft. Although fly ash is thought to be largely mineral matter, it still contains from 10 to 20 percent unburned organic matter. The minerals are mainly silicon, iron, aluminum, calcium, and ash sulfur. Wet collectors can be used to collect the fly ash; water can be re-used after settling out the fly ash, but they use a lot of water and create potential maintenance and explosion problems.

## FINAL COST RELATIVE TO OTHER TREATMENTS

The final costs of disposal using incineration that meet 1979 federal pollution standards appear to be some 30 percent higher than those arising from rail haul and sanitary landfill for the large coastal city (shown in Chapter 7).

## REFERENCE

Stern, A.C. 1962. *Air Pollution*, Vol. II. New York: Academic Press, p. 465.

# 9 PYROLYSIS

## DEFINITION AND PRODUCTS

Pyrolysis involves burning of refuse at about 900°C in the absence of air (oxygen). It is actually a process used in the chemical industry known as "destructive distillation."

According to the U.S. Bureau of Mines, a ton of refuse will yield the following products: 154-424 pounds of *solid residue*; 0.5-6.0 gallons of *tar*; 1-4 gallons of *light oil*; 97-133 gallons of *liquor*; 16-32 pounds of *ammonium sulfate;* and 7,380-18,058 cubic feet of *gas.*

The energy from the gas is more than sufficient to provide the heat for the pyrolysis.

To date, most of the development of this process has been only in the pilot-plant stage. One such pilot plant has been reported in the following description and data.

## PYROLYSIS PILOT PLANT AND TEST PROCEDURES

The experiments were conducted in the pilot plant shown schematically in Figure 9-1. The plant consists essentially of an electric furnace, cylindrical

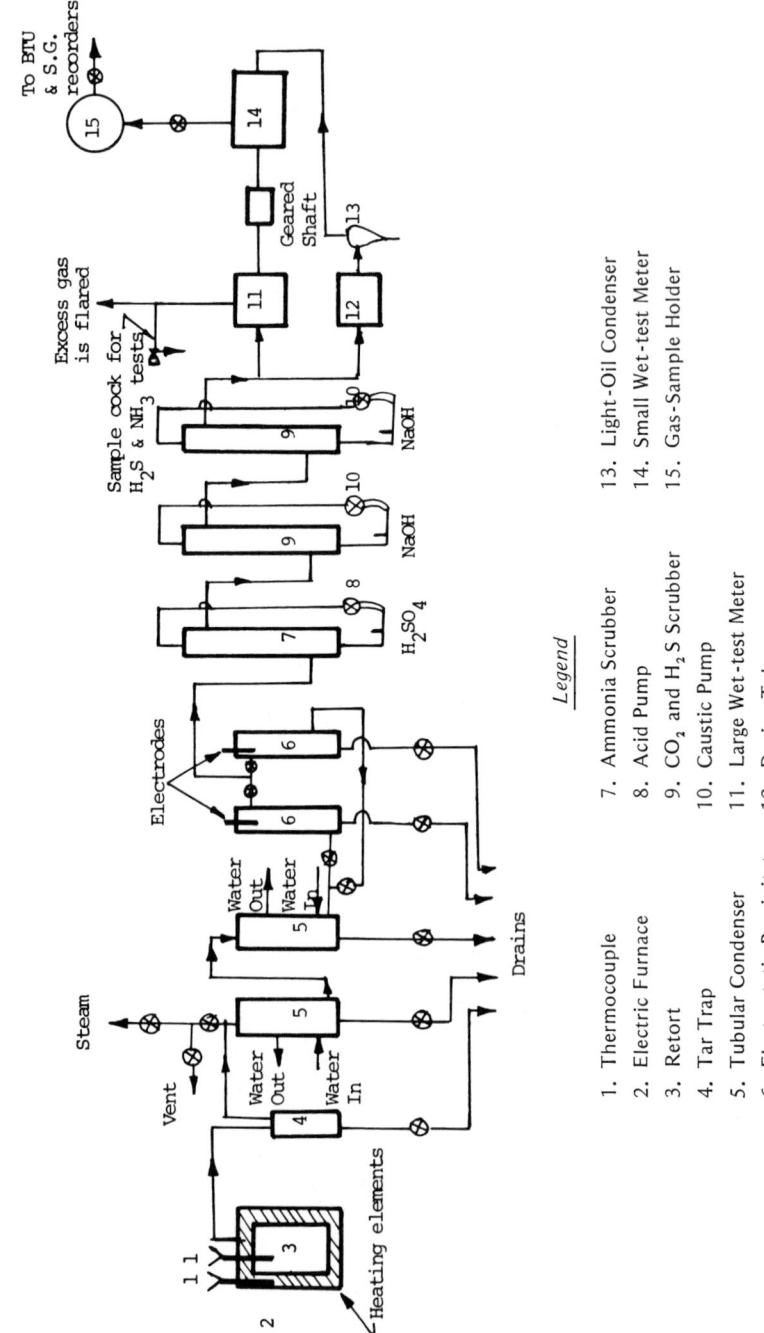

Figure 9-1. Flow Diagram of Pilot Plant Used To Pyrolyze Municipal and Industrial Refuse.

Legend

1. Thermocouple
2. Electric Furnace
3. Retort
4. Tar Trap
5. Tubular Condenser
6. Electrostatic Precipitator
7. Ammonia Scrubber
8. Acid Pump
9. $CO_2$ and $H_2S$ Scrubber
10. Caustic Pump
11. Large Wet-test Meter
12. Drying Tube
13. Light-Oil Condenser
14. Small Wet-test Meter
15. Gas-Sample Holder

steel retort, condensing and scrubbing train for product recovery, and gas-metering and sampling devices.

The electric furnace (2) (refer to Figure 9-1) is 26 inches inside diameter and 48 inches deep and is heated by nickel-chromium resistors spaced evenly in the furnace wall. The retort (3) is 18 inches in diameter and 26 inches deep and is made of 16-gage steel in the wall and 10-gage steel in top and bottom. Gases and vapors exit from the retort through a 2-inch-diameter offtake pipe and enter an air-cooled trap (4), where tar and heavy oils are collected. The gases and vapors are cooled to room temperature in two water-cooled condensers (5) connected in series, where additional heavy oil and liquor are collected. Final traces of heavy-oil mist are removed by one of the alternate electrostatic precipitators (6). The gas then passes successively through packed scrubbers, where ammonia is removed with sulfuric acid (7), and carbon dioxide and hydrogen sulfide are removed with caustic soda solution (9). The scrubbed gases pass to the large (11) and small (14) meters, which are geared together so that 99 percent of the gas passes through the large meter and is flared. The suction side of the small meter is cross-connected to the inlet side of the large meter so that 1 percent of the gas passes through a drying tube (12) and a condenser (13) immersed in acetone and solid carbon dioxide, where light oil is removed. Light oil recovered from the gas that passes through the small meter is calculated to the total gas yield. The gas from the condenser passes through the small meter (14) to the gas holder (15), and representative samples are taken from the holder for analyses. Steam is used to purge the condensers and piping at the conclusion of the test.

## YIELDS OF PRODUCTS FROM PYROLYSIS OF REFUSE

*Tar, heavy oil,* and *liquor* were collected from the air-cooled condenser, water-cooled condensers, and the electrostatic precipitator. *Light oil* was collected from the carbon dioxide-acetone condenser after the heavy oils had been removed and the gas had passed through the acid and caustic scrubbers. *Residue* remained in the retort after the test. Total gas, the sum of readings from both meters, was corrected to $60°F$ and 30 inches of mercury pressure.

A ton of municipal refuse pyrolyzed at *$900°C$* yielded *154 pounds* (7.7 percent of refuse charge) of *solid residue, 0.5 gallons of tar, 114 gallons of liquor, 25 pounds of ammonium sulfate,* and *17,741 cubic feet of gas.* Pyrolysis at $750°C$ increased tar production to 2.6 gallons per ton and light oil to 2.5 gallons per ton, but showed a marked decrease in gas production to 9,628 cubic feet per ton, which was slightly more than half that obtained at $900°C$. Pyrolysis starting at $500°C$ and increasing to $900°C$ yielded the largest amount of tar (4.8 gallons per ton), but the gas yield was 11,509 cubic feet, which was intermediate to that obtained at the other two temperatures.

*Pyrolsis at $900°C$* of a ton of *municipal refuse* containing *mainly plastic film* yielded 382 pounds (19.0 percent of refuse charge) of solid residue, 1.4 gallons of tar, 0.6 gallons of light oil, 97.4 gallons of liquor, 31.5 pounds of ammonium sulfate, and 18,058 cubic feet of gas. At $750°C$, gas production decreased to 7,380 cubic feet, less than half that obtained at $900°C$, and tar and light-oil production increased. Pyrolysis through a temperature range of $500°$ to $900°C$ yielded the largest quantity of tar, 5.6 gallons per ton, and 11,545 cubic feet of gas.

A ton of *industrial refuse* pyrolyzed at $900°C$ and containing mainly paper, rags, cardboard, and small amounts of metals yielded 618 pounds (10.9 percent of refuse charge) of solid residue, small amounts of tar, 1.4 gallons of light oil, 68.5 gallons of liquor, 22.9 pounds of ammonium sulfate, and 14,065 cubic feet of gas. Pyrolysis of industrial refuse at lower temperatures generally increased tar and light-oil yields but did not result in the marked decreases in gas production that were obtained when municipal refuse was pyrolyzed at the lower temperatures.

The solid residue, a lightweight, flaky, carbonaceous material, represents a 90-percent weight reduction for the municipal refuse and a 65-percent reduction for industrial refuse. The residue from municipal refuse has the highest fuel value and can range from 10 to 17 million Btu per ton.

As much as 53 percent of the raw product gas is carbon monoxide, methane, and ethylene, and the volume of gas produced is ample to provide the energy for pyrolysis of the refuse. The high hydrogen content of the gas may also have promise as a source of hydrogen or methane.

The tar from the pyrolysis of refuse will provide an additional 736,000 Btu per ton of refuse, and the light oil produced is a potential source of benzene and toluene.

The cities of San Diego, California, South Charleston, West Virginia, and Baltimore, Maryland, have modified pyrolysis systems for solid waste treatment (see Table 9-1).

The Midland Ross Corporation (*Wall Street Journal*, May 5, 1981, p. 9), is reputed to be the world's biggest supplier of industrial high temperature furnaces. It proposes a pyrolysis system shown in Figure 9-2.

Table 9-1. Pyrolysis Pilot Study Data.

| Location | Type of Plant | Capacity | Process Hi-lites | Pollution Control | Residue | Utilities Required | Economics | Products |
|---|---|---|---|---|---|---|---|---|
| San Diego | Pryolysis oil | 200 Tons/day | Coarse shredding air classification pulverizing ferrous metal, aluminum and glass separation | Pyrolysis gas cleaned of acids, particulates, and water. Process water, treated before emptying to sewer. | Heavy organic and miscellaneous inorganics to landfill 15–25% by weight | ~750 h.p. and electric utilities support | Estimated to break even depending upon marketing products | Pyrolysis oil 10,000 Btu/# CHAR 9,000 ferrous metal, aluminum, glass, CU/LD |
| South Charleston | Pryolysis gas | 200 Tons/day | High temperature partial oxidation with pure oxygen pryolysis with little or no feed preparation required | Pyrolysis gas cleaned of acids, particles and water. Process water treated and recirculated. | Inorganics are 10–15% by weight, metal, ash, glass, slag | 200 h.p. and electric service support 2400 kw. for $O_2$ plant, 5000 kw. for total plant. | Break even to small profit depending upon product market | Pyrolysis gas 300–320 Btu ft.$^3$ glass aggregates for building materials |
| Baltimore | Pyrolysis gas/steam | 1,000 Tons per day 135T/D (pilot plant) | Partial oxidation with air, pyrolysis using rotary-kiln reactor off gas burned, boiler to produce steam | Combined gases scrubbed to remove solubles and particulates | Glassy aggregate and CHAR separated by flotation. Ferrous separations by magnets | ~2500 h.p. plus elective service | $15 million operating net cost. $4.77/ton including sale of ferrous metals | 200,000 pounds per hour, steam glass, aggregates, ferrous metals |

Figure 9-2. Pyrolysis System.

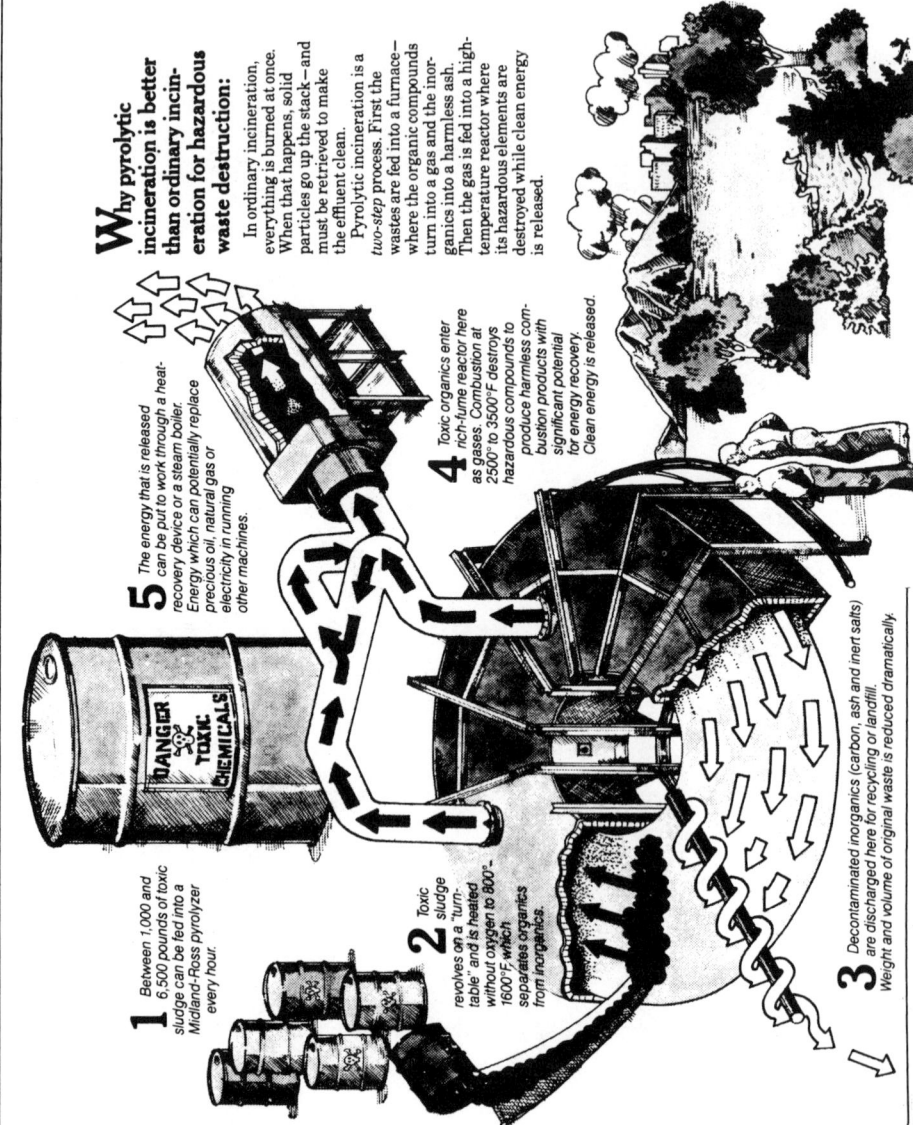

# 10 HAZARDOUS WASTES

## INTRODUCTION

The Environmental Protection Agency (EPA) defines the characteristics of a hazardous waste in terms of ignitability, corrosivity, reactivity, and Extraction Procedure toxicity. These are set forth specifically in Table 10-1.

An historic summary of federal legislation relating to the control of toxic substances in the United States is given in Table 10-2.

There is a great deal of concern by environmentalists and the general public about so-called hazardous wastes. Briefly, these are wastes which appear to be incompatible with normal disposal techniques and, at the same time, they have been shown to exhibit toxicity to some form of biological life. Some effects are generally well-known to the public, such as fires, explosions, and oils on beaches. Others, more recently revealed, are not so obvious to people, such as the leaching of metals and complex and potentially carcinogenic organic matter from industrial dumps and even sanitary landfills. Whether they originate from either source they warrant our special attention in this text just as they are receiving currently from the U.S. Environmental Protection Agency (See Chapter 10), Resource Conservation and Recovery Act of 1976, Subtitle C Section 3001-3011). Major hazards are described briefly here.

**Table 10-1.** Characteristics of Hazardous Waste.

1. *Ignitability* (261.21), EPA HW #D001
   - A liquid that has a flash point less than 60°C (140°F). *Exemption* — aqueous solution with less than 24 percent alcohol.
   - A waste that is not a liquid but is capable under standard temperature and pressure of causing fire through friction, absorption or moisture or spontaneous chemical changes, and, when ignited, burns so vigorously and persistently that it creates a hazard.
   - An ignitable compressed gas.
   - An oxidizer.

2. *Corrosivity* (261.22), EPA HW #D002
   - An aqueous waste that has a pH less than or equal to 2, or greater than or equal to 12.5.
   - A liquid that corrodes steel at a rate greater than 6.35mm (0.250 inches) per year at 55°C.

3. *Reactivity* (261.23), EPA HW #D003
   - Normally unstable, readily undergoes violent change without detonating.
   - Reacts violently with water.
   - Forms potentially explosive mixtures with water.
   - When mixed with water, generates toxic gases, vapors, or fumes in dangerous quantitites.
   - Cyanide or sulfide bearing waste which, when exposed to pH conditions between 2 and 12.5 can generate toxic gases, vapors, or fumes in dangerous quantities.
   - Capable of detonation or explosive reaction if subjected to a strong initiating source or heated.
   - Readily capable of detonation or explosive decomposition or reaction at standard temperature and pressure.
   - Forbidden explosives, Class A or Class B explosives.

4. *EP Toxicity* (261.24), EPA HW #D017
   A solid waste is toxic if, when extracted by the EP method, the leachate contains concentrations of constitutents equal to or greater than 100 times the Primary Drinking Water Standard. If the waste contains less than 0.5 percent filterable solids, then the filtrate is considered the extract.

Source: *Federal Register* (1980).

## FIRE AND EXPLOSIONS

A real danger of fire, especially those from affluent U.S. municipalities and industries, exists in untreated solid wastes largely because roughly 50 percent is paper products, wood, or plastics. There exists a possibility that these materials will be ignited by carelessly disposed-of smoking materials. However, there is always a potential for organic materials compacted together to undergo spontaneous combustion.

## SPONTANEOUS COMBUSTION

Spontaneous combustion occurs when organic materials decompose. At the beginning the action of aerobic (oxygen-loving) bacteria releases heat which is unable to escape causing the temperature of the mass to rise. Thermal insulation is usually provided by the inorganic or mineral materials.

A high concentration of uninsulated organic material has to be present before the temperature rises and burning commences.

Spontaneous combustion is a problem when storing agricultural materials such as hay or coal, and other finely divided fuels such as wood chips and sawdust. A recent example emphasizing the need to avoid large masses of any materials which have combustion potential is the fusing together of a pile of steel turnings which were awaiting reclamation in an open-air yard. The turnings were wet with cutting oil. The plant supervisor maintained that the inside of the pile became red-hot and subsequently cooled down to form a solid mass. The resulting pyramidical block was too thick to be cut with torches, and all other attempts to reduce it to moveable pieces failed.

Another contemporary problem has been experienced in grain storage bins. Spontaneous combustion from the fine grain dust particles has occurred in many midwestern areas of the United States during storage or transfer.

Little scientifically useful data is available to advise how to avoid spontaneous combustion in all types of solid wastes. A good practice is simply to avoid storing solid wastes in warm moist conditions for several days when the smallest dimension of the waste mass is greater

Table 10-2. U.S. Federal Laws and Agencies Affecting Toxic Substances Control.

| Statute | Year Enacted | Responsible Agency | Sources Covered |
|---|---|---|---|
| Food, Drug and Cosmetic Act | 1938 | FDA | Basic coverage of food, drugs, and cosmetics |
| Food additives amendment | 1958 | FDA | Food additives |
| Color additives amendment | 1960 | FDA | Color additives |
| New drugs amendment | 1962 | FDA | Drugs |
| New animal drugs amendment | 1968 | FDA | Animal drugs and feed additives |
| Medical device amendment | 1976 | FDA | Medical devices |
| Federal Insecticide, Fungicide and Rodenticide Act | 1948, amended 1972, 1978, 1980 | EPA | Pesticides |
| Federal Hazardous Substances Act (Formerly the Federal Hazardous Substances Labelling Act) | 1960, amended 1966 | CPSC | "Toxic" household products (equivalent to consumer products) |
| Wholesome Meat Act | 1967 | USDA | Food, feed, and color additives and pesticide residues in meat |
| Wholesome Poultry Products Act | 1968 | USDA | Food, feed, and color additives and pesticide residues in poultry products |
| Clean Air Act | 1970, amended 1977 | EPA | Hazardous air pollutants |
| Occupational Safety and Health Act | 1970 | OSHA | Work-place toxic chemicals |
| Poison Prevention Packaging Act | 1970 | CPSC | Packaging of dangerous children's |

| | | | |
|---|---|---|---|
| ...Transportation Act | ... | DOT | Transportation of toxic substances generally |
| Federal Railroad Safety Act | 1970 | EPA | Railroad safety |
| Clean Water Act (Formerly Federal Water Pollution Control Act) | 1972, amended 1977 | EPA | Toxic water pollutants |
| Marine Protection Research and Sanctuaries Act | 1972 | EPA | Ocean dumping |
| Consumer Product Safety Act | 1972 | CPSC | "Toxic" household products (equivalent to consumer products) |
| Ports and Waterways Safety Act | 1972 | DOT (Coastguard) | Shipment of toxic materials by water |
| Lead Based Paint Poison Prevention Act | 1973, amended 1976 | CPSC | Use of lead paint in federally assisted housing |
| Safe Drinking Water Act | 1974, amended 1977 | EPA | Drinking water contaminants |
| Resource Conservation and Recovery Act | 1976 | EPA | Hazardous wastes |
| Toxic Substances Control Act | 1976 | EPA | Regulate existing chemical hazards and to conduct premanufacturing of all new chemicals—not covered by other laws related to toxic substances |
| Comprehensive Environmental Response, Compensation and Liability Act (Superfund Act) | 1980 | EPA | Cleanup of spills of hazardous substances |

Source: *Federal Register* 1980.

than six to ten feet (2 to 3 meters). Also recommended is storage of smaller quantities, or keeping the refuse very dry or very wet to inhibit spontaneous combustion.

## EXPLOSIONS

Live ammunition and other explosive materials may be detonated during the handling and processing of mixed solid wastes. Mixtures or large quantities of apparently innocuous materials can be exploded in certain situations. In incinerator explosions, for example, explosions may occur from loading large amounts of plastic wastes. Probably in this case, gaseous pyrolysis products, when combined with oxygen, reached explosive proportions. Methane gas formed from anaerobic decomposition of organic waste can also cause explosions as discussed in Wilson 1979.

## GAS PRODUCTION

The principal gaseous product of anaerobic decomposition of organic wastes is methane ($CH_4$), although other gases such as carbon dioxide ($CO_2$), nitrogen ($N_2$), and hydrogen sulfide ($H_2S$), are also present. Wastes buried in landfills undergo predominently anaerobic decomposition, and the free gases find their way up through the fill to the surface. As discussed in Chapter 7, lateral movement of these gases—especially when restrained from rising vertically in the fill—may result in distant problems. In some cases, gases have been known to emanate in basements of houses and buildings. Deaths and injuries have resulted from asphyxiation, poisoning, or from the detonation of explosions of air-methane mixtures. Sanitary-landfill regulations require that ducts be incorporated into landfills to lead the evolved gases to be discharged at a safe location. Building codes for structures near landfills generally require particularly stringent standards for the construction of impermeable basements or sealed and ventilated underground services.

## LEACHATE PRODUCTION

During decomposition of organic wastes a liquid is normally produced which is termed a "leachate." In landfills, some of this liquid results from the weight applied by the fill above pressing out hydrolysis liquid products from refuse. Another principal component of leachate is rainwater which, during its passage through the landfill, dissolves many components of the fill. Leachate may emerge from a landfill with a biological oxygen demand of over 20,000 milligrams per liter, which is about 100 times stronger than raw sewage. Sanitary-landfill regulations call for the capture of all leachate and for its treatment by normal sewage-treatment methods or the equivalent. Studies have shown that the leachate concentration from a landfill might still be significant after twenty years. Wilson 1979 has described some of the danger points in handling solid wastes.

## WASTEBASKETS

Wastebasket refuse from any source is usually dry and reasonably clean and emptied every one or two days, so that few problems of fly-breeding and multiplication of undesirable bacteria exist. The paper contents are therefore usually highly flammable and hazardous to personnel. Wastebaskets contain ashes from cigarettes, cigars, and pipes, and frequently for the butts and pipe residues, the contents of ash trays, and the matches and spent cigarettes-lighter cartridges. Only some of these materials causes a smoldering condition and frequent fires; when these baskets are placed near curtains, drapes, and upholstered furniture, fire danger is imminent. Controls include limiting smoking within buildings, providing a large capacity of ashtrays near each receptacle, or using self-closing waste containers. Hazards to persons emptying wastebaskets comes predominantly from thoughtlessly discarded dangerous materials, such as partially filled bottles of strong acids and used hypodermic needles from animal experiments. Workers emptying waste refuse containers should be instructed to watch carefully for these dangerous objects and to empty them without using their hands to unload. Another approach is to hold the person who discards prohibited materials into the regular waste stream legally liable for damages resulting from these actions.

## REFUSE CHUTES

Handling of solid wastes is inadequately considered in the design stage of buildings. Refuse chutes, when they are incorporated, greatly reduce handling costs and at the same time reduce some of the dangers involved in solid-waste handling. There is the dangers of fire if a refuse chute is designed without appropriate traps to prevent the chute from acting as a chimney and if the materials and construction are insufficiently fire resistant. Building codes in most cities require that fire traps be used. Because of the difficulty in tracing the person using the refuse chute such systems encourage people to dispose of all kinds of prohibited, dangerous, or incriminating wastes.

## COMPACTORS

Compactors are often exposed to explosions of partially filled vapor propellant cans. Fortunately, the use of these cans is diminishing because of air-quality regulation. Fires are normally extinguished by compaction which removes oxygen for combustion. When fire inside a truck body occurs, standing instructions to the crew usually dumps the burning load in the middle of the highway where the fire department can easily extinguish it. Other dangers to operators from compactors originate from exploding T.V. tubes or certain light bulbs usually prevented by a ram and sometimes a screw which is closed off sealing the compacted refuse from the loading chamber before operation.

## INCINERATORS

Small incinerators manufactured for apartment houses and commercial establishments and described in Chapter 8 contains all the dangerous characteristics of an industrial processing plant. The mechanical handling of extremely varied materials, along with some danger of explosions resulting from explosives or from a high concentration of plastics, causes hazards mainly to workers but also potentially to people living in the area. Poisonous gases are given off during the combustion of some plastics, generally from the urethanes and the vinyls. Furnaces are normally maintained under negative pressure which protect the workers, but the surrounding air may be contaminated and therefore present a hazard to a wide number of people.

## SHREDDERS

The high strength materials used in the construction of shredders described in Chapter 2 enables them to contain all but the most violent explosions.

## RECLAMATION PLANTS

Reclamation when unorganized is extremely dangerous because a landfill area is unsafe and because it is difficult to reclaim desired items from their location in the dump, and from the danger of being in the direct path of vehicles discharging their loads. Fencing and policing landfills to prevent scavenging will alleviate most of these dangers. Modernization and mechanization of landfill reclamation plants will also ameliorate the dangers described above.

## LANDFILLS

Dangers from improper landfilling procedures just discussed fall into the categories of fires, gas production leading to fires of possibly long duration, asphyxiation, and explosions in buildings, leachate production leading to possible pollution of aquifers, rat infestation leading to spread of disease, and scavenging leading to a probability of injuries, such are the major hazards of landfills. Once fires begin in a land landfill, they are very difficult to extinguish totally. Fires reduce the density of the fill and lead to a sudden structural collapse of the surface, with obvious dangers to people and to any buildings in or on the area at the time.

## HAZARDOUS INDUSTRIAL WASTES

The food industry generates about 650,000 tons per year of organic sludge (Suler 1979). Again, this sludge is mostly organic, composed of cotton, wood, and synthetic fibers, dyes, sizing, and so on. The pulp and paper industry generates about 2 million tons per year of carbonaceous containing cellulose fibers and biomass sludges. Landspreading has been used for disposal of this material. The pharmaceutical industry generates about 200,000 tons per year of organic sludges. Wood wastes from the lumber, plywood, pulp, and related industries amount to many millions of tons per year. The leather tanning industry produces about 100,000 tons per year of sludge, much of it containing a high percentage of chromium from the use of chrome tanning agents. Feed loss generates about 50 million tons per year of manure, which is used for fertilizer and also disposed of by land-spreading. The petroleum industry produces approximately 850,000 tons per year of petroleum-related sludge. Much of it is disposed of by landfill. Suler (1979) identifies two arbitrary categories of organic wastes for the purpose of this discussion. Type 1 wastes can be considered those which are generally nontoxic, readily degradable, and often have characteristics or components which render them valuable in some way. He gives an example of a Type 1 waste food-processing sludge, such as from dairy waste, whey, or tomato pulp. These are not hazardous and can be composted for re-use. Suler defines Type 2 wastes as those wastes which contain toxic, hazardous, and recalcitrant compounds in sufficient quantity to limit or restrict the use of the waste in either its original or a converted form. He recommends treatment of Type 2 solid wastes to degrade the toxic or offensive compounds to nonhazardous form, thereby facilitating the ultimate disposal of the waste. The Toxic Substances Control Act and the Resource Conservation and Recovery Act legislative edicts severely restrict the handling of waste materials classified as toxic (see Table 10-3). While the advantage of degrading toxic or recalcitrant materials in a Type 2 waste is evident, the question is: How effective is composting in accomplishing this goal? Suler reports that more resistant compounds, which are mixed with and come in close contact with the more easily degradable substrates and the active micro-organisms, may also be degraded to a far greater extent than without the easily degradable substrate. Composting also seems to be somewhat resistant to the toxic effects of certain compounds. Suler identifies three types of reactions for Type 2 wastes in a compost system—positive, neutral, and negative. He defines a positive reaction as one in which the active organisms are able to degrade the compounds; neutral as one not degraded; and negative as one whose their presence inhibits the breakdown of other substrates in the compost. Preliminary results have shown some interesting interactions. He reports that a resistant crankcase oil mixture is not only difficult to degrade but its presence in a refuse mixture inhibits the degradation of the refuse. On the other hand, he found that a readily degradable oil, such as our synthetic oil, actually stimulates degradation of the refuse. Bench-scale compost studies indicates that waste army TNT wastes degrade to near complete mineralization or conversion to biomass. These composting of toxic solid wastes offer a potential for conversion from a toxic compound to a nontoxic, environmentally safe form which can be easily disposed of.

Studies by the National Canners Association with continuous thermophilic compost systems and batch-type systems have demonstrated the rapid degradation of diazion and parathion pesticides in the processes. DDT, unfortunately, was not apparently degraded by composting.

A recent report from Texaco describes that company's success in experimental trials using a Beltsville-type static pile system for the purpose of treating refinery sludge. They found a decrease in the amount of extractable oils and greases during composting. They used waste-activated sludge combined with small amounts of oily sludge. The resulting compost was used as a good soil additive. Suler believes that the utility of composting in treating these wastes depends on the ability to degrade them under optimum conditions. He is optimistic that composting will be an important part of the overall industrial waste-management picture.

Stearns Conrad Schmidt Engineers (1978) report a growing interest in nonstandard disposal techniques for hazardous wastes that can provide for a practical utilization of the waste materials. These have been instigated by several recent events: (1) contamination of underground water supplies; (2) increasing cost of landfill sites; (3) awareness that wastes may actually be useful to other industries; (4) increasing air-pollution control; (5) difficulty in finding economical landfill sites. Some illustrations of nonstandard (or re-use) techniques for specific wastes are given in Chapter 15.

Recently ("Reagan Plans to Ease Toxic Waste Rules . . ." 1981-84) there is some evidence that "environmental controls will vary depending on the 'degree of hazard' posed by the chemicals stored or disposed of at the sites and on the way they are handled."

A new detoxification process designed to eliminate the health and environmental threats posed by common insulating chemicals such as PCBs ("New Process . . ." 1981), has been approved by U.S. government. "The new process, involving chemical gear that can be moved

## 104 INDUSTRIAL SOLID WASTES

Table 10-3. Water Quality Limits for Toxic Pollutants for Three Uses.

| Toxic Chemical | Concentration of Toxic Material Considered Limit (ug/ℓ) | | | | |
|---|---|---|---|---|---|
| | Freshwater Aquatic Life | | Saltwater Aquatic Life | | Human Health |
| | Acute | Chronic | Acute | Chronic | |
| 1. Acenaphthene | 1,700 | — | 970 | 710 | 20 (est.) |
| 2. Acrolein | 68 | 21 | 55 | — | 320 |
| 3. Acrylonitrile | 7,550 | — | not available | | 058–.006 lifetime |
| 4. Aldrin–Dieldrin | .0019 (24 hr avg.) 2.5 maximum conc. | | 0.0019 (24 hr avg.) 0.71 (max. conc.) | | .0071 ng/ℓ–.71 ng/ℓ |
| 4A. Aldrin | 3.0 | — | 1.3 | — | .0074 ng/ℓ–.74 ng/ℓ |
| 5. Antimony | 9,000 | 1,600 | not available | | 146 |
| 6. Arsenic | 440 | 40 | 508 | — | .22 ng/ℓ–22 ng/ℓ |
| 7. Asbestos | not available | | not available | | 3,000–300,000 fibers/ℓ |
| 8. Benzene | 5,300 | — | 5,100 | 700 | .066–6.6 |
| 9. Benzidine | 2,500 | — | not available | | .01 ng/ℓ–1.2 ng/ℓ |
| 10. Beryllium | 130 | 5.3 | not available | | .37 ng/ℓ–37 ng/ℓ |
| 11. Cadmium | 3.0 (100 ppm hardness) max. .025 (100 ppm hardness) avg. | | 59 (maximum) 4.5 (avg.) | | 10 |
| 12. Carbon tetrachloride | 35,200 | — | 500,000 | — | .04–4.0 |

HAZARDOUS WASTES 105

| | | | |
|---|---|---|---|
| 13. Chlorodane | 2.4 max<br>0.0043 (24 hr. avg.) | .09 max<br>.0040 (24 hr. avg.) | .046–4.6 |
| 14. Chlorinated benzenes | 250    50 (fish 7.5 days) | 160    129 | hexachlorobenzene<br>.072 ng/ℓ – 7.2 ng/ℓ<br>tetrachlorobenzene<br>38 ug/ℓ – 48<br>pentachlorobenzene<br>74–85<br>monochlorobenzene<br>488 |
| 15. Chlorinated Ethanes | <u>Acute</u><br>118,000 (1, 2 dichlorethane)<br>18,000 (two trichloroethane)<br>9,320 (two tetrachlorethanes)<br>7,240 (pentachloroethane)<br>980 (hexachloroethane)<br><u>Chronic</u><br>20,000<br>9,400<br>2,400<br>1,100<br>540 | <u>Acute</u><br>113,000<br>31,200<br>9,020<br>390<br>940<br><u>Chronic</u><br>—<br>—<br>—<br>261<br>—<br>Same as above | .094–9.4 (1, 2 dichloro-<br>ethane)<br>18.4 mg/ℓ – 1.03 g/ℓ<br>.017–1.7 ug/ℓ<br>—<br>.19–19 ug/ℓ |

(*Table 10-3. continued overleaf*)

Table 10-3. continued

| Toxic Chemical | Concentration of Toxic Material Considered Limit (ug/ℓ) | | | | |
|---|---|---|---|---|---|
| | Freshwater Aquatic Life | | Saltwater Aquatic Life | | Human Health |
| | *Acute* | *Chronic* | *Acute* | *Chronic* | |
| 16. Chlorinated Naphthalenes | 1,600 | — | 7.5 | — | not available |
| 17. Chlorinated Phenols | 30 to 500,000 | 970 | 440 to 29,000 | — | 0.1 (3 monochlorphenol)<br>0.1 (4 monochlorphenol)<br>0.04 (2, 3 dichlorophenol)<br>0.5 (2, 5 dichlorophenol)<br>0.2 (2, 6 dichlorophenol)<br>0.3 (3, 4 dichlorophenol)<br>1.0 (2, 3, 4, 6 tetrachlorophenol)<br>2.6 mg/ℓ (2, 4, 5 tri-chlorophenols)<br>.12–12 ug/ℓ (2, 4, 6 trichlorophenol)<br>1,800 ug/ℓ (2 methyl 4 chlorophenol)<br>3,000 (3 methyl, 4 chlorophenol)<br>20 (3 methyl, 6 chlorophenol) |

HAZARDOUS WASTES 107

| | | | |
|---|---|---|---|
| 18. Chloroalkyl Ethers | 238,000 | not available | .00038 ng/ℓ to .038 ng/ℓ (for bischloro-methyl ether) .003 ug/ℓ–.3 ug/ℓ (for bis 3 chloroethyl ether) 34.7 ug/ℓ (for bis-2-chloro-isopropyl ether) |
| 19. Chloroform | 28,900 | not available | .019–1.9 ug/ℓ |
| 20. 2 Chlorophenol | 4,380 | not available | 0.1 ug/ℓ |
| | 2,000 (one fish species) | | |
| 21. Chromium | 21 (max) hexavalent .29 (avg. 24 hr.) 4700 ug/ℓ (100 ppm trivalent hardness) 44 ug/ℓ chronic toxicity | 1260 (max) Cr$^{vi}$ 18 (24 hr. avg.) 10,300 (Cr$^{+3}$) chronic toxicity | 170 mg/ℓ Cr$^{iii}$ 50 ug/ℓ (Cr$^{vi}$) |
| 22. Copper | 5.6 (24 hr. avg.) 22 ug/ℓ (100 ppm hardness) max | 4.0 (24 hr. avg.) 23 (max) | 1 mg/ℓ (for taste and odor) none other available |
| 23. Cyanide | 3.5 (24 hr. avg.) 52 (max.) | 2–30 ug/ℓ | 200 ug/ℓ |
| 24. DDT and Metabolites | 1.1 max .001 (24 hr. avg.) DDT 0.6 acute toxicity TDE 1050 acute toxicity DDE | 0.13 (max DDT) .001 (24 hr. avg.) 3.6 acute toxicity TDE 14 acute toxicity DDE | .0024 ng/ℓ to .24 ng/ℓ for DDT |

(*Table 10–3, continued overleaf*)

Table 10-3. continued

| Toxic Chemical | Concentration of Toxic Material Considered Limit (ug/ℓ) | | | | Human Health |
|---|---|---|---|---|---|
| | Freshwater Aquatic Life | | Saltwater Aquatic Life | | |
| | Acute | Chronic | Acute | Chronic | |
| 25. Dichlorobenzenes | 1,120 | 763 | 1970 | — | 400 |
| 26. Dichlorobenzidines | not available | | not available | | .00103–.103 |
| 27. Dichloroethylenes | 11,600 | — | 224,000 | — | .0033–.33 |
| 28. 2,4 Dichlorophenol | 2,020 | 365 | not available | | .3 ug/ℓ (for taste and odor) 3.09 mg/ℓ (toxicity) |
| 29. Dichloropropanes Dichloropropenes | 23,000 | 5,700 | 10,300 | 3,040 | 87 ug/ℓ |
| 30. 2,4 Dimethylphenol | 2,120 | — | not available | | 400 ug/ℓ taste and odor |
| 31. 2,4 Dinitrotoluene | 330 | 220 | 590 | 370 | .001–1.1 ug/ℓ |
| 32. 1,2-Diphenylhydrazine | 270 | — | not available | | 4–422 ng/ℓ |
| 33. Endosulfan | .056 (24 hr. avg.) .22 (maximum) | | .0087 (24 hr. avg.) .034 (maximum) | | 74 ug/ℓ |
| 34. Endrin | .0023 (24 hr. avg.) .18 (maximum) | | .0023 (24 hr. avg.) .037 (maximum) | | 1 ug/ℓ |
| 35. Ethylbenzene | 32,000 | — | 430 | — | 1.4 mg/ℓ |
| 36. Fluoranthene | 3,980 | — | 40 | 18 | 42 ug/ℓ |
| 37. Haloethers | 360 | 122 | not available | | not available |

HAZARDOUS WASTES 109

| | | | | |
|---|---|---|---|---|
| 38. | Halomethanes | 11,000 | 12,000 6,400 | .019–1.9 ug/ℓ |
| 39. | Heptachlor | .0038 (24 hr. avg.) .52 (maximum) | .0036 (24 hr. avg.) .053 (maximum) | .028–2.78 ng/ℓ |
| 40. | Hexachlorobutadiene | 90 | 32 | .045–4.47 ug/ℓ |
| 41. | Hexachlorocyclohexane (Lindane) BHC | .080 (24 hr. avg.) 2.0 (maximum) 100 | 0.16 | .92–9.2 ng/ℓ |
| | | | | 1.63–163 ng/ℓ |
| 42. | Hexachlorocyclopentadiene | 7.0 | 0.34 | 206 ug/ℓ |
| | | | | 1.0 ug/ℓ (taste and odor) |
| 43. | Isophorone | 117,000 | 7.0 | 5.2 mg/ℓ |
| 44. | Lead | 3.8 ug/ℓ (100 ppm hardness) (24 hr. avg.) 170 ug/ℓ (maximum) | 12,900 | 50 ug/ℓ |
| | | | 668  25 | |
| 45. | Mercury | .00057 ug/ℓ (24 hr. avg.) .0017 ug/ℓ (maximum) | .025 (24 hr. avg.) 3.7 (maximum) | 144 ng/ℓ |
| 46. | Naphthalene | 2,300  620 | 2,350 | not available |
| 47. | Nickel | 96 (100 ppm hardness and 24 hr. avg.) 1,800 (maximum) | 7.1 (24 hr. avg.) 140 (maximum) | 13.4 |
| 48. | Nitrobenzene | 27,000 | 6,680 | 19.8 mg/ℓ |
| | | | | 30 mg/ℓ (taste and odor) |
| 49. | Nitrophenols | 230  150 | 4,850 | 13.4 (for 2, 4 dinitro- cresol) |
| | | | | 70 (for dinitrophenol) |

(Table 10–3. continued overleaf)

Table 10-3. continued

| Toxic Chemical | Concentration of Toxic Material Considered Limit (ug/ℓ) | | | | Human Health |
|---|---|---|---|---|---|
| | Freshwater Aquatic Life | | Saltwater Aquatic Life | | |
| | Acute | Chronic | Acute | Chronic | |
| 50. Nitrosamines | 5,850 | — | 3,300,000 | — | .14–14 ng/ℓ (for n-nitrosodimethylamine) .08–8.0 ng/ℓ (for n-nitrosodiethylamine) .064–64 ng/ℓ (for n-nitrosodi-n butylamine) 490–49,000 ng/ℓ (for n-nitrosodiphenylamine) 1.60–160 ng/ℓ (for n-nitrosopyrolidine) |
| 51. Pentachlorophenol | 55 | 3.2 | 53 | 34 | 1.01 mg/ℓ 30 ug/ℓ (for taste and odor) |
| 52. Phenol | 10,200 | 2,560 | 5,800 | — | 3.5 mg/ℓ 0.3 mg/ℓ (for taste and odor) |
| 53. Phthalate Esters | 940 | 3 | 2,944 | 3.4 | 313 mg/ℓ (dimethylphthalate) 350 mg/ℓ (diethylphthlate) 34 mg/ℓ (dibutylphthlate) |

# HAZARDOUS WASTES

|  | Pollutant | Freshwater Aquatic Life | Saltwater Aquatic Life | Human Health |
|---|---|---|---|---|
|  |  |  |  | 15 mg/ℓ (di-2-ethyl-hexyl-phthalate) |
| 54. | Polychlorinated Biphenyls | 0.014 (24 hr. avg.) 2.0 (maximum) | 0.030 (24 hr. avg.) 10 (maximum) | .0079–.79 ng/ℓ |
| 55. | Polynuclear Aromatic Hydrocarbons (PAH's) | not available | 300 | .28–28 ng/ℓ |
| 56. | Selenium | 35 (24 hr. avg.) 260 (maximum) 760 (inorganic selenate) | 54 (24 hr. avg.) 410 (maximum) | 10 ug/ℓ |
| 57. | Silver | 4.1 (maximum 100 ppm hardness) 0.12 (average chronic) | 2.3 — | 50 ug/ℓ |
| 58. | Tetrachloroethylene | 5,280 840 | 10,200 450 | .08–8 ug/ℓ |
| 59. | Thallium | 1,400 40 | 2,130 — | 13 ug/ℓ |
| 60. | Toluene | 17,500 — | 6,300 5,000 | 14.3 mg/ℓ |
| 61. | Toxaphene | .013 (24 hr. avg.) 1.6 (maximum) | .070 (maximum) | .07–7.1 ng/ℓ |
| 62. | Trichloroethylene | 45,000 21,900 | 2,000 — | .27–27 ug/ℓ |
| 63. | Vinyl chloride | not available | not available | .20–20 ug/ℓ |
| 64. | Zinc | 47 ug/ℓ (24 hr. avg.) 320 (max. at 100 ppm hardness) | 58 (24 hr. avg.) 170 (maximum) | 5 mg/ℓ |

Source: Adapted From *Federal Register* 1980.

about on a special tractor-trailer truck, was said to offer on-site detoxification of polychlorinated biphenyls." The system, called PCBX, uses a chemical reagent to strip chlorine atoms from insulating liquids. Sunohio's people believe the "chlorine atom is what makes the PCB toxic and perhaps carcinogenic and what is left after the process is a non-toxic residue that can be used in landfills." They claim that the insulating fluid is not destroyed by the chemical process and can be returned for re-use in transformers. The cost would be about three dollars per gallon to strip PCB from insulating fluids.

The Vulcanus, a cargo ship owned by the Waste Management Company of Chicago, Illinois, has been converted to an ocean-going incinerator. It will burn 3.6 million gallons of highly toxic PBC wastes on the high seas in the Gulf of Mexico. EPA has estimated that about 750 million pounds of PCBs (half or more than the total manufactured since production began in 1929) are still in use today (Wilson 1979). Most of the material has been used as insulating fluid in heavy electrical equipment. All of it will either have to be retrieved and destroyed. Previously the only two licensed incinerators in the United States (one in Texas and one in Emelle, Alabama) burn the high level wastes, while the low level PCB wastes are buried in the Alabama site.

The advantages of the ship incinerator have been proclaimed to be: dispersion of acid wastes directly into the ocean without scrubbing; harmless byproducts (water, carbon dioxide, and aqueous hydrochloric acid) will fall out harmlessly into the ocean within a half-mile of the ship; and burning will safely be conducted far away from populated areas. The major problem of the ship incinerator appears to be involved with the collection and transportation of the toxic wastes from the origin and to the ship. Careful moving of the wastes, similar to the process of moving gasoline, will be used. Admittedly an accident could cause lasting environmental damage.

In Table 10-4 we list some representative hazardous substances found in industrial wastes, some of which are in the solid waste form.

EPA also gives the volumes of solid sludge wastes from various industries and the fractions of these found in the various geographic areas of the United States (see Table 10-5) (Office Solid Wastes, 1974).

On July 3, 1980, the governor of Florida signed into law (h.311) a "cradle-to-grave" program for disposal, treatment, and transportation of hazardous chemical wastes. Florida was (until this date) the last Southern state without hazardous waste legislation.

**Table 10-4.** Representative Hazardous Substances Within Industrial Waste Stream.

| | Hazardous Substances | | | | | | | | | | |
|---|---|---|---|---|---|---|---|---|---|---|---|
| Industry | As | Cd | Chlorinated Hydrocarbons[a] | Cr | Cu | Cyanides | Pb | Hg | Miscellaneous Organics[b] | Se | Zn |
| Mining and metallurgy | ✓ | ✓ | | ✓ | ✓ | ✓ | ✓ | ✓ | | ✓ | ✓ |
| Paint and dye | | ✓ | | ✓ | ✓ | ✓ | ✓ | ✓ | | ✓ | ✓ |
| Pesticide | ✓ | | ✓ | | | ✓ | ✓ | ✓ | ✓ | | |
| Electrical and electronic | | | ✓ | | ✓ | ✓ | ✓ | ✓ | ✓ | ✓ | |
| Printing and duplicating | ✓ | | | ✓ | ✓ | ✓ | ✓ | | | ✓ | |
| Electroplating and metal finishing | | ✓ | | ✓ | ✓ | ✓ | | | | | ✓ |
| Chemical manufacturing | | | ✓ | ✓ | ✓ | | ✓ | ✓ | ✓ | | |
| Explosives | ✓ | | | | | ✓ | ✓ | ✓ | | | |
| Rubber and plastics | | | ✓ | | | | | ✓ | ✓ | | ✓ |
| Battery | | ✓ | | | | | ✓ | ✓ | | | ✓ |
| Pharmaceutical | ✓ | | | | ✓ | | | ✓ | ✓ | | |
| Textile | | | | ✓ | | | | | ✓ | | |
| Petroleum and coal | ✓ | | ✓ | | | | ✓ | ✓ | ✓ | | |
| Pulp and paper | | | | | | | | | ✓ | | |
| Leather | | | | ✓ | | | | | ✓ | | |

a. Including polychlorinated biphenyls.
b. For example, acrolein, chlorophenol, dimethyl sulfate, dinitrobenzene, dimitrophenol, nitroaniline, and pentachlorophenol.

Source: U.S. EPA (1974).

Table 10-5. Solid Waste Materials Classified as Toxic.*

| Waste Stream Title | Standard Industrial Code | Percentage by Geographic Area† | | | | | | | | Volume (lb/yr) | Remarks |
|---|---|---|---|---|---|---|---|---|---|---|---|
| | | NE | MA | ENC | WNC | SA | ESC | WSC | M | W | | |
| Solid, slurry, or sludge: | | | | | | | | | | | | |
| Recovered arsenic from refinery flues (stored) | 1021 | — | — | — | — | — | — | — | — | 1.00 | $4 \times 10^7$ | Tacoma, Wash. |
| Sodium dichromate production wastes | 2819 | — | .150 | .243 | — | .437 | — | .170 | — | — | $3 \times 10^8$ | |
| Solvent-based paint sludge | 285 | .044 | .243 | .269 | .072 | .103 | .041 | .069 | .012 | .147 | $4 \times 10^7$ | |
| Water-based paint sludge | 285 | .044 | .243 | .269 | .072 | .103 | .041 | .069 | .012 | .147 | $3 \times 10^7$ | |
| Tetraethyl and tetramethyl lead production wastes | 2869 | — | — | — | — | — | — | .63 | — | .37 | $3 \times 10^5$ | |
| Urea production wastes | 2873 | — | .05 | .09 | .18 | .09 | .15 | .29 | — | .14 | $2 \times 10^5$ | Dry basis |
| Benzoic herbicide contaminated containers | 2879 | — | — | .655 | .154 | .006 | .017 | — | .009 | .160 | $2 \times 10^4$ | |
| Calcium arsenate contaminated containers | 2879 | .03 | .02 | .08 | .07 | .16 | .16 | .35 | .09 | .03 | $6 \times 10^3$ | |
| Carbonate pesticide contaminated containers | 2879 | .0008 | .016 | .382 | .070 | .022 | .108 | .321 | .020 | .060 | $5 \times 10^4$ | |
| Chlorinated aliphatic pesticide contaminated containers | 2879 | .381 | — | .076 | .418 | — | .105 | .010 | — | .010 | $1 \times 10^4$ | |
| Dinitro pesticide contaminated containers | 2879 | .496 | .168 | .023 | .017 | .228 | — | .003 | .006 | .165 | $2 \times 10^4$ | |
| Lead arsenate contaminated containers | 2879 | .03 | .02 | .08 | .07 | .17 | .17 | .35 | .03 | .08 | $1 \times 10^4$ | |
| Mercury fungicide contaminated containers | 2879 | .02 | .03 | .04 | .03 | .28 | .32 | .05 | .01 | .22 | $5 \times 10^2$ | |
| Miscellaneous organic insecticide contaminated containers | 2879 | .148 | .084 | .054 | .039 | .197 | .143 | .148 | .017 | .170 | $4 \times 10^4$ | |
| Organic arsenic contaminated containers | 2879 | — | .007 | — | — | .011 | .764 | .218 | — | — | $5 \times 10^3$ | |
| Organic fungicide contaminated containers | 2879 | .048 | .125 | .047 | .028 | .441 | .001 | .036 | .007 | .266 | $8 \times 10^4$ | |
| Organophosphorus contaminated containers | 2879 | .043 | .050 | .018 | .125 | .139 | .192 | .175 | .049 | .208 | $1 \times 10^5$ | |
| Phenoxy contaminated containers | 2879 | .035 | .033 | .196 | .321 | .031 | .030 | .067 | .141 | .146 | $2 \times 10^5$ | |
| Phenyl urea contaminated containers | 2879 | .106 | .085 | .106 | .033 | .106 | .424 | .042 | .024 | .095 | $9 \times 10^3$ | |
| Polychlorinated hydrocarbon contaminated containers | 2879 | .017 | .107 | .019 | .138 | .306 | .211 | .133 | .024 | .044 | $2 \times 10^5$ | |
| Triazine contaminated containers | 2879 | .147 | .121 | .320 | .372 | .013 | .003 | .011 | .002 | .011 | $6 \times 10^4$ | |

HAZARDOUS WASTES    115

| | | NE† | MA | ENC | WNC | SA | ESC | WSC | M | W | |
|---|---|---|---|---|---|---|---|---|---|---|---|
| Miscellaneous organic pesticide contaminated containers | 2879 | .014 | .162 | .385 | .068 | .162 | .123 | .041 | .014 | .034 | $1 \times 10^4$ |
| Petroleum refining still bottoms | 2911 | .006 | .086 | .159 | .055 | .025 | .025 | .477 | .033 | .134 | $2 \times 10^4$ |
| Petroleum waste brine sludges | 2911 | .002 | .06 | .09 | .011 | .12 | .10 | .55 | .022 | .045 | $4 \times 10^6$ |
| Iron manufacturing waste sludge | 331 | .05 | .05 | .56 | .02 | .12 | .03 | .09 | .05 | .03 | $6 \times 10^6$ |
| Arsenic trioxide from smelting industry | 333 | — | .03 | .015 | .07 | .005 | .01 | .10 | .70 | .07 | $2 \times 10^7$ |
| Selenium production wastes | 3339 | — | .75 | — | — | — | — | — | .25 | — | $2 \times 10^4$ |
| Duplicating equipment manufacturing wastes | 3555 | — | 1.00 | — | — | — | — | — | — | — | $7 \times 10^5$ |
| Refrigeration equipment manufacturing wastes | 3585 | .013 | .232 | .408 | .096 | .040 | .069 | .086 | .011 | .045 | $2 \times 10^8$ |
| Battery manufacturing waste sludge | 3691 | .117 | .043 | — | .118 | .117 | — | — | .118 | — | $5 \times 10^7$ |
| Arsenic trichloride recovered from coal | 49 | .05 | .23 | .07 | .05 | .33 | .25 | — | .07 | — | $6 \times 10^6$ |
| Military paris green (stored) | 9711 | — | — | 1.00 | — | — | — | — | — | — | $3 \times 10^4$ |
| Stored military mercury compounds | 9711 | .47 | — | — | .51 | — | — | — | — | .02 | $2 \times 10^2$ |
| Subtotal | | | | | | | | | | | $\overline{7 \times 10^8}$ |

*This is an updated version of the table that appeared in the first edition of this report.

†NE = New England: Connecticut, Maine, Massachusetts, New Hampshire, Rhode Island, and Vermont; MA = Mid Atlantic: New Jersey, New York, and Pennsylvania; ENC = East North Central: Illinois, Indiana, Michigan, Ohio, and Wisconsin; WNC = West North Central: Iowa, Kansas, Minnesota, Missouri, Nebraska, North Dakota, and South Dakota; SA = South Atlantic: Delaware, District of Columbia, Florida, Georgia, Maryland, North Carolina, South Carolina, Virginia, and West Virginia; ESC = East South Central: Alabama, Kentucky, Mississippi, and Tennessee; WSC = West South Central: Arkansas, Louisiana, Oklahoma, and Texas; M = Mountain: Arizona, Colorado, Idaho, Montana, Nevada, New Mexico, Utah, and Wyoming; W = West (Pacific): Alaska, California, Hawaii, Oregon, and Washington. Upstate New York

Source: U.S. EPA (1974).

116    INDUSTRIAL SOLID WASTES

In carrying out the RCRA of 1976 (see Chapter 11) EPA specified rules for an extraction procedure toxicity test to identify a hazardous waste (*Federal Register* 1980). EPA methods for toxicity analysis only of various wastes are given here in detail as they appear in the *Federal Register*. The importance of verifying toxicity cannot be stressed enough.

## APPENDIX II: EP TOXICITY TEST PROCEDURE

### A. Extraction Procedure (EP)

1. A representative sample of the waste to be tested (minimum size 100 grams) should be obtained using the methods specified in Appendix I or any other methods capable of yielding a representative sample within the meaning of Part 280. [For detailed guidance on conducting the various aspects of the EP see "Test Methods for the Evaluation of Solid Waste, Physical/Chemical Methods," SW-846, U.S. Environmental Protection Agency Office of Solid Waste, Washington, D.C. 20460."]

2. The sample should be separated into its component liquid and solid phases using the method described in "Separation Procedure" below. If the solid residue obtained using this method totals less than 0.5% of the original weight of the waste, the residue can be discarded and the operator should treat the liquid phase as the extract and proceed immediately to Step 8.

3. The solid material obtained from the Separation Procedure should be evaluated for its particle size. If the solid material has a surface area per gram of material equal to, or greater than, 3.1 $cm^3$ or passes through a 9.5 mm (0.375 inch) standard sieve, the operator should proceed to Step 4. If the surface area is smaller or the particle size larger than specified above, the solid material should be prepared for extraction by crushing, cutting or grinding the material so that it passes through a 9.5 mm (0.375 inch) sieve or, if the material is in a single piece, by subjecting the material to the "Structural Integrity Procedure" described below.

4. The solid material obtained in Step 3 should be weighed and placed in an extractor with 16 times its weight of deionized water. Do not allow the material to dry prior to weighing. For purposes of this test, an acceptable extractor is one which will impart sufficient agitation to the mixture to not only prevent stratification of the sample and extraction fluid but also insure that all sample surfaces are continuously brought into contact with well mixed extraction fluid.

5. After the solid material and deionized water are placed in the extractor, the operator should begin agitation and measure the pH of the solution in the extractor. If the pH is greater than 5.0, the pH of the solution should be decreased to 5.0 ± 0.2 by adding 0.5 N acetic acid. If the pH is equal to or less

than 5.0, no acetic acid should be added. The pH of the solution should be monitored, as described below, during the course of the extraction and if the pH rises above 5.2, 0.5N acetic acid should be added to bring the pH down to 5.0 ± 0.2. However, in no event shall the aggregate amount of acid added to the solution exceed 4 ml of acid per gram of solid. The mixture should be agitaged for 24 hours and maintained at 20° - 40°C (68° - 104°F) during this time. It is recommended that the operator monitor and adjust the pH during the course of the extraction with a device such as the Type 45-A pH Controller manufactured by Chemtrix, Inc., Hillsboro, Oregon 97123 or its equivalent, in conjunction with a metering pump and reservoir of 0.5M acetic acid. If such a system is not available, the following manual procedure shall be employed:

(a) A pH meter should be calibrated in accordance with the manufacturer's specifications.

(b) The pH of the solution should be checked and, if necessary, 0.5N acetic acid should be manually added to the extractor until the pH reaches 5.0 ± 0.2. The pH of the solution should be adjusted at 15, 30 and 60 minute intervals, moving to the next longer interval if the pH does not have to be adjusted more than 0.5N pH units.

(c) The adjustment procedure should be continued for at least 6 hours.

(d) If at the end of the 24-hour extraction period, the pH of the solution is not below 5.2 and the maximum amount of acid (4 ml per gram of solids) has not been added, the pH should be adjusted to 5.0 ± 0.2 and the extraction continued for an additional four hours, during which the pH should be adjusted at one hour intervals.

6. At the end of the 24 hour extraction period, deionized water should be added to the extractor in an amount determined by the following equation:

$V = (20)(W) - 16(W) - A$
$V$ = ml deionized water to be added
$W$ = weight in grams of solid charged to extractor
$A$ = ml of 2.5N acetic acid added during extraction

7. The material in the extractor should be separated into its component liquid and solid phases as described under "Separation Procedure."

8. The liquids resulting from Steps 2 and 7 should be combined. This combined liquid (or the waste itself if it had less than 0.5 percent solids, as noted in Step 2) is the extract and should be analyzed for the presence of any of the contaminants specified in Table 1 of §261.24 using the Analytical Procedures designated below.

## Separation Procedure

Equipment: A filter holder, designed for filtration media having a nominal pore size of 0.45 micrometers and capable of applying a 5.3 kg/cm$^3$ (75 psi) hydrostatic pressure to the solution being filtered shall be used. For mixtures containing nonabsorptive solids, where separation can be affected without imposing a 5.3 kg/cm$^3$ pressure differential, vacuum filters employing a 0.45 micrometers filter media can be used. (For further guidance on filtration equipment or procedures see "Test Methods for Evaluating Solid Waste, Physical/Chemical Methods.")

Procedure:

(i) Following manufacturer's directions, the filter unit should be assembled with a filter bed consisting of a 0.45 micrometer filter membrane. For difficult or slow to filter mixtures a prefilter bed consisting of the following prefilters in increasing pore size (0.65 micrometer membrane, fine glass fiber prefilter, and coarse glass fiber prefilter) can be used.

(ii) The waste should be poured into the filtration unit.

(iii) The reservoir should be slowly pressurized until liquid begins to flow from the filtrate outlet at which point the pressure in the filter should be immediately lowered to 10-15 psig. Filtration should be continued until liquid flow ceases.

(iv) The pressure should be increased stepwise in 10 psi increments to 75 psig and filtration continued until flow ceases or the pressurizing gas begins to exit from the filtrate outlet.

(v) The filter unit should be depressurized, the solid material removed and weighed and then transferred to the extraction apparatus, or, in the case of final filtration prior to analysis, discarded. Do not allow the material retained on the filter pad to dry prior to weighing.

(vi) The liquid phase should be stored at 4°C for subsequent use in Step 8.

## B. Structural Integrity Procedure

Equipment: A Structural Integrity Tester having a 3.18 cm (1.25 in.) diameter hammer weighing 0.33 kg (0.73 lbs.) and having a free fall of 15.24 cm (6 in.) shall be used. This device is available from Associated Design and Manufacturing Company, Alexandria, VA., 22314, as Part No. 125, or it may be fabricated to meet the specifications shown in Figure 1.

Procedure:

1. The sample holder should be filled with the material to be tested. If the sample of waste is a large monolithic block, a portion should be cut from the block having the dimensions of a 3.3 cm (1.3 in.) diameter x 7.1 cm (2.8 in.) cylinder. For a fixated waste, samples may be cast in the form of a 3.3 cm (1.3 in.) diameter x 7.1 cm (2.8 in.) cylinder for purposes of conducting this test, in such cases, the waste may be allowed to cure for 30 days prior to further testing.

2. The sample holder should be placed into the Structural Integrity Tester, then the hammer should be raised to its maximum height and dropped. This should be repeated fifteen times.

3. The material should be removed from the sample holder, weighed, and transferred to the extraction apparatus for extraction.

## Analytical Procedures for Analyzing Extract Contaminants

The test methods for analyzing the extract are as follows:

(1) For arsenic, barium, cadmium, chromium, lead, mercury, selenium or silver: "Methods for Analysis of Water and Wastes." Environmental Monitoring and Support Laboratory, Office of Research and Development, U.S. Environmental Protection Agency, Cincinnati, Ohio 45283 (EPA-800/4-79-020, March 1979).

(2) For Endrin: Lindane: Methoxychlor: Toxaphene: 2, 4-D: 2, 4-5-TP Silver: in "Methods for Benzidine, Chlorinated Organic Compounds, Pentachlorophenol and Pesticides in Water and Wastewater," September 1978, U.S. Environmental Protection Agency, Environmental Monitoring and Support Laboratory, Cincinatti, Ohio 42588,

as standardized in "Test Methods for the Evaluation of Solid Waste, Physical/Chemical Methods."

For all analyses, the method of standard addition shall be used for the quantification of species concentration. This method is described in "Test Methods for the Evaluation of Solid Waste." (It is also described in "Methods for Analysis of Water and Wastes.")

## INDUSTRIAL INSURANCE REQUIREMENTS

A growing significant problem facing the American industry of the 1980s is that of providing security for its continual survival against pollution damages. This can only be done through a combination of effective and successful waste treatment and adequate insurance against potential failures.

The RCRA of 1976 dealing with solid wastes focused industry's attention on its legal obligations to protect the environment from wastes—and especially hazardous ones. Industry, in addition to life insurance and then health insurance, must also now consider environmental disaster insurance. Any company that stores, treats, or disposes of solvents, acids, alkalis, or other hazardous materials must have some insurance ("Insurers Fret . . ." 1981).

The new rules (RCRA—'76) are said to cover nearly 60 million metric tons of acids, corrosive metals, toxic chemicals, and other hazardous materials produced by the United States alone each year.

Besides the cradle-to-grave tracking required, industry is also responsible for its disposal practices and for the facilities on its properties.

The rules proposed by the Environmental Protection Agency in January 1981 require coverage of $3 million for each case of long-term damage up to a maximum of $6 million a year, excluding legal defense costs. For sudden spills and the like, the required coverage is $1 million for each case and up to $2 million a year. Many insurance brokers are of the opinion that EPA's proposal limits are not sufficient to cover a major accident.

Uncertainty of how the courts will interpret a slow release of toxic material over an elongated period rather than a sudden discharge is causing concern among chemical manufacturers.

The risk is of uncertain dimensions because the long-term health effects of many chemicals have not been determined. In addition, the records of past dumpings, discharges, and seepages have been poor. The insurance companies have little precedence of past cases upon which to base their risk.

Most assessments of current effectiveness of disposal facilities and likelihood of accidental spills must be made by competent consulting engineers. This often proves to be an important part—and an often costly one—to any insurance plan or claim. Because of the difficulty of judging risk, most insurers are not yet ready to provide coverage. At this writing (1981) three U.S. insurance companies are providing coverage. Some industries are providing their own insurance because they believe the federal $6 million limit may be a drop in the bucket compared to real costs of up to $30 million for a significant suit involving a spill.

## REFERENCES

Office Solid Waste Management Programs EPA "Report to Congress Disposal Hazardous Wastes" U.S.E.P.A. Report S.W-115, 1974.

*Federal Register* 45, 96. 1980. (May 19): 33127-33128.

"Insurers Fret Over Covering Pollution Costs." 1981. *Wall Street Journal* (July 17): 29.

"New Processes to Detoxify PCB's Approved by U.S." 1981. *Miami Herald* (May 29).

"Reagan Plans to Ease Toxic Waste Rules in Major Concession to Chemical Industry." 1981. *Wall Street Journal* (May 6): 4.

Wilson, David G. 1979. "Chapter 6: Health Hazards of Solid Waste Treatment." In *Dangerous Properties of Handling Industrial Materials*, 5th ed., edited by Irving Sax. (New York: Van Nostrand).

# 11 LEGAL ASPECTS OF SOLID WASTES

## INTRODUCTION

The rapid geometric increase in municipal and industrial growth, and therefore generation of solid wastes, have necessitated new laws to require environmental protection from these wastes. The two primary pieces of legislation enacted in the United States were the *Solid Waste Act*, 42 U.S.C., 3251 of 1970, and the *Resource Conservation and Recovery Act* (RCRA) of 1976. Florida also finally enacted a law (H311, July 3, 1981) regulating the discharge of hazardous types of solid wastes.

## RESOURCE RECOVERY ACT OF 1970

This act was really an amendment to the Solid Waste Disposal Act of 1965. The latter act (1965) was the first recognition in United States that a problem generally considered to be local in nature was actually one of national scope. Although it left collection and disposal to local governments, the act provided for a national research and development program for new and improved methods of disposal. It also provided for federal technical and financial aid to state and local governments. It organized solid waste specialists into a new location under the Department of Health, Education and Welfare. Grants were given for demonstration projects as well as research, training of personnel, and for cooperative planning between states, as well as for regional management systems. The 1970 legislation provided for demonstration, research, and training grants to enhance resource recovery and to encourage implementation in the building of disposal facilities. It called for a study of resource recovery that would develop and provide guidelines, model ordinances, and data to be used by all agencies. It also directed attention to the hazardous waste problem by calling for a plan for locating national disposal sites for these wastes. Several sections of this act offered significant directives for the solutions to the increasing solid-waste dilemma. Section 205 directs the federal government to carry out comprehensive investigations on waste reduction and recovery of re-usable materials and energy. Among these investigations changes in existing products and ways of packaging them, incentives for recylcing, how recycling will affect subsidies and tax systems, and how recovery of valuable materials will affect final disposal costs. Despite the fact that the act required the secretary to report at least annually to the people (Bryson 1974), the first report was not issued until February 1973. Several significant problems were revealed in this report, however, which are worth mentioning here. They enable the reader to understand better the development of legislation in the solid waste field. The decrease in use of secondary materials was decried in the report. But more important was the revelation that artificial subsidies for virgin materials existed as well as "loaded" freight rates favoring the use of new rather than re-used materials. They reported, however, that when using secondary materials, less environmental damages occurred compared to that happening from production with primary resources. They did not recommend, however, federal subsidies at that time for recycling. It also took two and a half years for the EPA to propose solid waste guidelines (April 27, 1973). These were only directed to incineration and sanitary landfill techniques. Bryson (1974) also reported the absence of action specified by the act regarding training of personnel, model ordinances, and a plan for national disposal sites for hazardous wastes. Bryson also puts the blame on "large waste-generating and virgin materials industries and the decision of the Nixon Administration largely to withdraw the Federal Government from efforts to assist with solid waste management" for these inactions. Ultimately, the lack of progress in solving the solid-waste problems led to the development and eventually the enactment of the subsequent act of 1976. The problem would not go away, solve itself, or be "shoved under the rug." Eventually it emerged in the political arena again in the form of more strict legislation.

## RESOURCE CONSERVATION AND RECOVERY ACT OF 1976

This act, known as Public Law 94-580, was enacted by the 94th Congress as an amendment to the Solid Waste Act of 1970. Under Title II, "Solid Waste Disposal," it provided for the following major matters as subtitles:

Subtitle A—General Provisions
Subtitle B—Office of Solid Waste; Authorities of the Administrator
Subtitle C—Hazardous Waste Management
Subtitle D—State or Regional Solid Waste Plans
Subtitle E—Duties of Secretary of Commerce in Resource and Recovery
Subtitle F—Federal Responsibilities
Subtitle G—Miscellaneous Provisions
Subtitle H—Research, Development, Demonstration, and Information

The significant aspects of each of these subtitles are given below to inform the reader about the broad coverage and important details only of the act.

### Subtitle A

The continuing concentration of our population, economic and population growth, improvements in standard of living, continuing technological progress, and improvements in manufacturing, packaging, and marketing have all contributed to a rising tide of scrap, discarded, and waste materials. These have caused communities serious financial, management, intergovernmental, and technical problems in the disposal of these wastes. The problem, although primarily a local one, has become national in scope so as to necessitate federal action.

With respect to our environment and health, Congress found that:

1. Most solid wastes are placed in open dumps or sanitary landfills, thus "polluting" a "national resource" despite the increasing value of the land.
2. Disposal of solid waste and hazardous waste in or on the land without careful planning and management can present a danger to human health and the environment.
3. As a result of other state and federal laws protecting both the air and water environments, greater amounts of solid waste in the forms of sludge have been created. Also, inadequate and environmentally unsound practices for the disposal or use of solid waste have created greater amounts of all types of environmental and health problems.
4. Open dumping is particularly harmful to health by contaminating drinking-water supplies, air, and the land.
5. Hazardous waste presents, in addition, special dangers to health and requires a greater degree of regulation than nonhazardous solid waste.
6. Alternatives to existing methods of land disposal must be developed since many cities in the United States will be running out of suitable waste-disposal sites within five years unless immediate action is taken.

Congress found simultaneously three important facts in respect to materials:

1. Millions of tons of recoverable material which could be used are needlessly buried each year.
2. Methods are available to separate usable materials from solid waste.
3. The recovery and conservation of such materials can reduce the U.S. dependence on foreign resources and reduce the deficit in its balance of payments.

Congress concluded with these three findings regarding energy:

1. Solid waste represents a potential source of solid fuel, oil, or gas that can be converted into energy.
2. The need exists to develop alternative energy sources.
3. Technology exists to produce usable energy from solid waste.

Because of these findings the Congress enacted this legislation with the following objectives:

1. Technical and financial assistance will be provided to state and local governments and interstate agencies for the development of solid-waste management plans which will promote improved techniques of management, collection, separation, and recovery of solid wastes, and the environmentally safe disposal of nonrecoverable residues.

2. Training grants will be provided in occupations involving the design, operation, and maintenance of solid-waste disposal systems.

3. Future open dumping will be prohibited on the land, and existing dumps will be converted to nondangerous facilities.

4. Regulations will be enforced regarding the treatment, storage, transportation, and disposal of hazardous wastes which have adverse effects on health and the environment.

5. The promulgation of guidelines will be provided for solid waste collection, transport, separation, recovery, and disposal practices and systems.

6. A national research and development program will be promoted for improved management, resource conservation, and methods of solid-waste disposal.

7. Demonstration, construction, and application of solid-waste systems will be promoted which preserve and enhance the environmental quality.

8. Cooperation will be established among federal, state, and local governments and private industry to recover valuable materials and energy from solid waste.

Nothing in the act shall be construed to apply to any activity or substance which is subject to the Federal Water Pollution Control Act, the Safe Drinking Water Act, the Marine Protection, Research and Sanctuaries Act of 1972, or the Atomic Energy Act of 1954 except to the extent that such application is not inconsistent with the requirements of such acts.

Congress has specified for the administrator during the first year to establish guidelines which set forth technical and economic description of levels of performance attainable by various available solid-waste management practices. Within two years these guidelines should include levels of performance as well as appropriate methods and degrees of control that yield (1) protection of public health and welfare, (2) protection of water quality from leachates, (3) protection of surface water quality from runoff, (4) protection of ambient air quality, (5) disease and vector control, (6) safety, and (7) aesthetics. The guidelines should also provide minimum criteria to be used by the states for solid-waste management practices of open dumping such as adequate location, design, and construction of facilities.

### Subtitle B

Congress has decreed that the administrator shall establish within the Environmental Protection Agency an Office of Solid Waste to be headed by a deputy assistant administrator of the Environmental Protection Agency. Thirty-five million dollars were appropriated for fiscal year ending September 30, 1977; $38 million for 1978; and $42 million for 1979. Not less than 20 percent of these amounts were designated for the purpose of Resource Recovery and Conservation Panels (teams of personnel from all aspects of solid waste).

### Subtitle C

Within eighteen months the administrator shall develop and promulgate criteria for identifying the characteristics of hazardous wastes. In preparing these criteria, consideration should be given to toxicity, persistence, degradability in nature, potential for accumulation in tissue as well as well as flammability, corrosiveness, and other hazardous characteristics.

Also within these eighteen months the administrator shall promulgate standards requiring: (1) record-keeping of quantities and constituents, and disposition in the environment; (2) labeling for all containers being stored, transported, or disposed of; (3) use of appropriate containers; (4) information on general chemical composition; (5) use of a manifest system designating treatment, storage, or disposal facilities; and (6) reports submitted to the administrator designating quantities and disposition of hazardous wastes.

Congress was also deeply concerned in this act about transporting hazardous materials for re-use or for their ultimate designation. As a result, they include within this eighteen-month period a stipulation that the administrator should promulgate regulations establishing standards for transporting hazardous waste. These shall include record-keeping, proper labeling, compliance with the manifest system, and transporting only to agreed and accepted ultimate licensed disposal locations. He should also promulgate standards applicable to owners and operators of facilities for the treatment, storage, or disposal of hazardous waste. These standards shall include: (1) record-keeping; (2) satisfactory reporting, monitoring, and inspection and compliance with the manifest system; (3) treatment storage, or disposal of all such waste; (4) location, design, and construction of such waste treatment systems; (5) contingency plans for effective action to minimize unanticipated damage from any treatment, storage, or disposal of any such hazardous waste; (6) maintenance of operation of such facilities including qualification of ownership, continuity of operation, training for personnel, and financial responsibility; and (7) compliance with requirements respecting permits for treatment, storage, or disposal.

Congress required each person owning or operating a facility to have a permit. The permit issuance depends upon data supplied about composition, quantities, and concentration of any hazardous waste and the time, frequency, or rate of which such waste is proposed to be disposed of, treated, transported, or stored as well as the site at which it will be done. During this eighteen-month period the administrator should promulgate guidelines to assist states in the development of state hazardous-waste programs. Ultimately Congress proposes that the states enact and administer their own regulatory programs which must first be approved by the administrator. The act also provides for authority to enter any facility and to inspect and obtain samples of any hazardous wastes. If any analysis is made of such samples, an equal portion must be left with the owner or operator and a copy of the results furnished promptly to him and also to the public.

Enforcement of the hazardous solid-waste portion of the act is provided by the administrator's notice to the violator of his failure to comply with the requirements. If the violation persists after thirty days, either a specified time period or immediate civil action may be specified. Failure to comply liables the violator to a civil penalty of not more than $25,000 for each day of continued noncompliance, and the administrator may suspend or revoke any permit issued to the violator. A public hearing may be called by the administrator at the request of the violator to argue the merits of compliance.

The act also provides for criminal prosecution of any person who knowingly transports any hazardous waste without having obtained a permit or who makes any false statement or representation in any application, label, manifest, record, report, permit, or other document used to show compliance. The violator will be fined not more than $25,000 for each day of violation, or be sentenced to imprisonment not to exceed one year, or both. A second conviction under this part of the act results in double this penalty.

### Subtitle D

To implement the objectives of the act provisions are made to give federal technical and financial assistance to states or regional authorities for comprehensive planning. Guidelines for regulations for these plans will be issued by the administrator within eighteen months of the enactment date. The guidelines should consider (1) the varying regional, geologic, hydrologic, climatic, and other conditions which may affect drinking-water supplies from leachate contamination; (2) characteristics and conditions of collection, storage, processing, and disposal as well as materials and site locations; (3) methods for closing or upgrading open dumps to prevent potential health hazards; (4) population existence and predicted growth; (5) physical and hydrologic conditions (6) type and location of transportation; (7) industrial profiles; (8) constituents and rates of solid-waste generation; (9) political, economic, organizational, financial, and management problems; (10) resource-recovery facilities; and (11) new and added markets available for recovered material.

The act also sets forth details of minimum requirements for approval of the state plans such as identifying responsibilities of various authorities involved, prohibition of open dumps, utilization of resource recovery or sanitary landfills for all solid wastes regardless of origin, closing or upgrading all existing open dumps, state regulatory powers, permission of local governments to enter into contract for resource recovery, and positive plans for resource conservation or recovery in combination with sanitary landfills in such a manner that is environmentally sound.

Within the eighteen-month period the administrator must promulgate regulations containing criteria for determining which facilities should be classified as sanitary landfills and which should be classified as open dumps. Open dumps must be closed or upgraded to a sanitary landfill within a stated period of time.

Within six months after a state plan has been submitted, the administrator will approve or disapprove the plan. Congress has authorized for appropriation of $30 million for 1978 and $40 million for 1979 to make grants to the states for the development and implementation of these plans. Of this total $15 million has been authorized for each year for studies, consultation, surveys of market needs, marketing of recovered resources, technology assessments, legal, construction feasibility, source separation projects, and so on. Also, $2.5 million of the total is to be used each year for conversion, improvement, or consolidation of existing solid-waste disposal facilities, or for the construction of new facilities. Finally, $25 million of the total was authorized for rural community assistance (for municipalities smaller than 5,000 persons or counties of less than 10,000, or fewer than 20 persons per square mile. The amount of any such grant will not exceed 75 percent of the project cost.

### Subtitle E

To put this act into operation, Congress has specified that the Secretary of Commerce, in protecting the environment, shall have specific obligations to provide accurate specifications for recovered materials, to stimulate development of markets for recovered materials, to promote proven technology and a forum for the exchange of technical and economic data relating to resource-recovery facilities.

The secretary will work through the National Bureau of Standards, and within two years will publish guidelines for the development of specifications for the classification of materials recovered from waste which were destined for disposal. Also within this period he will: (1) identify the geographical locations of existing or potential markets for recovered materials; (2) identify the economic and technical barriers to the use of recovered materials; and (3) encourage the development of new uses for recovered materials. He is also authorized to evaluate commercial feasibility of resource-recovery facilities and to

130    INDUSTRIAL SOLID WASTES

publish the results of such evaluation, and to develop a data base for purposes of assisting persons in choosing such a system.

### Subtitle F

The federal government and its agencies has the legal obligation and responsibility of: (1) procuring recovered materials for re-use within reason; (2) making use of recovered material—derived fuel as a primary or supplementary fuel to generate heat, mechanical, or electrical energy; (3) requiring that vendors certify the percentage of the total material utilized for the performance of the contract which is recovered materials. All federal agencies having functions relating to solid or hazardous waste shall cooperate to the maximum legal extent with the EPA administrator in carrying out his function under this act.

### Subtitle H

The administrator, in consultation with all other interested governmental agencies, will conduct, encourage, cooperate with, and render financial and other assistance to all parties for research, investigation, experiments, training, demonstrations, surveys, public education programs relating to thirteen areas of solid-waste problems: (1) adverse health and welfare effects and methods for their alleviation; (2) operation and financing of management programs; (3) planning, implementation, and operation of resource-recovery and resource-conservation systems and hazardous-waste management systems, including marketing of recovered resources; (4) production of usable forms of recovered resources, including fuel; (5) reduction of amounts of solid waste and unsalvagable material; (6) development and application of new methods of collecting and disposing and processing and recovering materials and energy; (7) identification of components and potential materials and energy recoverable from such components; (8) small-scale and low-technology solid-waste management systems; (9) methods of improving performance characteristics of resources recovered from solid waste and their relationship with marketing of them; (10) improvements in land disposal practices to reduce adverse environmental effects; (11) methods for the sound disposal or recovery of resources, including energy, from industrial and municipal sludges; (12) methods of hazardous-waste treatment that are environmentally safe; (13) any adverse effects of air quality particularly with regard to the emission of heavy metals which result from burning solid waste.

The act also directs the administrator to undertake a study and publish a report on resource recovery from glass and plastic waste, including a scientific, technological, and economic investigation of potential solutions to implement such recovery. This will include the composition of the solid-waste stream and anticipated future changes in that composition. Priorities are given to promising techniques of energy recovery such as waterwall furnace incinerators, dry shredded fuel systems, pyrolysis, densified refuse-driven fuel systems, anaerobic digestion, and fuel and feedstock preparation systems. Research studies are called for of small-scale and low-technology waste-management, including household, office, and multiple-dwelling unit systems. Other research studies required include front-end source separation systems, mining wastes, sludges, and tires. In the front-end research Congress is primarily concerned with the compatibility of such source-separation systems with high-technology resource-recovery systems. In mining-waste problems Congress is concerned with the adverse effects of solid wastes from active and abandoned surface and underground mines on the environment. Recommended studies should include: (1) sources and volumes of discarded material generated per year from mining; (2) present disposal practices; (3) potential dangers to human health and the environment from leachate and dust; (4) alternatives to current disposal methods, (5) the cost of those alternatives in terms of the impact on mine product costs; and (6) potential for use of discarded material as a secondary source of the mine product. Congress is also concerned with identifying all types of sludges in addition to those originating from pollution treatment plants such as industrial sludges from extraction of oil from shale, liquification and gasification of coal, and coal pipeline operations. With sludges, they want to know the effects of pollution-abatement legislation on the creation of large volumes of sludge; the amounts originating in each state and in each industry; methods of disposal of such sludge, including the cost and efficiency of these methods; alternative methods for the use of sludge, including agricultural and energy applications; and methods to reclaim areas which have been used for the disposal of sludge or which have been damaged by sludge.

The act also establishes a Resource Conservation Committee composed of the administrator as chairman, secretaries of the U.S. Departments of Commerce, Labor, the Treasury, the Interior, Energy, and chairman of the Council on Environmental Quality and Council of Economic Advisors, as well as a member of the important Office of Management and Budget. The committee shall conduct full and complete investigation and study of all aspects of the economic, social, and environmental consequences of resource conservation. These studies may even include pilot-scale projects. Two million dollars have been allocated for this committee's use. In addition, a further study of airport landfills is authorized to alleviate the hazards to aviation from birds congregating and feeding on landfills in the area. Eight million dollars have been authorized for all research and studies for each fiscal year of 1978 and 1979. Informational, library, and model code systems are also to be established.

The administrator is authorized to make grants for construction of new or improved solid-waste disposal facilities. The federal share will not exceed 75 percent of the project cost which involves more than one governmental entity. Thirty-five million dollars for fiscal year 1978 were authorized for construction costs.

The Hazardous Waste Bill of 1980 ("The Costly Cleanup" 1981) was passed by the Florida Legislature in 1980 but is not expected to be enforced until at least 1982. Briefly, the new program will extend government control over industry's production, use, and disposal of potentially hazardous waste by means of a cradle-to-grave tracking system. Every business that handles these wastes must file reports outlining where the chemicals came from and where they are going. The program will also set technical standards for permissible waste disposal and tax industries that generate hazardous solid wastes. Revenues would feed a cleanup fund for today's accidental spills or yesterday's leaky dumps. The tax amounts to 1 percent of the cost of disposing of the wastes. By 1984 the tax will rise to 4 percent of the cost of disposal.

## REFERENCES

Bryson, John E. 1974. "Solid Waste and Resource Recovery." In *Federal Environmental Law*, edited by E. L. Dolgin and T.G.P. Guilbert, p. 1296. St. Paul, MN: Environmental Law Institute West Publishing Co.

Public Law 94-580, October 21, 1976. 94th Congress 90 stat. 2828-2841 Resource Conservation and Recovery Act of 1976, pp. 387-433.

Solid Waste Disposal Act of 1965, Public Law No. 89-272, Title 2, 79 Stat. 997.

"The Costly Cleanup." 1981. *The Miami Herald* (June 15): 1.

# 12 HEALTH HAZARDS IN CONNECTION WITH SOLID WASTES

## INTRODUCTION

Solid wastes are apt to contain almost any unwanted, used, or unused constituent produced in society. They have been found to contain, for example, matter as unhealthy as live ammunition, hospital or laboratory disease cultures, and various organic and inorganic toxic, concentrated, semi-solid chemicals.

A constituent of solid wastes can be hazardous to health in many ways. For example, the Environmental Protection Agency asks eleven questions, and the answers of *yes* to any one will place the particular solid waste on the hazardous list. Figure 12-1 presents these eleven queries leading to a health-hazardous definition. Containers of liquids of a toxic nature must be classified as a health-hazard solid waste. Many are dangerous at the point of discharge, such as hospital wastes described in Chapter 14. Some may become a health hazard after reaching the disposal site. Most municipalities require that hospital wastes be wrapped and contained in plastic bags; however, these bags are apt to be broken during transmission to or compaction at the landfill or during handling at the incinerator.

Air near landfills, composters, or incinerators is known to contain higher pathogens than ambient air. The air surrounding untreated refuse dumping is especially dangerous for it may contain many con-

**Figure 12-1.** Health Hazards of Solid Waste—A Question and Answer Decision.

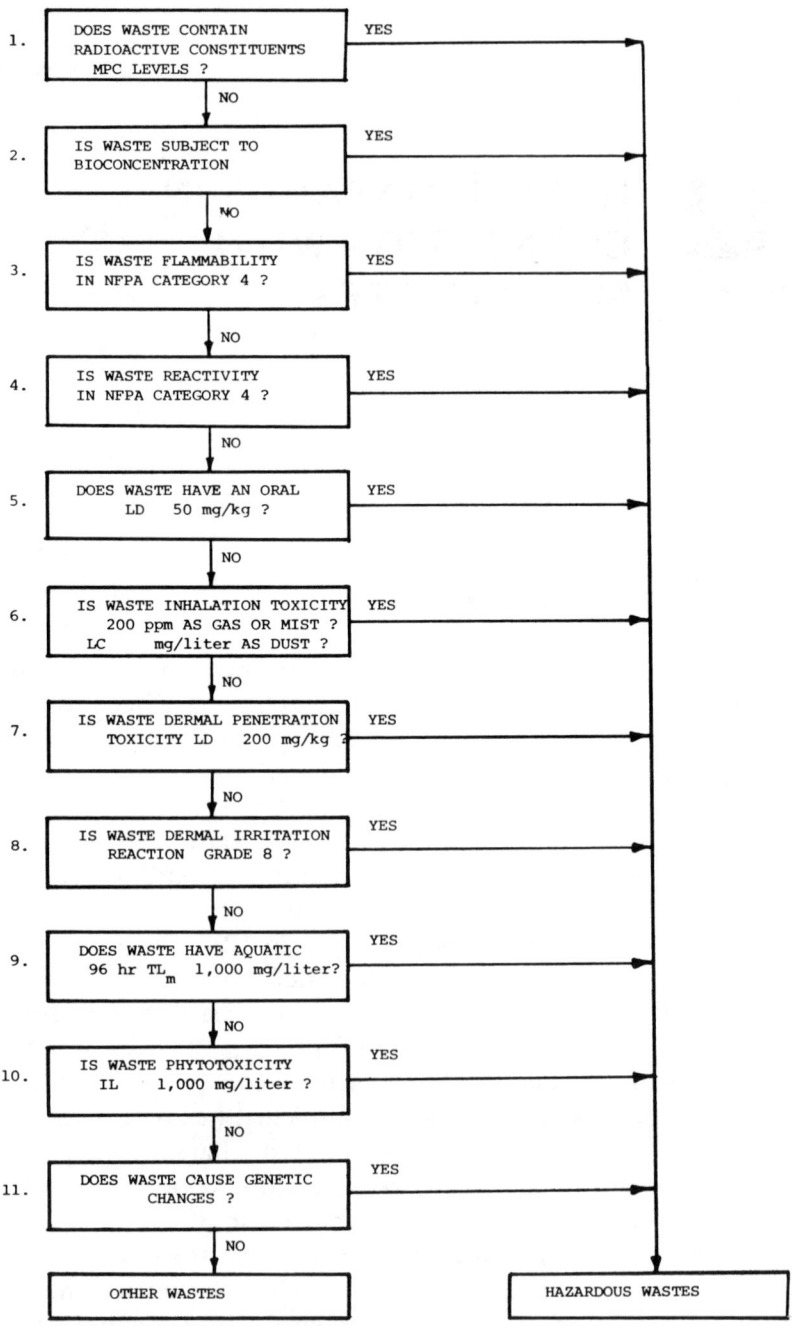

Source: Office Solid Waste Management Programs, EPA (1974).

taminants. Ideally, this air should be collected and used in processing to prevent contact with workers. Workers should also wear gas masks to ensure their health and safety.

## ASBESTOS DUST

Asbestos dust originates both in industries (lagging for boilers, tanks, and pipes and scrap brake linings) and in normal municipal refuses (demolition and used-auto brake linings) and can cause cancer of the lung in workers who breathe the dust. The dustiness of the processing can influence whether the disease occurs over a relatively long period of exposure or during a short concentrated exposure. Protection includes wet stripping of lagging, using protective screens around areas, or workers' use of positive-pressure respiratory equipment.

Asbestos is the first of dozens of potentially toxic substances now becoming the subject of considerable litigation. According to a recent disclosure (reported in the *Wall Street Journal*, March 18, 1981, p. 10), asbestosis is a disease which has manifestations similar to emphysema. It usually develops after about ten years of exposure to asbestos. Lung cancer can take twenty years or more to develop, and mesethelioma, a cancer of the chest lining, may not show up until forty years of exposure.

American Re-Insurance estimates (according to the *Wall Street Journal*, June 16, 1981, p. 54) that 12,000 asbestos settlement suits are outstanding against 725 companies and that asbestos settlements now run to $1.4 billion annually.

Wetting down of asbestos products before demolotion is an acceptable healthful practice. However, care must be exercised in carrying away the demolition material. Material is placed in large plastic bags during storage and transportation. All deposited material should be covered immediately with at least nine inches of compacted soil to prevent dusting of the buried asbestos.

The specific health hazards of some particularly troublesome metals and organic solids are listed and described in Table 12-1.

## HEALTH AND PROTECTION OF WORKERS

In 1965 the average accident frequency of *156 disabling* injuries per million man hours was four and one-quarter times the highest rate

Table 12-1. Troublesome Metals and Organic Solids.

| Contaminant | Source | Health Hazard |
| --- | --- | --- |
| *Inorganic* | | |
| 1. Lead | Paint chips, contaminated air fallout, demolition debris, leachates | Brain damage especially in children; lowers infection resistance |
| 2. Mercury | Industrial and agricultural wastes | Severe mental and motor symptoms, neurological symptoms, kidney damage; gastrointestinal and pulmonary symptoms; genetic damage |
| 3. Beryllium | Rocket fuel, low density metal as alloy | Respiratory tract irritation, dermatitis conjunctivitis, potential mortality |
| 4. Manganese | Alloying element in steels | Inhalation causes manganic pneumonia |
| *Organic* | | |
| 5. Benzopyrene and Benzofluoranthene | Formed from heating (incineration) > 500°C; Soils, root vegetables grown in contaminated soils. | Carcinogenic to animals |
| 6. Polybrominated biphenyls (PBB) | Flame retardants for fibers and thermoplastics | Toxic to liver, kidney and thyroid. It also severely affects the nervous and reproductive systems. It is a teratogenic agent— persistent and bioaccumulates 20–30,000 times. |
| 7. PCB | Electrical equipment cooling systems, carbon copy paper re-used in paperboard as used in food packages | Causes liver malfunction; collects in fatty tissue; severe skin eruptions |

Table 12-1. continued

| Contaminant | Source | Health Hazard |
|---|---|---|
| 8. Fiber Polymers | Open burning of fiberpolymers such as orlon, foamed plastics, etc. | Poison gas such as HCn, HCe |

(36.71 for coal mining) of any major industry reported by the National Safety Council. Illness data reported in the same survey showed that sanitation workers suffered average or below-average illness rates for industry in general.

The following are some: *occupational mycoses*—infections from fungi—that affect workers who handle certain types of wastes:

*Histo-plasmosis*—contact with chicken feces.

*Cryptococcosis*—contact with pigeon feces. It can affect demolition workers in particular.

Some deaths have been attributed to *Mycosis* and more specifically *Coccidividomycosis, Sporotrichosis,* and *Histoplasmosis.*

## PUBLIC HEALTH HAZARDS

Any previously sterile organic matter such as food refuse can become infested with disease-producing micro-organisms in a short time especially in warm temperatures. These disease organisms can then be spread to other animals and humans mainly by *flies, mosquitoes,* and *rodents.* The specific diseases caused by each of these three vectors are given in Table 12-2.

Refuse-containing organic matter, when kept sealed in one way or another, will not allow spreading of disease by these vectors. However, this is more easily stated than actually accomplished. Compaction to a density of 1,000 pounds per cubic yard will reduce the spaces between the particles of refuse and hence prevent insect penetration. Another possible treatment is that of processing the refuse immediately or at least within three days after receipt so as to prevent hatching of insect eggs and propagation of disease organisms.

Table 12-2. Specific Diseases Caused by Flies, Mosquitoes, and Rodents.

| Fly-Borne Diseases | Mosquito-Borne Diseases | Rodent-Borne Diseases |
|---|---|---|
| Typhoid | Dengue | Echinostomiasis |
| Bacillary dysentery | Encephalitis | Hemorrhagic septicemia |
| Amoebic dysentery | Filariasis | Histoplasmosis |
| Diarrheas | Malaria | Lymphocytic choriomeningitis |
| Asiatic cholera | Yellow fever | Plague |
| Helminth/infections (worm) | Tularemia | Rat-bite fever |
| Myiasis | Lymphocytic choriomeningitis | Rat-mite dermatitis |
| Loiasis | Melioidosis | Rat-tapeworm infection |
| Onchocerciasis | Rift-Valley fever | Rocky Mountain spotted fever |
| Ozzard's filariasis |  | Salivary-gland virus infection |
| Leishmaniasis |  | Salmonellosis |
| African sleeping sickness (trypanosomiasis) |  | Schistosomiasis |
| Yaws |  | Bilharziasis |
| Tularemia |  | Sporotrichosis |
| Bartonellosis |  | Swine erysipelas |
| Cararrhal conjunctivitis |  | Trichinosis |
| Sandfly fever |  | Leptospirosis |
|  |  | Leishmaniasis |
|  |  | Relapsing fever |
|  |  | Tularemia |
|  |  | Rickettsial pox |
|  |  | Murine typhus |

Also shredding of refuse increases the surface area and allows an increase in density after compaction. This prevents disease transmission since insects and rodents are unable to exist in the refuse. The refuse then begins to decompose aerobically, generating enough heat to discourage vermin. Fires set in landfills of shredded and compacted refuse tend to smolder rather than burn.

Health problems are also associated with the combined discharged of refuse with domestic sewage sludge. Potential health conditions are related to the following:

1. *Heavy metals* such as lead and cadmium are found in sewage sludge. These metals can cause hypertension, cardiovascularism, and respiratory illness. They are also found in municipal solid waste.

2. *Pathogens* from sewage sludge have resulted in increases in hepatitis, salmonelloses, and bovine cysticerosis.
3. Other *bacterial, viral, protozoa,* and *helminthsic diseases* are shown in Table 13-1.

## REFERENCE

Office of Solid Waste Management Programs, U.S. Environmental Protection Agency. 1974. "Report to Congress—Disposal of Hazardous Wastes," U.S. E.P.A. Report SW-155.

# 13 AGRICULTURAL WASTE

## INTRODUCTION

Fifty percent (50%) of the U.S. solid waste production (2,280 million tons in 1968) comes from agricultural operations. Historically, most of this waste has been returned to the land by farmers. The most common disposal practice has been composting, but more agriculturists are looking at incineration of these wastes because of the possible environmental impact. Up until now the economics have not been conducive to the reduction of these wastes by incineration.

Agricultural waste consists of two major categories: crop solid wastes and animal solid wastes. Figure 13-1 depicts the general sources of both types of wastes. Middlebrooks (1979) gives an excellent presentation of the entire subject of agricultural wastes. In this chapter we have concentrated our findings on the solid-waste data available.

## SEWAGE CONTAMINATED SLUDGE AGRICULTURAL WASTE

If sewage sludge-refuse mixtures are used for agricultural irrigation, no curing takes place, such as in composting, largely because it often

# INDUSTRIAL SOLID WASTES

Figure 13-1. Agricultural Waste Sources.

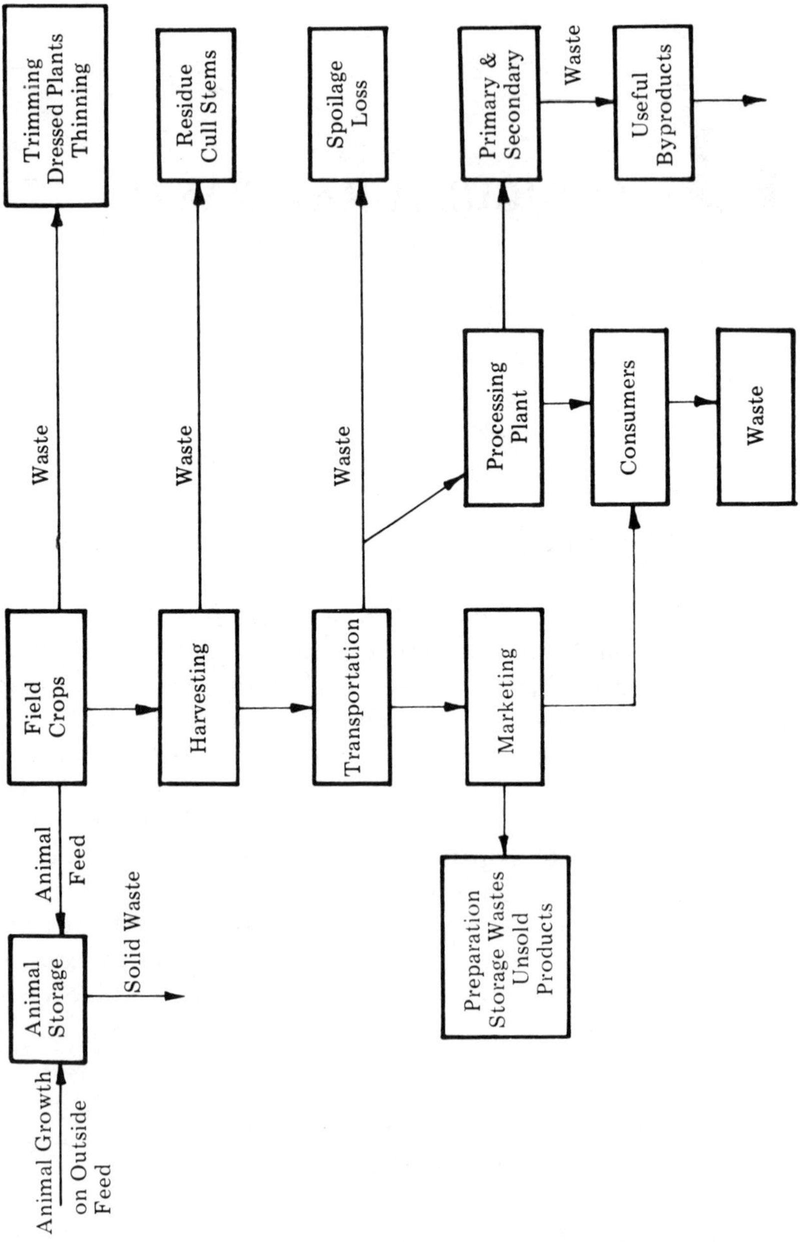

becomes anaerobic. Viruses such as poliomyelitis have been found to build up and accumulate in the vegetables grown in the soil. Parasite—such as hookworm—disease can also be spread by this practice. Walking barefoot in these fields can result in picking up the hookworms or larvae through the skin which are then transmitted through the liver to the lungs. This gives rise to pulmonary diseases evidenced by coughing. Hookworms also cause anemia. Some of these diseases are shown in Table 13-1.

Metals in refuse and sewage sludge, when finding their way to rivers, bays, estuaries, and coastal zones, are picked up by fishes and transmitted to eaters of fish, causing human disease to the nervous system.

Shellfish are also often contaminated by human sludge organisms. Hepatitis especially is transmitted in this way by eating contaminated shellfish.

A large portion of the waste is generated in the growing and harvesting of field crops. An example in this area is in the sugar cane industry. For years the crushed cane stalks (called bagasse) have been burned in "milk bottle" type incinerators to generate steam at the sugar mills. The cane is burned on a hearth in a multiple-chamber, vertical, in-line incinerator. Air is injected at the flame port between the primary and secondary combustion chambers.

Much of the waste from processing and marketing of agricultural products is incinerated at the plant sites.

The wastes from produce trimmings at the retail outlets are often incinerated along with other packaging wastes in commercial-sized units.

Table 13-1. Diseases Associated with Human Fecal Waste.

| | |
|---|---|
| Bacterial infections | Protozoal infections |
|   Typhoid fever |   Entamoeba histolytica |
|   Paratyphoid fevers | Helminthiasis |
|   Cholera |   Fish tapeworm |
|   Shigellosis (bacillary dysentery) |   Beef tapeworm |
| Viral infections |   Pork tapeworm |
|   Poliomyelitis |   Pinworm |
|   Coxsackie infection |   Roundworm |
|   Infectious hepatitis |   Whipworm |
|   (very many other enteric viruses exist) |   Hookworm |

## CONTAMINANTS AND TREATMENT

The principal constituents in agricultural wastes and common means for disposal are shown in Table 13-2.

Some actual weights of solid wastes are given in Table 13-3 for certain types of field crop and animal agricultural wastes.

When cotton crop is grown and harvested, hulls, sticks and stems, and leaf and dirt constitute the solid wastes that require disposal. Pounds per bale of each are given in Table 13-4.

On the other hand, total animal wastes in the United States for the year 1965 are given for each seven major animal types in Table 13-5.

Farm animals are often compared to man in relation to the amount of solid waste, bacteria, and Biochemical Oxygen Demand each produces. Table 13-6 relates chickens, swine, and cattle to the BOD of man.

The contributions of ducks, sheep, chickens, cows, turkeys, and pigs to solid wastes are also given in Table 13-7.

The defecation solids of cattle, poultry, swine, and sheep must be disposed of as solid wastes. Quantities, BOD, and fertilizer nutrients of the solids of these animals are given in Table 13-8.

The nutrient chemical characteristics of animal solid wastes are important for re-use possibilities and are given in Tables 13-9 and 13-10.

As animal solid wastes are dried, the offensive odors diminish. Figure 13-2 demonstrates this with chicken solid manure.

Table 13-2. Principal Agricultural Waste Components.

| Source | Waste | Composition | Means of Treatment or Disposal |
|---|---|---|---|
| Farms, ranches, livestock feeders, and growers | Household refuse | Same as household solid wastes | Landfill |
| Farms, ranches, livestock feeders, and growers | Crop residue | Cornstalks, tree prunings, pea vines, sugarcane stalks (bagasse), green drop, cull fruit, cull vegetables, rice barley, wheat and oats stubble, rice hulls. Fertilizer and insecticide residue | Plowed back into the land, incineration, stock feed |
| | Animal manure (paunch manure) | Lignaceous and fibrous organic matter, nitrogen, phosphorus, potassium, volatile acids, proteins, fats, carbohydrates | Fertilizer composting, stock feed |
| | Poultry manure | Same as animal manure | Fertilizer composting, lagooning |

Table 13-3. Agricultural Wastes.

|  | Source Units[a] | Multiplier | Ton/Year |
|---|---|---|---|
| Apricots | 310 acres | 1.5420 ton/acre year[b] | 478 |
| Cherries | 53 acres | 1.4265 ton/acre year[b] | 76 |
| Vegetables, berries, and seed crops | 3,614 acres | 3.0 ton/acre year[c] | 10,842 |
| Greenhouse and nursery | 125 acres | 25.0 ton/acre year[d] | 3,125 |
| Feedlot cattle manure | 840 head | 13.21 ton/head year[e] | 11,096 |
| Total agricultural waste |  |  | 25,617 |

a. Basic crop acreage and livestock inventory are from Alameda County Agriculture Commission Report (1967). Data were allocated to subcounty areas with help of Alameda County Agriculture Extension Office.

b. Apricot and cherry orchard waste multipliers determined by estimate of trees per acre from University of California Agricultural Extension Service and per tree waste multipliers from FMC Corporation Santa Clara Study.

c. State of California Department of Public Health, *Status of Solid Waste Management in California,* 1968.

d. Professor Harry C. Kohn, Department of Environmental Horticulture, Davis Campus of the University of California (verbal communication).

e. The manure multiplier for cattle is from Taiganides, E. Paul, Table 4, *Agricultural Solid Wastes.* Three of the cattle manure multipliers in his table averaged to 72.4 lb./day.

Source: Golueke and McGauhey (1970: 50).

Table 13-4. Average Amount and Type of Trash in Seed Cotton Harvested by Various Methods.

| | Pounds per Bale | | |
|---|---|---|---|
| Types | Machine Picked | Machine Stripped | Machine Scrapped |
| Hulls | 29 | 397 | 329 |
| Sticks and stems | 9 | 50 | 143 |
| Leaf and dirt | 43 | 78 | 398 |
| Total | 81 | 525 | 870 |

Source: *Control and Disposal of Cotton-Ginning Wastes* (1967: 30).

Table 13-5. Production of Wastes by Livestock in the United States, 1965.

| Livestock | Population, Millions | Annual Production of Solid Wastes, Million Tons | Annual Production of Liquid Wastes, Million Tons |
|---|---|---|---|
| Cattle | 107 | 1004.0 | 390.0 |
| Horses | 3 | 17.5 | 4.4 |
| Hogs | 53 | 57.3 | 33.9 |
| Sheep | 26 | 11.8 | 7.1 |
| Chickens | 375 | 27.4 | — |
| Turkeys | 104 | 19.0 | — |
| Ducks | 11 | 1.6 | — |
| Total | | 1138.6 | 435.4 |

Source: U.S. Department of Agriculture (1968).

Table 13-6. Population Equivalents[a] of Animal Wastes ($BOD_5$[b] Basis).

| Animal | Reference |
|---|---|
| Chickens (4-5 lb.) | |
| 11.8 | 1 |
| 11.3 | 2 |
| 12 | 3 |
| 11 | 4 |
| 10-20 | 5 |
| Swine (100 lb.) | |
| 0.6 | 1 |
| 0.33 | 6 |
| Cattle (1,000 lb.) | |
| Dairy 0.13 | 7 |
| Beef 0.17 | 7 |

a. Equivalent animals per capita, that is, 11.8 chickens contribute the BOD equivalent to one person per day.

b. BOD (Biological Oxygen Demand) is a measure of the water pollution potential, of an organic waste. It corresponds to the amount of oxygen required by the bacteria which consume the organic waste.

Source: *Pollution Implications of Animal Wastes — A Forward Oriented Review* (1968: 44).

**Table 13-7.** Per Capita Animal Contribution of Indicator Micro-Organisms.

| Animals | Average Weight, Feces/Day Wet Weight, g | Moisture, % | Average Indicator Density/g of Feces | | Average Contribution/ Capita Day | | Ratio |
|---|---|---|---|---|---|---|---|
| | | | Fecal Coliform, Million | Fecal Streptococci, Million | Fecal Coliform, Million | Fecal Streptococci, Million | |
| Man | 150 | 77 | 13.0 | 3.0 | 2,000 | 450 | 4.4 |
| Duck | 336 | 61 | 33.0 | 54.0 | 11,000 | 18,000 | 0.6 |
| Sheep | 1,130 | 74 | 16.0 | 38.0 | 18,000 | 43,000 | 0.4 |
| Chicken | 182 | 72 | 1.3 | 3.4 | 240 | 620 | 0.4 |
| Cow | 23,600 | 83 | 0.23 | 1.3 | 5,400 | 31,000 | 0.2 |
| Turkey | 448 | 62 | 0.29 | 2.8 | 130 | 1,300 | 0.1 |
| Pig | 2,700 | 67 | 3.3 | 84.0 | 8,900 | 230,000 | 0.4 |

Source: Geldreich (1966); Kenner, Clark, and Kabler (1961).

**Figure 13-2.** Odor Offensiveness of Chicken Manure as a Function of Moisture Content.

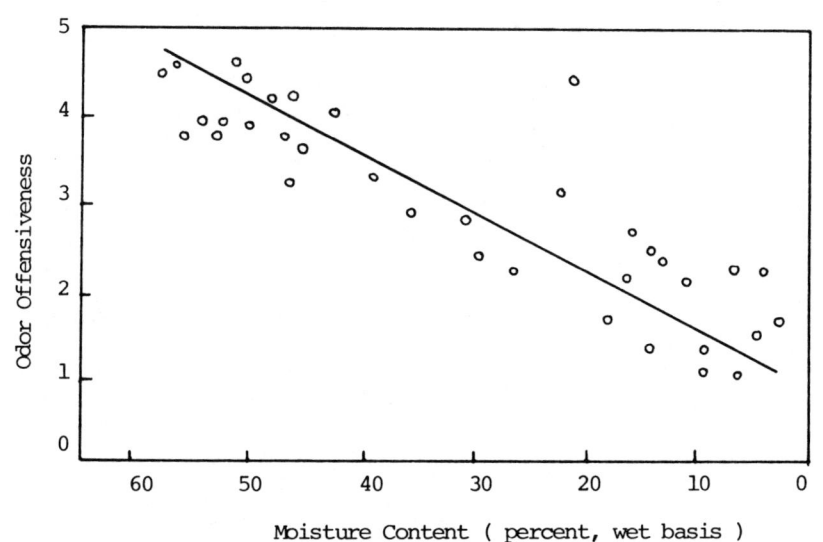

```
5 - Very strong offensive odor
4 - Strong odor
3 - Definite odor
2 - Faint odor
1 - Very faint odor
0 - No offensive odor
```

Source: Sobel (1969: 349).

Table 13-8. Animal Waste Defecation (Per 1,000 Pounds of Live weight).

| | Manure Production | | | | | | BOD Production | | | Fertilizer Nutrients | | | |
|---|---|---|---|---|---|---|---|---|---|---|---|---|---|
| | TS | | | VS | | | lb BOD[f]/lb TS | lb BOD/lb VS | BOD:COD,[g] % | N, % db | $P_2O_5$, % db | K, % db | Ref. |
| | lb/d | % wb | lb/d | % db | lb/d | lb/d | | | | | | | |

*Dairy Cattle*

| lb/d | % wb | lb/d | % db | lb/d | lb/d | lb BOD/lb TS | lb BOD/lb VS | BOD:COD % | N | $P_2O_5$ | K | Ref. |
|---|---|---|---|---|---|---|---|---|---|---|---|---|
| 72 | 12.5 | 9.0 | 80[a] | 7.2[a] | 1.84 | — | — | — | — | — | — | 1 |
| — | — | 10.4 | 80.3 | — | — | — | 0.183 | 18.3 | 3.7 | 1.1 | 3.0 | 3 |
| — | — | — | 80[a] | — | — | 0.102 | 0.129 | 8.2[a] | — | — | — | 5 |
| — | — | — | 71.5[a] | — | — | 0.278 | 0.388[a] | 12.7[a] | 2.8 | 1.04 | 0.34 | 6 |
| 105 | 9.0 | 9.4 | — | — | — | — | — | — | — | — | — | b |
| — | — | — | — | — | — | — | 0.232 | — | — | — | — | 8 |
| — | — | 6.8 | 85[a] | 5.7 | 1.32 | — | — | 22.8[a] | 5.5[a] | 1.1[c] | 1.7[c] | 10 |
| Average 88 | | 9.0[a] | | | | | 0.233[c] | 16.0[c] | 4.0[c] | | | |

*Beef Cattle*

| — | — | — | 73.2[a] | — | — | 0.195[a] | 0.267[a] | 13.3 | 12.5[a] | 1.52 | 0.44 | 6 |
|---|---|---|---|---|---|---|---|---|---|---|---|---|
| — | — | 3.6 | 86.5[a] | — | 1.02 | — | — | 31.3[a] | 7.2[a] | — | — | 10 |
| — | — | — | — | — | — | — | 0.236 | — | — | — | — | 8 |
| Average | | | 80.0[c] | | | | 0.252[c] | | 9.8[c] | | | |

*Poultry, Hens*

| 64[a] | 27.2 | 17.4[a] | 70.3 | 12.2[a] | — | — | 0.338 | 29.8[a] | 23.3[a] | — | — | 2 |
|---|---|---|---|---|---|---|---|---|---|---|---|---|
| — | — | 16.5[a] | 77.5 | 12.8[a] | — | — | 0.288 | 26.0 | 5.4 | 4.6 | 2.1 | 3 |
| 54 | 24.1 | 18.4 | 73.8 | 13.6 | — | — | 0.381 | — | 6.9 | — | — | c |
| 50[c] | | 17.4[c] | 74.0[c] | | | | 0.338 | 28.0[c] | 11.5[c] | | | |

## Swine

| | | | | | | | | | | | | | |
|---|---|---|---|---|---|---|---|---|---|---|---|---|---|
| 52 | 10.5 | 5.5 | 81.3 | 4.5 | 3.1 | — | 0.57 | — | 38.3 | 3.35 | — | — | d |
| — | — | — | 78.5 | 6.3 | — | — | — | 0.696 | 26.7[a] | 4.0 | 3.1 | 1.4 | 3 |
| 49[a] | — | — | — | — | — | — | — | 0.320 | 36.2[a] | — | — | — | 4 |
| — | — | — | — | — | — | — | — | — | — | — | — | — | 5 |
| — | 15.4 | — | 85.0 | — | — | — | 0.262 | 0.302 | 19.3 | 5.9[a] | — | — | 7 |
| — | — | — | — | — | — | — | 0.450[a] | 0.382[a] | 30.8[a] | 7.0 | — | — | 8 |
| — | — | — | — | — | — | — | — | — | — | — | — | — | e |
| — | — | — | — | — | — | — | — | — | — | — | 1.9 | 1.4 | 9 |
| — | — | — | — | — | — | — | — | — | — | 5.6 | 2.5 | 1.4 | |
| | | | | | | | | | | | | | |
| — | — | — | — | — | — | — | — | — | — | — | — | — | 5 |
| — | — | — | — | — | — | — | — | — | — | — | — | — | b |
| — | — | — | — | — | — | — | — | — | — | — | — | — | e |

a. Indicates value was calculated on the basis of data cited in the reference.
b. W.B. Roller, personal communication, Ohio Agric. Res. Center, 1968.
c. E.P. Taiganides, personal communication, Agric. Eng. Dept., Ohio State University, 1963.
d. J.C. Converse, personal communication, Agric. Eng. Dept., University of Illinois, 1970.
e. E.P. Taiganides, personal communication, Agric. Eng. Dept., Ohio State University, 1967.
f. BOD (Biological Oxygen Demand) is a measure of the water pollution potential of an organic waste. It corresponds to the amount of oxygen required by the bacteria which consume the organic waste.
g. COD (Chemical Oxygen Demand) is a measure of the water pollution potential of an organic waste. It corresponds to the amount of oxygen required for complete oxidation of the organic matter.

Source: Middlebrooks (1979).

*Note:* lb/d = pounds per day; wb = wet basis; db = dry basis; TS = Total solids; VS = Volatile solids.

Table 13-9. Characteristics of Animal Manures.

| Animal | Moisture, % | Lb/Ton Manure | | | | | | | | |
|---|---|---|---|---|---|---|---|---|---|---|
| | | N | P | K | S | Ca | Fe | Mg | Volatile Solids | Fat |
| Dairy cattle | 79 | 11.2 | 2.0 | 10.0 | 1.0 | 5.6 | 0.08 | 2.2 | 322 | 7 |
| Fattening cattle | 80 | 14.0 | 4.0 | 9.0 | 1.7 | 2.4 | 0.08 | 2.0 | 395 | 7 |
| Hog | 75 | 10.0 | 2.8 | 7.6 | 2.7 | 11.4 | 0.56 | 1.6 | 399 | 9 |
| Horse | 60 | 13.8 | 2.0 | 12.0 | 1.4 | 15.7 | 0.27 | 2.8 | 386 | 6 |
| Sheep | 65 | 28.0 | 4.2 | 20.0 | 1.8 | 11.7 | 0.32 | 3.7 | 567 | 14 |

Source: Benne et al. (1961).

Table 13-10. Nutrients in Animal Wastes.

| Parameter, % Slurry | Dairy Cattle | Beef Cattle | Hogs | Poultry | Reference |
|---|---|---|---|---|---|
| Nitrogen (as N) | 0.13–0.42 | 0.24–0.60 | 0.3–0.9 | 0.41–1.7 | 1 |
| Phosphorus (as P) | 0.06–0.09 | 0.09–0.25 | 0.2–0.6 | 0.30–1.5 | 1 |
| Potassium (as K) | 0.13–0.30 | 0.14–0.28 | 0.2–0.4 | 0.13–1.25 | 1 |

| Parameter, % Dry Basis | Cattle | Hogs | Sheep | Horse | Hen | Reference |
|---|---|---|---|---|---|---|
| N | 0.3–1.3 | 0.2–0.9 | 0.9 | 0.66 | 1.8–5.9 | 2 |
| $P_2O_5$ | 0.15–0.5 | 0.14–0.83 | 0.34 | 0.23 | 1.0–6.6 | 2 |
| $K_2O$ | 0.13–0.92 | 0.18–0.52 | 1.0 | 0.68 | 0.8–3.3 | 2 |

Source: Middlebrooks (1979).

## REFERENCES

Benne, E. J., et al. 1961. "Animal Manures—What Are They Worth Today?" *Agricultural Experiment Station, Michigan State University Bulletin.*

*Control and Disposal of Connon-Ginning Wastes.* 1967. Washington, D.C.: U.S. Department of Health, Education and Welfare, Public Health Service.

Geldreich, E. E. 1966. *Sanitary Significance of Fecal Coliforms in the Environment.* Washington, D.C.: U.S. Department of the Interior, Federal Water Pollution Control Administration.

Kenner, B. A.; H. F. Clark; and P. W. Kabler. 1961. "Fecal Streptococci: I. Cultivation and Enumeration of Streptococci in Surface Waters." *Applied Microbiology* 9:15.

Middlebrooks, E. J. 1979. "Animal Wastes." In *Industrial Pollution Control*, Vol. I., *Agro-Industries*, Chapter 2B, pp. 83–101. New York: Wiley Interscience Co.

*Pollution Implications of Animal Wastes—A Forward-Oriented Review.* 1968. Washington, D.C.: U.S. Department of the Interior, Federal Water Pollution Control Administration.

Sobel, A. T. 1969. "Removal of Water from Animal Manures." In *Animal Waste Management,* Conference on Agricultural Waste Management, Cornell University, January 13–15.

U.S. Department of Agriculture. 1968. *Wastes in Relation to Agriculture and Forestry.* Washington, D.C.: U.S. Government Printing Office.

# 14  HOSPITAL SOLID WASTES

INTRODUCTION

Today few will deny that hospitals are an industry as vital to our society as any other provider of goods and services. However, hospitals are so intimately involved with people that their problems have generally been considered municipal. In addition, they are usually located within the city to provide easy access for its people. Furthermore, their wastes are usually collected and treated along with other municipal solid wastes. Therefore, in this text we are still considering hospital wastes outside the realm of specific industrial wastes. Hospitals are generally built in congested areas and on limited land area. Under such conditions they have great difficulty in handling their own solid wastes. Hospitals also produce a much greater quantity of solid wastes than most institutions. Because of their danger to public health due to their contact with patients who have contagious disease, these hospital wastes represent a special problem.

A directive, for example, from the Florida Department of Environmental Regulation ("Hospital Wastes to Orlando. . . ." 1981) rejected a Metropolitan Dade County plan to dump 100 tons of infectious hospital wastes in a county landfill. The state and county officials warned that the wastes—including syringes, blood-soaked bandages, and human tissue—posed a potential health threat to Dade County residents. When we also consider solid wastes from animal experiments, autopsies, and surgeries, hospital wastes are varied as well as dangerous.

## CHARACTERISTICS

Hospital solid wastes are reported to be 1.5 percent of the total solid wastes generated in United States in 1969 ("A Comprehensive Assessment . . ." 1969). This amounted to an estimated 55 million tons per year (*Handbook on Hospital Solid Waste Management* 1973). Another interesting design parameter commonly used is ten pounds per day per hospital patient (*Handbook* 1973); however, these quantities vary from eight to twenty pounds per patient per day ("Hospital Solid Waste Disposal . . ." 1971). Obviously a considerable difference exists between the generation and collection quantities of solid wastes because of incomplete collection or burning of hospital wastes. Because of the predicted continued increase by hospitals in use of disposable items, their solid wastes are expected to increase twice as fast as total U.S. solid wastes (*Handbook* 1973).

In order to handle all the various types and quantities of hospital solid wastes, classification systems have been proposed (see Table 14-1).

The *Handbook* (1973) suggests classifying hospital solid wastes into four simple types as shown in Table 14-2. The *Handbook* authors emphasize that contaminated waste should go into one particular storage system rather than end up in the domestic waste system because an employee of the hospital does not consider the waste to be hazardous enough to be labeled pathological. Each of these four waste types presents its own special problems as given in Table 14-3. The various operations of the hospital produce various types (shown in Table 14-2) of solid wastes as given in Figure 14-1.

The University of Minnesota study ("Hospital Solid Waste Disposal . . ." 1971) revealed that at least 87 percent of hospital solid waste by weight belonged in the domestic waste category of which 78 percent is combustible. Pathological solid wastes (1.4 percent) and contaminated wastes (8 percent) and special waste (more than 1 percent) comprise the remainder of hospital solid waste. As is obvious, hazardous hospital solid waste represents a very small percentage of the total. Segregation and special treatment is a "natural" solution.

Table 14-1. Classification Schemes for Hospital Solid Wastes as Summarized from Various Sources.

| Author | Year Presented | Classes of Hospital Solid Waste |
|---|---|---|
| McKenna | 1963 | Garbage, rubbish, ashes, pathological waste, critical waste (for example, radioactive), and pathogencontaminated waste. |
| Falick | 1965 | Garbage, rubbish, organic waste ("animal carcasses, organs, solid organic waste and pathological waste"), other waste (rubber, plastics, metals, glass. etc.), and special waste ("radioactive and other waste representing a high degree of danger if handled and, (which) therefore, must be disposed of by separate means, by law"). |
| Black | 1967 | Garbage, "paper, trash, and other dry combustibles," treatment room waste, surgery waste, autopsy waste, and noncumbustibles. |
| Holbrook | 1968 | Garbage and similar material capable of being ground for discharge to a sewage system, "wet waste incompatible with the sewer and difficult to burn," combustible waste, "pathological and radioactive wastes, which require special treatment because of contamination and which are usually regulated by statute," and noncombustible waste. |
| Groce | 1968 | Garbage, rubbish, pathological waste (including body tissue, fluids and bones), and infectious waste. |

Source: "Hospital Solid Waste Disposal in Community Facilities," 1971.

The *Handbook* (1973) also derived density values for each of its four classifications in order to arrive at volume requirements. They are given in Table 14-4.

It is important to recognize that the volumes, characteristics, and densities of classified hospital wastes may vary considerably from those given in the tables. These factors must be ascertained specifically for each hospital in question. It is then necessary to collect, store, and transport these wastes either for further storage or to processing equipment, and then dispose of residue in a sanitary manner with a minimum of environmental impact.

Table 14-2. General Classification System.

| Waste Type | Description |
| --- | --- |
| Domestic | Kitchen waste, office waste, and packing waste |
| Pathological | Human tissue, animal carcasses, organs, animal bedding, and fecal matter |
| Contaminated | All items discarded after patient use. Floor cleanings |
| Special | Hazardous wastes other than pathological, radioactive wastes, discarded acids |

Source: *Handbook on Hospital Solid Waste Management* (1973).

As indicated in this and other chapters in this text, we expect an increase in the use of plastic materials in United States during the 1980s. With respect to hospital solid wastes there are special concerns with the plastics beyond their presumed nonbiodegradability. They are inflammable and can burst or melt when stored near a hot location such as a furnace. When charged into the furnace, they may adhere upon melting to any surface. Cleaning incinerators becomes more difficult. and the heat-exchange capability of these units decreases as a result of burning plastics. Some plastics, such as polyvinylchloride, yield acids such as hydrochloric when burning which are corrosive to the materials surrounding the incinerator. The *Handbook* (1973) recommends that domestic hospital waste be stored, transported, and disposed of in clear plastics while, on the other hand, opaque plastic bags can be used for hazardous or objectionable solids. For collection of these wastes from the food-handling area, grinders are recommended (*Handbook* 1973) at the "dirty" or cleanup area to speed up the return of trays and dishes for the next serving. Two, one-and-a-half horsepower grinders are needed when 1,000 or more meals are served daily.

Solid wastes must be transported to storage or disposal within the hospital. In Table 14-5 some advantages and disadvantages are given for the four major internal transportation systems.

Table 14-3. Problems Associated with Waste Type.

| Classification | Problems |
|---|---|
| 1. Domestic | Organic kitchen waste encourage disease vectors and unpleasant odors. Packing waste may provide habitation for vectors. Packing waste also represents a significant fire hazard. |
| 2. Pathological | Certain biological waste is hazardous due to contamination with pathogens. Tubercular lungs are particularly hazardous due to possibility of airborne dissemination of pathogens.<br><br>The American Hospital Association has identified as particularly hazardous the organism Mycobacterium tuberculosis, bacteria of the pasteurella, brucella and psittacosis groups and also certain viruses.<br><br>Autopsy and surgical waste present pathogenic hazard and also public insult if not correctly handled. |
| 3. Contaminated | Blood, pus and spittle can contaminate the air with a variety of microbiological agents. Contaminated casts and dressings are potentially equally hazardous.<br><br>Disposable gowns, sheets, boots, animal bedding utensils and floor sweeping, are all potentially contaminated.<br><br>Sharp wastes such as scalpels and hypodermic needles are potentially disease and accident hazards. Discarded hypodermic needles may also be stolen for drug abuse. |
| 4. Special Wastes | Radionuclides used in diagnosis and treatment have short half-lives. Special storage should be provided to permit decay to a satisfactory level before disposal. The most common radioactive wastes are carbon 14, chromium 51, gold 198, iodine 131, iron 50, phosphorus 22, and sodium 24.<br><br>Highly explosive or toxic chemical wastes require special handling and disposal procedures. |

Source: *Handbook on Hospital Solid Waste Management* (1973).

## 162 INDUSTRIAL SOLID WASTES

**Figure 14-1.** Operations of Hospitals That Produce Solid Wastes.

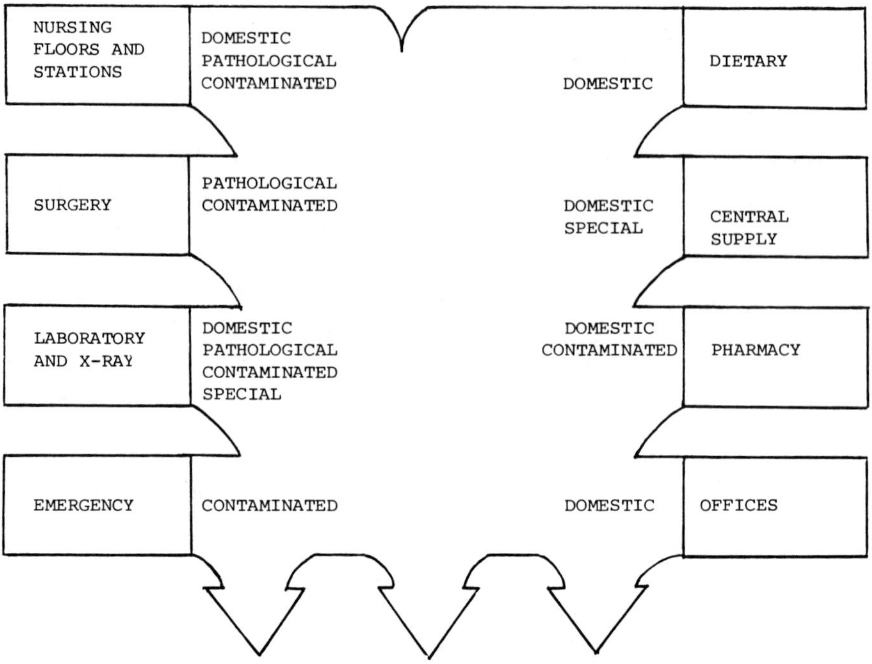

```
┌─────────────┐
│ NURSING     │ DOMESTIC                                      ┌─────────┐
│ FLOORS AND  │ PATHOLOGICAL                                  │ DIETARY │
│ STATIONS    │ CONTAMINATED           DOMESTIC               │         │
└─────────────┘                                               └─────────┘

┌─────────────┐ PATHOLOGICAL                                  ┌─────────┐
│ SURGERY     │ CONTAMINATED           DOMESTIC               │ CENTRAL │
│             │                        SPECIAL                │ SUPPLY  │
└─────────────┘                                               └─────────┘

┌─────────────┐ DOMESTIC
│ LABORATORY  │ PATHOLOGICAL           DOMESTIC               ┌─────────┐
│ AND X-RAY   │ CONTAMINATED           CONTAMINATED           │ PHARMACY│
│             │ SPECIAL                                       │         │
└─────────────┘                                               └─────────┘

┌─────────────┐                                               ┌─────────┐
│ EMERGENCY   │ CONTAMINATED           DOMESTIC               │ OFFICES │
└─────────────┘                                               └─────────┘
```

Source: *Handbook on Hospital Solid Waste Management* (1973).

**Table 14-4.** Hospital Solid Wastes Classification by Origin.

*Estimated Densities of Major Categories*

| Type of Waste | Density of Refuse (lbs/ft$^3$) | |
|---|---|---|
| Domestic | Garbage | 50 |
|  | Noncombustibles | 10 |
|  | Combustibles | 5 |
| Pathological |  | 3.75 |
| Contaminated |  | 5 |
| Special | Variable | |

Source: *Handbook on Hospital Solid Waste Management* (1973).

Table 14–5. Advantages and Disadvantages of Hospital Internal Transportation Systems for Solid Waste.

| | Initial Cost | Manpower Usage | Mechanical Complexity | Air Pollution Potential | Water Pollution Potential | Fire Hazard | Capable of Vertical and Horizontal Transport | Capable Alternate Uses |
|---|---|---|---|---|---|---|---|---|
| Standard cart | Low | High | Low | Low | None | Moderate | No except with elevator | Yes |
| Gravity chute | Low | Low | Low | High | None | High | No | Usually none |
| Pneumatic chute | High | Low | High | Low when filtered | None | Low | Yes | Yes |
| Wet pulper | — | — | — | — | Moderate | — | — | — |

Source: *Handbook on Hospital Solid Waste Management* (1973).

## TREATMENT OF HOSPITAL WASTES

Despite the disadvantages listed in Chapter 8, incineration is the most generally recommended treatment system for hospital solid wastes. Most municipal ordinances require incineration of pathological wastes; thus, these as well as much of the other solid wastes are burned. Most have a BTU value of about 6,500 BTUs per pound except for pathological waste which is much more watery and must be dried prior to burning—usually in a specially designed incinerator. Such a system is described (*Handbook* 1973) as a controlled air incinerator which emits only 0.05 grains of particulate matter per cubic foot per second along with 12 percent $CO_2$ and no odors or visible emissions. Wet scrubbers are also recommended to attain the present 0.03 grains per cubic foot per second of particulate standard and also to scrub out potentially hazardous gases.

The *Handbook* (1973) gives some 1973 capital costs for wet scrubbers and incinerators recommended for hospitals (see Tables 14-6 and 14-7).

When using incineration, separation of the various hospital solid wastes is not necessary. They can be fed automatically into the incinerator, thus eliminating the need for excessive handling. Incineration removes the fear of contaminated solid waste entering the environment. The *Handbook* (1973) provides data—Table 14-8—showing the advantages and disadvantages of hospital waste-handling systems. Some expected lives of major hospital-waste treatment equipment are also given in Table 14-9. It is significant to note that the refractories of incinerators must be repaired or replaced about every three years.

Development of a total hospital solid-waste system depends upon the least-cost-most-effective one for a particular hospital. It must include both the system within the hospital and that for the collective wastes after they leave the hospital. The system within the hospital normally contains a gravity or pneumatic collection for both contaminated and domestic wastes followed by a central compactor or incinerator for disposal outside the hospital. Since the pathological wastes must be incinerated, the choice usually is between burning all wastes, just pathological wastes, or both pathological and other contaminated wastes. State and local ordinances and economics usually dictate the choice selected. A Naval hospital survey was reported by the *Handbook* (1973) which compares collection and treatment system costs (Figure 14-2).

The *Handbook* (1973) compares all ultimate (final) disposal costs of hospital solid wastes to that of sanitary landfill. Unfortunately, most of the municipal landfills which accept hospital solid wastes have been found to be lacking either in design or operation. They will have to be updated and altered in operation in order to ensure this method as a safe one for hospital wastes. On the inside, hospitals, which are very sensitive to changes in labor costs, will have to continue to evaluate collection and compaction systems requiring more labor to handle more quantity.

## REFERENCES

"A Comprehensive Assessment of Solid Waste Problems, Practices, and Needs." 1969. Washington, D.C.: Office of Science and Technology.

*Handbook on Hospital Solid Waste Management.* 1973 Westport, CT.: Technomic Publishing Co.

Greene, J., and J.D. Salant. 1981. "Haul Hospital Wastes to Orlando, Feds Order After Illegal Dumping." *Miami Herald* (January 23), p. 1C.

"Hospital Solid Waste Disposal in Community Facilities." 1971. University of Minnesota.

Table 14-6. Equipment Cost for Wet Scrubbers.

| Equipment Type | Purchase Cost ($/CFM) | Smallest Particle Collected (microns) | Pressure Drop (in $H_2O$) | Power Used $\frac{KW}{1000\ CFM}$ | Remarks |
|---|---|---|---|---|---|
| 1. Spray tower | 0.1–0.2 | 10 | 0.1–0.5 | 0.1–0.2 | Common, low water use |
| 2. Jet | 0.4–1.0 | 2 | — | 2–10 | Pressure gain, high velocity liquid jet |
| 3. Venturi | 0.4–1.2 | 1 | 10–15 | 2–10 | High velocity gas stream |
| 4. Cyclonic | 0.3–1.0 | 5 | 2–8 | 0.6–2 | Modified dry collector |
| 5. Inertial | 0.4–1.0 | 2 | 2–15 | 0.8–8 | Abrasion problem |
| 6. Packed | 0.3–0.6 | 5 | 0.5–10 | 0.6–2 | Channeling problem |
| 7. Mechanical | 0.4–1.2 | 2 | — | 2–10 | Abrasion problem |

Source: *Handbook on Hospital Solid Waste Management* (1973).

Table 14-7. Cost of Incinerators.

| Hospital Size | System | Cost of System Installed (Incinerator, Gas Washer, ID Fan, Burners, Controls, Breeching, etc.)[a] | Chimney Installed[b] | Miscellaneous | Total Approx. |
|---|---|---|---|---|---|
| 100 Bed | One 1000 lb/hr | $85,000 | $12,000 | $10,000 | $107,000 |
| 250 Bed | One 2000 lb/hr | $165,000 | $14,000 | $10,000 | $189,000 |
| 400 Bed | Two 1500 lb/hr | $240,000 | $16,000 | $15,000 | $271,000 |
| 600 Bed | Two 2000 lb/hr | $315,000 | $18,000 | $15,000 | $348,000 |

a. These prices do not include an automatic stoker. An automatic feed system is not included.
b. The chimney prices are based on an estimated cost per foot, installed, for each respective size. The prices listed above are for 40 foot chimneys, thus:

30 in. dia. single flue = $300/ft
42 in. dia. single flue = $350/ft
35 in. dia. double flue = $400/ft
42 in. dia. double flue = $450/ft

Table 14-8. Advantages and Disadvantages of Waste-Handling Systems.

| System | Advantages | Disadvantages |
| --- | --- | --- |
| Central compactor(s) | May be chute fed. No heat generated. No air-pollution problems. No water-pollution problems. Simple to operate. Minimum volume for disposal. Minor changes in present house-keeping practices. Minimum maintenance. Can dispose of waste by landfill or in some instances incineration. Minimum handling of wastes reduces health hazards. Requires relatively unskilled labor. Can be operated 24 hours per day. | Chute fed does not allow separation of wastes. Need adjunct incineration facility for pathological wastes. Increase in disposables will be directly reflected in increased waste quantities. Syringes and medicines may be retrievable. May limit yourself to the number of contractors that will be able to handle the system. Weight restrictions precludes basement or manual movement of compacted waste. Should not be used for wastes of a pathological origin. |
| Central incineration | Maximum volume reduction. Sterile residue when operated properly. No auxiliary system required. No storage time required. Immediate consumption of combustibles. Precludes re-use of medicines, syringes, and hypodermic needles. Complies with most codes for disposal of pathological wastes. Psychologically amenable method of disposal. Minimum installation space. | Quantities of glass, cans, and other non-combustibles are not reduced in volume and must be handled separately. Complex operation considering environmental factors (air and water pollution control). Requires trained personnel to properly operate the system. Limited to operation during daylight hours. Non-combustibles and ash pose difficult handling problems. Auxiliary fuel necessary. |

| | | |
|---|---|---|
| Wet pulping | Reduction in volume (3 or 4 to 1). Will dispose of most materials except pathological waste. Vertical and horizontal transportation system are available utilizing this system. Quick and easy operation for disposal of waste on each floor. Completely eliminates potential hazard of medicines. System can accept kitchen and food handling wastes. | Increase in weight of material handling (80% moisture). Increased water consumption at hospital. Plastics are not particularly well handled presently by this system. Offensive cleaning problems. Rags, cords, and strings not easily handled. Uncertain destruction of needles and/or syringes. Dehydrated pulp not amenable to incineration chute system is negated except in central installation. More easily installed in new construction. |
| Combination system (individual compactors installed on each floor. | Compactors readily accessible at nursing stations. Presents easily handleable packages. Interchangeability during possible breakdowns (reduces cost). Minimum operator skill required. No need for storage. May be palletized. Minimum installation space in any one area. No centralized space consumption. | Use of existing chute system is eliminated. May cause difficulty if incineration is considered for ultimate disposal. Present equipment basically designed for home use which may not be easily extrapolated to hospital wastes. Should not be used for wastes of pathological origin. Needs manual pick-up. |

Source: *Handbook on Hospital Solid Waste Management* (1973).

**Figure 14-2.** Economic Evaluation of Hospitals.

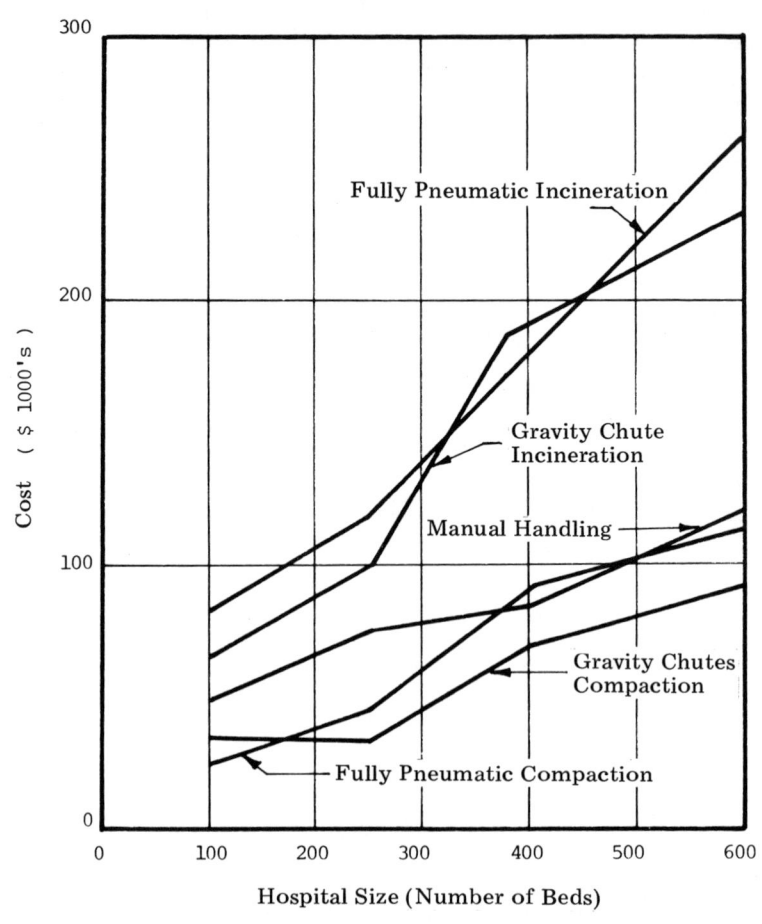

Source: *Handbook on Hospital Solid Waste Management* (1973).

Table 14-9. Service Life of Solid-Waste Processing Equipment.

| Equipment | Life |
|---|---|
| Food Grinders | 10 years |
| Pulping | 2 to 5 years |
| Gravity chutes | 25 years or more |
| Pneumatic exhausters | 20 years or more |
| Compaction | No record established at this time; however, based on similar hydraulic equipment in industry—20 years or more |
| Incinerators | refractors—3 years<br>casing—indefinite<br>washers—indeterminate depending upon water conditions and particulates ineffluent |

Source: *Handbook on Hospital Solid Waste Management* (1973).

# 15 INDUSTRIAL WASTES

## INTRODUCTION

In its *News Highlights* (1980), the American Cyanamid Co. chairman acknowledged that "public concern with the environment in the 1960s led to a flood of new government regulations air and water pollution. This, in turn, led in the 1970s to increased spending by many manufacturing companies in three areas: new pollution-control plant and equipment, increased costs for operating and maintaining that plant and equipment, and additional manpower costs in the environmental area."

The chairman continued with a point of significant interest to those of us concerned with the industrial solid waste problem: "While the bulk of Cyanamid's spending on environmental needs in the 1970s was to bring the company's U.S. plants into compliance with federal water pollution control regulations, *spending in the years ahead, will be concentrated on meeting new federal requirements for solid waste disposal*" (emphasis added).

### Background

Industrial solid wastes arise from a variety of sources (Grove and others 1969). Basically they can be divided into three categories: (1) wastes arising from the raw materials used in industry—mining and metallurgical wastes; (2) wastes resulting from industrial processes—metal scrap, office refuse, chemical waste, boiler house cinders, and so on; and (3) wastes created by worn-out products—scrapped automobiles, refrigerators, stoves, and other domestic and industrial appliances. These wastes vary in quantity, physical and chemical composition and properties, and in methods used for their disposal and potential re-use. The major portion of industrial refuse consists of solid waste coming from factories, processing plants, and other manufacturing enterprises. The disposal of such refuse may be an obligation of the industry producing them; or, especially for smaller industries, it may be the obligation of the municipality. Because of this mixture of solid waste in some communities, it is sometimes difficult to differentiate between municipal and industrial waste. This confusion is more pronounced when the industrial waste also contains organic matter, because it is common to speak of industrial refuse as an inorganic material. However, some industries—for instance, food-processing and canning—do produce waste that is essentially organic in nature. Organic and inorganic wastes, when combined, create very serious disposal problems.

Industrial solid wastes vary in many physical and chemical properties. They are acidic or basic, saline, inert, and may even be toxic. As a result of one or more of these properties, few solid industrial wastes can support useful vegetation until years after their initial deposition. The precise qualities and quantities of solid waste and particularly industrial solid waste are not available in textbooks for even those few texts dealing with solid-waste management rapidly become out of date. The available information on municipal refuse is widely scattered in journals, magazines, and conference proceedings, while industrial refuse disposal receives relatively scant attention. Most pertinent references up to 1969 were given by Grove and others (1969).

### General

No comprehensive statements can be made about the type, amounts, or the character of solid wastes produced by industry (Grove et al. 1969). Several general statements, however, can be made:

> Usually the wastes produced in any *food-processing operation* will closely resemble the garbage produced in residential areas. In addition, waste materials

from the *paper and plastics industries* are similar to the paper and plastic packaging materials found in domestic rubbish. On the other hand, the *metal-processing* industry will obviously generate metallic wastes, but in addition will also create large quantities of slags, processing chemicals, and other residues, many of which are produced in air-pollution control and water-pollution control activities.

*Construction and demolition wastes* are quite heterogeneous in character and may be composed of discarded building materials which are wasted during the construction of new structures and facilities, or they may consist of the wood, steel, bricks, concrete, and other construction materials. Generally this material is nonbiodegradable and is often used for fill material.

Of the 5.32 pounds per person per day national average of solid wastes *collected* (8–10 pounds per capita per day generated estimated) (Meyers 1968) industry produced 0.6 pounds per capita per day of the collected and (2–3 pounds per capita per day of the generated quantity).

Weston (1970) gives in his consultant report the industrial solid-waste quantities by Standard Industrial Code numbers (see Table 15-1).

The Office of Science and Technology (1969: 7) gives industrial solid wastes generated in 1967 as 110 million tons per year or about 3 percent of the total quantity generated.

Industrial wastes produced in a plant can be classified into several categories: (1) garbage (cafeteria wastes); (2) refuse (office and packing discards); and (3) process wastes—not saleable and not reworkable—from process cleaning operations.

The first two categories (garbage and refuse) can be disposed of in much the same manner as municipal refuse. The process wastes, however, usually requires a specially designed installation for disposal without causing pollution. Types of industrial wastes considered which require ultimate disposal include: monomers, polymers, and resins; solvents; waste oils—mineral and cutting; oil sludges; paint sludges; oil water emulsions; chlorinated hydrocarbons and solvents; trimmings; phenols, cresols, and tars; combustible chemicals; amines; fats, vegetable oils, and greases; and miscellaneous.

Although many in this list are in semi-liquid state, they may be classified almost as solid wastes because they are not discharged with liquid industrial wastes through sewer systems and are usually confined in solid containers.

Table 15-1. Solid Industrial Waste Quantities.

| SIC Code | Industry | Waste Production Rate (Tons/Employee/Year) |
|---|---|---|
| 2010 | Meat processing | 6.2 |
| 2033 | Cannery | 55.6 |
| 2037 | Frozen foods | 18.3 |
| other 203 | Preserved foods | 12.9 |
| other 20 | Food processing | 5.8 |
| 22 | Textile mill products | 0.26 |
| 23 | Apparel | 0.31 |
| 2421 | Saw mill and planing mills | 162.0 |
| other 24 | Wood products | 10.3 |
| 25 | Furniture | 0.52 |
| 26 | Paper and allied products | 2.00 |
| 27 | Printing and publishing | 0.49 |
| 2810 | Basic chemicals | 10.00 |
| other 28 | Chemicals and allied products | 0.63 |
| 2900 | Petroleum | 14.8 |
| 3000 | Rubber and plastic | 2.6 |
| 3100 | Leather | 0.17 |
| 3200 | Stone and clay | 2.4 |
| 3300 | Primary metals | 24.0 |
| 3400 | Fabricated metals | 1.7 |
| 3500 | Non-electrical machinery | 2.6 |
| 3600 | Electrical machinery | 1.7 |
| 3700 | Transportation equipment | 1.3 |
| 3800 | Professional and scientific institutions | 0.12 |
| 3900 | Miscellaneous manufacturing | 0.14 |

Source: Weston 1970.

In 1971, it is estimated that in the United States alone there were discarded (Bavin and Parker 1974): 70 billion metal cans, 38 billion glass containers, more than 7 million television sets, nearly 3.7 million tons of plastics, more than 100 million tons of paper products, and several million junk automobiles. Baum and Parker (1974) blamed the manner in which we live for contributing to these mountains fo waste and to the problems of disposal.

Solid waste is generated by society in the following approximate proportions: 44 percent from private families or households; 30 percent from construction activities and industry; and the remaining 26 percent from packaging—all of which contribute about 25 percent of the approximately 200 million tons of solid waste reported as collected in 1970. The food industry is considered the largest user of packages, followed by beverage and chemical product companies. Packaging materials in 1966 were estimated at about 52 million tons and were expected to rise to about 74 million tons of waste in 1976). Difficulties of disposing of such materials may become more complex because dissimilar materials are often combined in the same package (Baum and Parker 1974).

In 1970, the United States spent $5.7 billion on solid waste disposal. The CEQ (Council of Environmental Quality) estimated that the country would spend $7.8 billion in 1975—over 35 percent more than in 1970. For the years between 1970 and 1975, CEQ placed the total cost of solid waste disposal for the period at $43.5 billion. The Office of Solid Waste Management Programs (OSWMP) had budgets for dealing with solid waste of about $17 million in 1971, about $35 million in 1972, and about $15 million in 1973. The 1974 fiscal year budget in this area was estimated to be held at about $15 million.

In choosing a method for solid waste disposal, Baum and Parker (1974) propose some factors to be taken into consideration: (1) make the best possible use of natural resources, either recovering energy or product; (2) avoid the creation of some other form of pollution which may generate more hazard than is present in the original waste; (3) obtain the best balance of economy and efficiency. New methods of waste management vary widely because no two communities have the same problems and so no single technique will work in every case.

Two methods—incineration and landfill—account for about 98 percent of the collected solid waste disposed of in the United States. Other disposal methods presently are not used widely enough to be compared with sanitary landfill and incineration was able to handle the quantities of refuse generated in urban areas. However, management of solid wastes by other methods such as pyrolysis, anaerobic digestion, composting, and recycling is under diligent study, in some respects as a direct result of the Resource Recovery Act of 1970. This act put strong emphasis on recycling of wastes to recover products or energy and authorized funds for demonstration grants for recycling resources. The major purpose of the act was to encourage

178  INDUSTRIAL SOLID WASTES

Table 15-2. Principal Industrial Waste Components.

| Source | Waste | Characteristics | Composition | Means of Treatment or Disposal |
|---|---|---|---|---|
| Food and kindred product industries | Fruit, vegetable and citrus | | Hull, rinds, cores, seeds, vines, leaves, tops, roots, trimmings, pulps, peelings, hydrochloric acid (used in processing) | Screening, lagooning, soil absorption, spray irrigation reclamation |
| Canning | Cobs, shells, | High in suspended solids (liquid waste), colloidal and dissolved organic matter | | |
| Vegetable oil refining | | | "Still pitch"—tarry residue, fatty acids, sodium hydroxide, trichlorethylene | Reclamation |
| Dairy | Dilutions of whole milk, separated milk, buttermilk whey | High in dissolved organic matter, mainly protein, fat, lactose | $N$, $CaO$, $K_2O$, $P_2O_5$, $Fe$, $Cl$, $SiO_2$ | Aeration, trickling filter, activated sludge |
| Slaughtering of animals, rendering of bones and fats, | Manure, paunch manure, blood, flesh, fat particles, hair, | | $N$, $NH_3$, $NH_2$, $NO_3$, $NaCl$ | Reclamation, screening, trickling filters, chlorination |

| | | | |
|---|---|---|---|
| | residues in condensates, grease, wash water | bones, oil, grease | |
| Breweries and distilleries | Spent grain, spent hops, yeast, alkalis, amyl alcohol, dissolved organic solids containing nitrogen and fermented starches | High in dissolved organic solids, containing nitrogen and fermented starches or their products | Amyl alcohol (from processing) | Recovery, centrifugation and evaporation, trickling filtration, stock feeds, fertilizer |
| Pharmaceutical | Micro-organisms, organic chemicals | High in suspended and dissolved organic matter, including vitamins | Aniline, phenols | Evaporation, incineration, stock feeds |
| Textile mill products | Textiles, such as cotton, wool, and silk | Highly alkaline, colored, high BOD and temperature, high suspended solids | $H_2SO_4$, NaOH, aniline, chlorine, starch, malt, tin and iron salts, dyes, bleach, fibers, minerals | Neutralization, precipitation, trickling filtration, aeration, recovery |
| Cooking of fibers, desizing of fabrics | | Same as textile mill products | For complete list of chemicals used in textile industry, see source | |

(*Table 15–2. continued overleaf*)

Table 15-2. continued

| Source | Waste | Characteristics | Composition | Means of Treatment or Disposal |
|---|---|---|---|---|
| | Rayon, other man-made materials such as Acrilan, Dynel, Orlon, Nylon | Acidic, alkaline, inorganic | Sulfides and polysulfides, colloidal sulfur, NaOH, $H_2SO_4$, $ZnSO_4$, HCl, $NaHSO_4$, $H_2S$, $CaSO_4$, acrylonitrile, phenol, $HNO_4$, ammonia, adiponitrile, hexamethylenediamine, sodium carbonate, alcohols, ketones | Reclamation, neutralization, trickling filtration, lagooning |
| Laundry | | High turbidity and alkalinity | Spent soaps, synthetic detergents, bleaches, dirt, grease | Screening, precipitation, flotation, absorption |
| Lumber and wood products (forest, mills, factories) | Pulp and paper | High or low pH; colored; high suspended, colloidal, and dissolved solids; inorganic fillers | Sawmill usage (sawdust, shavings, wood chips), wood flour, soda, sulfate, sulfite | Reclamation, incineration, soil conditioning |
| | | Organic, inorganic, toxic, suspended and dissolved solids of | Sodium lignate, sodium resinate, complex organo-sulfur com- | Reclamation, settling, lagooning, biological treatment aeration |

| | | | |
|---|---|---|---|
| Chemical plants (general) | | lignin, resins, soda, ash, fiber, adhesives, ink, fats, soaps, tallow pounds, some fiber in relatively dilute solutions, sulfites, mercaptans, sulfides, disulfides, sulfates, terpenes, carbohydrates, $CaO$, $SO_2$, N, $PO_4$ | Reclamation, lagooning and all other known methods of treatment |
| | Toxic | Acrylonitrile, aniline,[a] amyl alcohol, carbon disulfide, carbon tetrachloride, chlorine, hydrogen cyanide, hydrochloric acid, phenol, sulfuric acid, toluene, xylene, dinitrobenzene, dimethyl sulfate, ethylene, chlorohydrin, benzene, metallic compounds of lead, arsenic, mercury | |
| | Fumes or dust | Arsenic | |
| | Particulate clouds and dusts | Mn, V, Cd, Be, Fe, Zn, and their oxides | |

(*Table 15–2. continued overleaf*)

Table 15-2. continued

| Source | Waste | Characteristics | Composition | Means of Treatment or Disposal |
|---|---|---|---|---|
| | Weed killer | | 2-4-D | Sewage |
| | Cyanide waste | | Cyanides | Ponding |
| | Plastics, synthetic resins | Toxic to aquatic life | Acrolein, acrylonitrile, formaldehyde, phenols, trichloroethylene | Reclamation, incineration |
| Aircraft manufacturing industry | Cd and hexavalent Cr | Traces of metals | Cd and hexavalent Cr | Leaching pits |
| Waste treatment plants | Well-designed sludge | Blackish, amorphous, nonplastic material | Mg, Ca, Zn, Cr, Sn, Mn, Fe, Cu, Pb | Anaerobic decomposition of organic waste solids |
| Petroleum industry | Spent chemical | Liquid wastes with oil, acid and alkaline solutions, inorganic salts, organic acids and phenols, and so on | Clays, $H_2SO_4$, $H_3PO_4$ | Streams |
| Drilling | | Oil, brine, chemicals | Sodium, calcium, magnesium, chlorine, $SO_4$, bromine | Separation, evaporation, lagooning |

| | | | |
|---|---|---|---|
| Storage | Muds, salt, oils, natural gas | | Separation, evaporation, lagooning |
| Distillation | Acid sludges, miscellaneous oils | Insoluble organic and inorganic salts, sulfur compounds, sulfonic and naphthenic acids, insoluble mercaptides, oil-water emulsions, soaps, waxy emulsions, metal oxides, phenolic compounds | $Na_2CO_3$, $(NH_4)_2S$, $Na_2S$, sulfates, acid sulfates, $H_2S$, NaOH, $NH_4OH$, $Ca(OH)_2$, $(NH_4)_2SO_4$, $NH_4Cl$, phenols | Settling, filtration, reclamation, evaporation |
| Treating | | See "Distillation" | See "Distillation"; also lead, copper, calcium | Reclamation, settling, filtration, evaporation, neutralization |
| Recovery | | See "Distillation"; also organic esters | See "Distillation"; also iron | See "Treating" |
| Leather and leather products | Tanneries | Organic and inorganic, high BOD-lime sludge, hair, fleshing, tan liquor, bleach liquor, salt, blood, dirt, chrome | Chromium, sulfuric acid, nitrogen, $CaCO_3$, $P_2O_5$, $K_2O$, Fe | Sedimentation, lagooning |

(*Table 15-2. continued overleaf*)

## Table 15–2. continued

| Source | Waste | Characteristics | Composition | Means of Treatment or Disposal |
|---|---|---|---|---|
| Energy-producing industry | Fly ash | Hollow spheres of fused or partially fused silicate glass or as small solid spheres of fused silicates, iron oxides or silica, unburned carbon and mineral | Silicates, iron oxide, silica | Sold for use in concrete, landfills, and so on |
| Pulverized coal-fired plants; stoker-fired, cyclone-fired plants: wet-bottom pulverized coal-fired plants | | | | |
| Electrical industry | Ash | Dust | Silicates and aluminates of Fe, Cu, Mg with small percentages of Na, K | |
| Metal finishing industry | Pickling and washing liquors | Toxic, waste waters | Cu and Cu alloys | Sewage |
| | Acid wastes | Harmful to aquatic life, salts of metals | Cu, Ni, Zn, Cr, Fe | Sewage |

INDUSTRIAL WASTES 185

| Industry | Origin of waste | Major characteristics | Major treatment |
|---|---|---|---|
| Rubber and miscellaneous plastic products | Rubber | High BOD, odor, high suspended solids, variable pH, high chlorides | Aeration, chlorination, Sulfonation, biological treatment |
| | Washing of latex: coagulated rubber; exuded impurities from crude rubber; rejects, cuttings, mold flashings, trims, excess extrusions | | |
| | Scraps from molding, extrusion, rejects, trimming, finishing | Sulfuric acid, trichlorethylene, xylene, amyl alcohol, aniline, benzene, chromium formaldehyde | |
| Explosives, washing TNT and guncotton for purification, washing and pickling of cartridges | | TNT, colored, acid, odorous, and contains organic acids and alcohol from powder and cotton, metals, acid, oils and soaps | Dilution, neutralization, lagooning, flotation, precipitation, aeration, chlorination |
| | | $H_2SO_4$, $HNO_3$, $NO_2$, $SO_3$, picric acid, TNT isomers, copper, zinc, nitrogen, toluene | |
| Phosphate and phosphorus | Washing, screening, floating rock, condenser bleed-off | Clays, slimes, tallows, low pH, high suspended solids | Settling, clarification, (mechanical) lagooning |
| | | Phosphorus, silica, fluoride | |

(*Table 15–2. continued overleaf*)

Table 15-2. continued

| Source | Waste | Characteristics | Composition | Means of Treatment or Disposal |
|---|---|---|---|---|
| Fertilizers | | | Nitrogen, phosphorus, potassium, sulfuric acid, traces of other chemicals | |
| Coke byproducts | Slag from ovens, ammonia still waste, spent acids and phenols | Suspended solids, volatile suspended solids, organic and $NH_3$-N, phenol, cyanide acids, alkalis | Ammonia, benzene, $H_2SO_4$ phenol | Discharged to sewers, dumped, incineration |
| Industrial, not otherwise identified | Inorganic industrial waste or stabilization | Metals and compounds thereof | Na, K, Ca, chlorides, sulfates, bicarbonates, nitrates, phosphates, fluorides, borates, chromates, and so on Pb, V, As, Be and compounds thereof | |
| | Metallic fumes and dusts | | | |
| | Industrial wastes | Mineral fines | Chromates, heavy metals | Underground aquifers |
| | Laboratory wastes | | Metallic ions, phenolics, cyanides, oils, synthetic fibers, phar- | Landfill or dump |

| Industrial wastes | | Toxic metals | |
|---|---|---|---|
| | | | chemicals Pb, Be |
| Insecticides | Washing and purification of products | High organic matter, toxic acidic | Carbon, hydrogen, chlorine, carbon disulfide, carbon tetrachloride |
| | | Chlorinated hydrocarbons: toxaphene, benzene, hexachloride, DDT, aldrin, dieldrin, lindane, chlordane, methoxychlor, heptachlor | |
| | | Organic phosphorus compounds: parathion, Malathion, phosdrin, tetraethyl pyrophosphate | Phosphorus, oxygen, carbon, hydrogen, carbon disulfide, carbon tetrachloride |
| | | Other organic compounds | Carbonates, dinitrophenols, organic sulfur compounds, organic mercurials, rotenone, pyrethrum, nicotine, strychnine |
| | | Inorganic substances | Copper sulfate, arsenate of lead, compounds of chlorine and fluorine, zinc phosphide, thallium sulfate, sodium fluoracetate |
| | | | See Chemical plants (general) |

a. Most common and troublesome toxics.

Source: *Solid Waste/Disease Relationships* (1967: 161).

Table 15-3. Relative Quantities of Solid Waste Per Year Generated from Durable and Nondurable Industries for 1967.

| Industry | Employment,[a] July 1967 | Multipliers,[b] Ton/Employee/ Year | Waste, Ton/Year |
|---|---|---|---|
| Nondurables | | | |
| Food products | | | |
| Seasonal foods | 2,200 | 5.56570 | 12,245 |
| Other foods | 11,482 | 4.81655 | 55,304 |
| Total food products | 13,682 | | |
| Paper, printing, and publishing | 6,478 | 12.87060 | 83,376 |
| Chemicals | 1,900 | 8.21075 | 15,600 |
| Other nondurables | | | |
| Textiles and apparel | 2,193 | .52575 | 1,153 |
| Rubber and plastics | 1,835 | 1.54810 | 2,841 |
| Leather | 355 | 2.49365 | 885 |
| Total other nondurables | 4,383 | | |
| Durables | | | |
| Stone, clay, glass, and concrete | 3,708 | 18.11425 | 67,168 |
| Primary and fabricated metals | 15,250 | 6.73000 | 102,632 |
| Electrical and non-electrical machinery | 12,478 | 3.58040 | 44,676 |
| Other durables | | | |
| Lumber and wood products | 1,033 | 21.68805 | 22,404 |
| Furniture and fixtures | 1,562 | 20.15545 | 31,483 |
| Transportation equipment | 2,768 | 3.39330 | 9,393 |
| Instruments | 915 | 2.51700 | 2,303 |
| Total other durables | 6,278 | | |
| Other manufacturing | 2,500 | 2.49365 | 6,234 |
| Total manufacturing employment | 66,657 | | 457,697 |

a. Basic employment data are from the State of California Department of Employment Community Labor Market Survey and are limited to the special study area in California. Data were adjusted to exclude Union City which is not in the study area. Employment in the categories "Other Durables" and "Other Nondurables" was distributed to the relevant SIC groups by using the same proportions as existed in the 1965 employment data from ABAG.

b. Multipliers for the manufacturing industries were developed and reported in Table VI, *Comprehensive Studies of Solid Waste Management*, Second Annual Report.

Source: Golueke and McGauhey (1970:53).

Table 15-4. Large Firm Multipliers.

| Standard Industrial Classification | Employment[a] | Annual Wastes, Vol., $Yd^3$ [b] | Annual Wastes per Employee, $Yd^3$ [c] |
|---|---|---|---|
| Ordnance and accessories | 29,356 | 131,404 | 4.476 |
| Canning and preserving[d] | 11,389 | 102,238 | 8.977 |
| Other food processing | 2,012 | 17,545 | 8.720 |
| Tobacco | NA | NA | NA |
| Textiles | NA | NA | NA |
| Apparel | 601 | 1,248 | 2.077 |
| Lumber and wood products | NA | NA | NA |
| Furniture and fixtures | NA | NA | NA |
| Paper and allied products | 250 | 9,360 | 37.440 |
| Printing, publishing, and allied | 968 | 7,020 | 7.252 |
| Chemicals and allied | NA | NA | NA |
| Petroleum refining | NA | NA | NA |
| Rubber and pastics | 481 | 9,069 | 18.854 |
| Leather | NA | NA | NA |
| Stone, clay, glass, and concrete | 1,258 | 6,617 | 5.260 |
| Primary metals | NA | NA | NA |
| Fabricated metal products | 3,565 | 47,078 | 13.206 |
| Non-electrical machinery | 8,872 | 101,153 | 11.401 |
| Electrical machinery | 7,807 | 57,252 | 7.333 |
| Transportation equipment | 4,100 | 100,776 | 24.580 |
| Instruments | NA | NA | NA |
| Miscellaneous manufacturing industries | NA | NA | NA |

a. Data on employment were obtained for those large firms which were surveyed and included in the wastes calculation from the research department of the Association of Metropolitan San Jose (Greater San Jose Chamber of Commerce).

b. FMC report, 1968. *Solid Waste Disposal System Analysis (Preliminary Report).* Tables 10 and 11.

c. Annual wastes, vol. $yd^3$/employment.

d. For canning and preserving, no individual firm data were available. The industry total developed for the county as a whole was divided by the total employment in the industry (specially tabulated) to arrive at the multiplier.

NA = not available.

Source: Golueke and McGauhey (1969:221).

Table 15-5. Small Firm Multipliers.

| Standard Industrial Classification | Weekly Wastes, Vol. per Firm, $Yd^{3a}$ | Annual Wastes, Vol. per Firm, $Yd^{3b}$ | Average Employment per Firm$^c$ | Annual Waste, Vol. per Employee, $Yd^{3d}$ |
|---|---|---|---|---|
| Ordnance and accessories | 2.500 | 130.00 | NA | NA |
| Canning and preserving | | (Not surveyed) | | |
| Other food processing | 10.875 | 565.50 | 26.979 | 20.961 |
| Tobacco | NA | NA | | |
| Textiles | NA | NA | | |
| Apparel | 4.000 | 208.00 | 5.882 | 35.360 |
| Lumber and wood products | 16.083 | 836.33 | 17.247 | 48.492 |
| Furniture and fixtures | 23.000 | 1,196.00 | 13.767 | 86.877 |
| Paper and allied products | 44.650 | 2,321.80 | 35.479 | 65.442 |
| Printing, publishing, and allied | 6.448 | 335.29 | 13.289 | 25.230 |
| Chemicals and allied | 6.506 | 338.31 | 18.439 | 18.348 |
| Petroleum refining | NA | NA | NA | NA |
| Rubber and plastics | 5.275 | 274.30 | 9.596 | 28.583 |
| Leather | NA | NA | NA | NA |
| Stone, clay, glass, and concrete | 9.415 | 489.60 | 16.747 | 29.235 |
| Primary metals | 2.000 | 104.00 | 23.409 | 4.443 |
| Fabricated metal products | 5.284 | 274.75 | 12.951 | 21.214 |
| Non-electrical machinery | 4.450 | 231.40 | 12.921 | 17.909 |
| Electrical machinery | 6.733 | 350.13 | 21.036 | 16.645 |

| | | | |
|---|---|---|---|
| Transportation equipment | 4.550 | 236.60 | 16.490 | 14.348 |
| Instruments | 3.600 | 187.20 | 20.933 | 8.943 |
| Manufacturing industries | 1.250 | 65.00 | 10.931 | 5.946 |

a. Data obtained and calculated for each SIC on the basis of small firm questionnaire responses supplied by FMC.
b. Weekly average in first column multiplied by 52.
c. Average size of small firm estimated from the distribution of firms by employment size, supplied by the California Department of Employment (Research and Statistics). San Francisco Office.
d. Annual wastes/average employment per firm.
NA = not available.

Source: Golueke and McGauhey (1969).

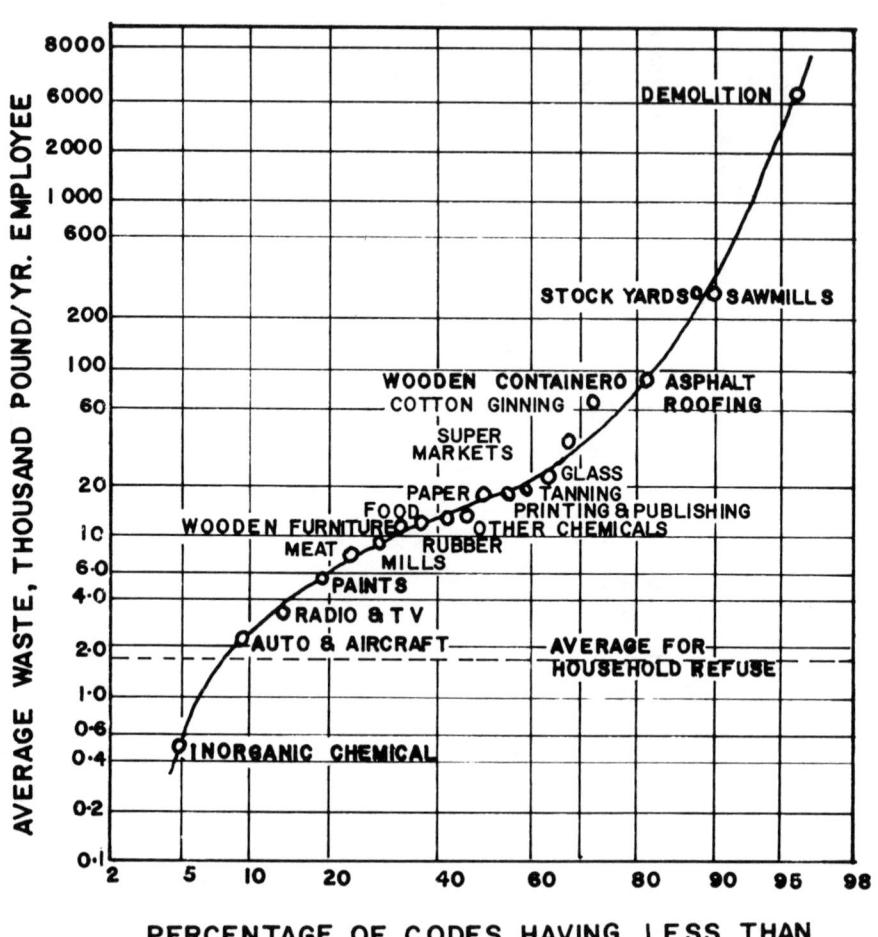

Figure 15-1. Distribution of Waste for Disposal: Employee Ratios in Twenty-one SIC Code Groups.

Source: *Technical Economic Study of Solid Wastes Disposal Needs and Practices Industrial Inventory* (1969: 8).

INDUSTRIAL WASTES 193

**Figure 15-2.** Distribution of Waste for Disposal Among Twenty-one SIC Code Groups.[a]

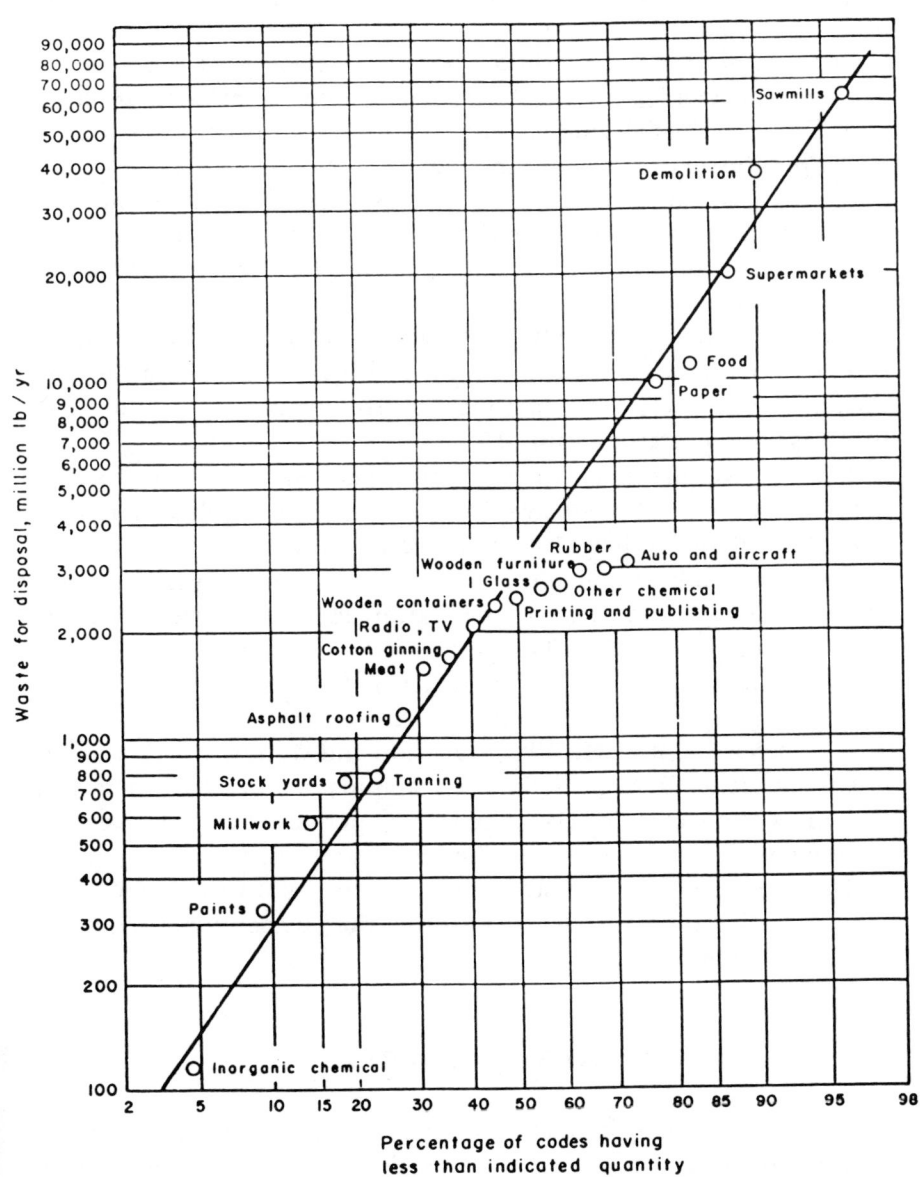

a. Standard Industrial Classification.

Source: *Technical Economic Study of Solid Waste Disposal Needs and Practices Industrial Inventory* (1969: 7).

resource recovery at all levels. This reclamation effort, although considerable by some standards, represents a recovery of less than 2 percent of the total at-the-dump solid wastes generated by the household, industrial, and commercial sources.

Current and novel methods for the recycling of solid wastes, whether the yield is in the form of products or energy, are important. There is special emphasis on the recycling of plastics. Connecticut is progressing both from the technical and marketing viewpoints. Energy relationships in recylcing are significant from the standpoints of energy required to produce the original material and of the energy released when such materials are disposed of by pyrolysis or incineration methods.

The Department of Health, Education and Welfare made a survey (*Solid Waste/Disease Relationships* 1967) of U.S. industries' solid waste, composition, and methods of treatment or disposal. Their findings are reported in Table 15-2.

The relative quantities of solid waste per year generated from durable and nondurable industries for 1967 were compiled in 1970 by Golueke and McGauhey (1970). These are shown in Table 15-3.

A similar study was made for the volume of solid wastes for most of the Standard Industrial Classification large industries on a total and a per-employer basis. This is shown in Table 15-4.

The same was done for the small SIC industries, and the results are given in Table 15-5.

The Department of Health, Education and Welfare, *Technical Economic Study* (1969) constructed a graph plotting frequency distribution of the weight of solids generated per year per employee. This is shown in Figure 15-1 for all the SIC code groups.

The same frequency distribution relationship was drawn for the total quantity produced per year for all the SIC code industries. This is shown in Figure 15-2.

## MEAT INDUSTRY IN GENERAL

In the meat-packing industry the major solid wastes treatment systems are recovery and re-use systems. Seven basic conservation methods have been recommended in this industry: (1) dry-cleaning of pens and floors before washing down; (2) collecting the killing blood in a separate blood tank, and squeezing the blood flow to the blood sewer during cleanup; (3) retaining casing slimes in casing washing operations, and then drying these slimes and adding them to animal feed products; (4) screening paunch manure and hog stomach contents with either rotary or vibrating screens; (5) separating grease from all grease-bearing waste by means of efficient gravity or air flotation separators; (6) evaporating tank water, the liquid residue from wet rendering; after evaporation to about 35 percent moisture, the liquid "stock" can be mixed with tankage and dried; (7) evaporating blood water if blood is coagulated rather than dried directly.

Paunch manure should be segregated and disposed of by dumping on land as a potential soil builder (conditioner) or in city plants by being trucked away with the garbage.

Blood, casing slimes from the stripping operations, and tank water can be concentrated and dried, then used in feeds and fertilizers. Grease can be rendered to produce both edible and inedible foods. Hair which is removed by mechanical scrapers is washed and sold unprocessed or is further processed in boiling water, dried, and baled for sale. It may also be hydrolyzed by steam-rendering by adding lye. This yields a powdered material after drying. Further processing of meat-packing plant byproducts such as tanning, wool pulling, manufacturing glues, soaps, and fertilizers are normally done in individual plants. Glue is a collagenous matter extracted from head, feet, bones, tendons, and especially hide trimmings, by rendering them in water at different temperatures. The inedible tallow is used in soap manufacturing. Animal feeds of fairly high grade are usually produced when this dry rendering of the solids is done.

## METAL-PLATING SOLID WASTES

The dry sludge resulting as a waste product from metal-finishing waste treatment and usually containing mainly metal oxides and metal hydroxides can be hauled to a dump site. However, if these metals become solubilized due to the acid conditions which develop in the dumps, they may contaminate underground water supplies,

### Metal-Fabricating and Plating Wastes

*Origin.* The wastes originating from the metal-fabricating and plating plants are mostly liquid effluents containing heavy metals like copper, chromium, cyanide, zinc, lead, nickel and so on. The origin of these solid wastes in these types of plants is from the heavy volume of sludge produced at the effluent treatment plants.

*Characteristics.* The characteristics of the solid waste as sludge from the effluent treatment plants is usually of high metallic content and strongly acidic or alkaline in nature. The pH and the metal content varies depending on the type of metal plating being used and the method of treatment used to precipitate the metals in the form of sludge.

*Disposal.* The sludge produced in the effluent treatment operations is usually dewatered by vacuum filters or centrifuges. Due to the increasing costs of raw materials and overall production, there has been a general tendency to recover the metals from the waste produced. Although it is perhaps more convenient and economical to recover the metals from the liquid streams, methods also are being developed to recover the metal from the sludges. One of the methods studied has been with the use of oxalic acid to separate iron and chromium from copper, nickel, and zinc in an electroplating mixed sludge. Another method employed has been the recovery of chromium from waste sludges by calcination process. Studies have also been made to evaluate the feasibility of mixing electroplating sludges with other organic material and subjecting the mixture to high temperatures. As a result, the nonleaching aggregate produced may be suitable for landfilling. In spite of considerable efforts being spent in developing techniques for metal recovery from metal-plating sludges, the economic viability of some of the processes, though technologically attractive, has been the general constraint in their widespread applications. Although highly undesirable due to pollutional effects of leaching of heavy metals into ground and surface waters, landfilling has still been the commonly practiced method of disposal of metal-plating sludges.

One new process to prevent metallic hydroxide sludge resolublization is known as the Soliroc Process (Rousseaux and Craig 1981). Here a chemical reaction occurs between metals present in an acid solution and a siliceous reagent. This synthesizes a monomer which then polymerizes to form an insoluble bulk mass claimed to be stable in landfills. Specifically, hexavalent chromium is reduced to the trivalent state during which acid is added to maintain a pH of 1.5. The siliceous reagent is then slowly added into the solution. Then the pH is elevated to about 11 to begin polymerization. This causes a sludge to develop which, when enhanced by lime or cement, coagulates and hardens before placing the solid waste into a landfill.

On the other hand, Mehta (1981) states that trying to immobilize the heavy metals in a sludge by any one of the fixation techniques would add to the cost of the final disposal. Mehta contends that RCRA rules and severe restrictions may still be encountered by land disposal. Recovery and recycling is the better practice, but it is an uneconomical one in terms of capital and operating costs, is an incompatible operation, and, individually, the quantities generated are not large enough to warrant a recovery system. However, Mehta does recommend a centralized waste treatment facility to handle metal-plating wastes from many platers. The precipitated metal hydroxide sludges could be treated further to recover valuable metals by direct recycling to a smelter or a metal alloy industry or by leaching with acids or other solvents to recover the metals from them. The dissolved metals in lean solutions could be concentrated by cementation or solvent extraction.

## STEEL MILLS

### General

The iron and steel industry is one of the major industries and is considered as a "backbone" industry for any major industrial country. Today, it produces in excess of 700 million tons of steel each year, ("The Iron and Steel Industry and the Environment," 1980), and, as with other basic manufacturing industries, consumes large quantities of resources and in turn discharges atmospheric, water-borne, and solid wastes. The solid waste generated in steelmaking is in a major way dependent upon the process that is used. In recent years the iron and steel industry is undergoing major changes in its processes. It is moving increasingly away from old, established, energy-hungry, and pollution-generating practices toward more energy-efficient and pollution-free ones. These changes were initiated partly be economic necessity, after the explosion of the oil prices, and partly by increasing environmental pressures expressed in recent years in the major industrialized countries and particularly the United States.

There are a considerable number of stages in the entire iron- and steelmaking process, where effluents and emissions of a polluted or polluting nature occur. In 1979, 747 million metric tons of crude steel were produced, of which 55 percent was produced by the basic oxygen steelmaking process (BO), which is a relatively new technology. The electric-arc furnace (BAF) which mainly consumes scrap as recycled material accounted for 21 percent of the world crude steel production, while 24 percent was still made with the old open-hearth furnace process (OHF). Most steel-producing countries of the Western world have discarded the OHF entirely.

At a rough estimation ("The Iron and Steel Industry" 1980), in 1979, some 940 million metric tons of iron ores and 460 million metric tons of coal were consumed by the industry. In addition, there were approximately 400 million metric tons of scrap used, some generated within the steel plants, some generated in the steel-manufacturing process, but most of it from consumer waste material—shipbuilding, motor cars, and consumer durables of many kinds.

### Origin

In the manufacture of steel, one of the essential pre-operations is the manufacture of coke from coal in coke ovens. Coke is used as a source of pure carbon in the blast-furnace operation for reduction of oxidized form of iron present in the raw iron ore to produce the pure iron. The processing of coal to coke produces solid wastes, the characteristics and treatment of which are described subsequently. The hot briquetting operation before the blast-furnace process agglomerates many kinds of steel plant waste and produces briquettes which are used as blast-furnace charge. The major contribution of solid wastes from the steel plant emerges from the blast-furnace operation used for the manufacture of pig iron from raw iron ore using limestone and coke as reacting compounds. Slag is produced in the process which contributes the largest volume of solid waste produced in the steel plant. The other sources of solid waste are from the treatment plants used for the processing of liquid effluents emerging from the various manufacturing operations.

### Characteristics

The solid wastes generated at the coke oven plant consist of tar, light oil emulsion, and so on. Ammonium sulfate is produced during the processing of coke oven gas containing ammonia passed through sulfuric acid solution. The treatment of coke oven effluents produce tar sludges, "dirty" phenolates, and blow-down sludge. The slag produced in the blast furnace operation consists of mixture of limestone, silica, and iron ore and is generally porous in nature. Slag, scrap, and dust solid wastes are characterized in the following sections.

*Slag.* In blast furnaces, where iron ores and coke are charged together at the top of the furnace, the gangue materials are separated from the molten iron at the bottom and removed as slag. For each metric ton of iron some 300 kilograms of slag are produced ("The Iron and Steel Industry" 1980). If this figure is multiplied by the total steel world production, the total amount of blast-furnace slag produced is 150 million metric tons. By selective cooling of the molten slag, three types of material are produced: air-cooled slag, foamed slag, and granulated slag.

1. *Air-cooled slag* is formed when allowed to solidify by slow cooling either in a ladle or pit, whereby it develops a crystalline structure and becomes a rockline material.
2. *Foamed slag* was first produced by Carl Schol in Germany in 1911. A method used nowadays to produce it discharges slag contained in a large ladle rapidly into a pond, the bottom and sides of which contain water jets (Gutt et al. 1974). The disposition of the water jets is arranged to generate steam inside the molten slag and to produce a uniformly foamed, gray-colored mass.
3. *Granulated slag* is produced when the molton slag is cooled rapidly by means of high-pressure water jets, crystallization does not occur, and the material solidifies to give a slag glass.

In 1971 from total of seventy-one blast furnaces in Britain alone, the output of slag was 9 million tons.

Blast-furnace slag is utilized as roadstone, as railway ballast, as raw material for the manufacture of Portland blast-furnace cement, as an aggregate for concrete, as a filter medium for sewage-disposal plants, and as an agricultural fertilizer.

Electric arc furnace slag is reported to vary from 2 to 10 percent of the weight of scrap melted.

Rolling mill scale from steel mills average about 2 percent of the rolled steel output.

*Scrap.* The other side of the coin in the steel industry as far as solid wastes are concerned is the utilization of scrap for the production of steel. In addition to new metallic inputs, a large quantity of ferrous metal also "circulates" in the industry; that is, it is consumed every year in the initial production steps (the furnaces), but then it reappears in the final production steps (rolling, fabrication, and conversion) and is once more consumed.

This circulating material is called "home scrap" if it appears within the steel industry itself, and "prompt scrap" if it appears in industrial operations outside the steel industry. These types of scrap are contrasted to "obsolete scrap," which means metal derived from products or structures that have completed their useful economic life and are ready for recycling. According to the EPA (1972), the U.S. industry in 1967 sent to final consumers products weighing 95.9 million tons, and lost an estimated 11.8 million tons of metals in processing; that is, total metal consumption was 107.7 million tons. This was supplied by iron ores of 86.1 million tons (80 percent), 21.6 million tons of obsolete scrap (20 percent), and by imported scrap of 229,000 tons. In this period, however, 63.7 million tons of additional scrap circulated the system and were once more available at the end of the year.

*Dust.* Dust is taken out of emission gases as part of air-pollution control; sources are ore storage and handling, pelletizing, and steel mills.

### Disposal

The treatment of slag is a very important aspect of steel-plant solid-waste disposal. The large volume of slag produced for every ton of steel makes the problem more difficult. One of the methods of treatment is known as the granulation process wherein the molten slag runs into a pit of water into which high-pressure water streams are sprayed, which break the liquid slag into granules because of the sudden cooling and spraying action. This process however, generates a large amount of steam and $H_2S$ gas which is not desirable from air-pollution considerations. An alternative method of treating slag is by an air-cooled slag process in which the molten slag is poured into sand pits in the ground for slow cooling. Limited amounts of water are added to accelerate cooling.

Molten slag forms a solid mass and is taken out of the pit by power shovels. One of the common uses of blast-furnace slag is for the manufacture of pozzolona cement. Often bricks or hollow blocks are manufactured from crushed slag powder for use in construction of buildings and other structures. The granular form of slag obtained is sometimes disposed of by landfilling. The tar sludges produced in the coke-making process are further refined and sold for manufacture of various byproducts. Sometimes part of the tar sludge is incinerated. The ammonium sulfate produced by passing coke oven gas through sulfuric acid is crystallized and dried for sale.

### Additional Re-use and Disposal Techniques

Scale from rolling mills can be sent to sinter plant for recovery. Sand from foundry can be put on a slag pile or landfilled. Barden fines from air washers at sinter plant, skip hoist charging area, blast-furnaces gas washers, open hearth, and BOF can be sent to sinter plant for recovery as sinter (to be re-used in making molten steel). Fly ash from coal-fired furnaces can be used in cement or cement block additives. Coal and coke from coke plants can be collected and burned. Steel-mill slag re-use depends upon the particle size. The National Slag Association (n.d.) reports the main two uses:

1. As re-use in construction, it can be a pozzolanic additive to Portland cement concrete, a mineral wool-base material, a lightweight aggregate, a railroad ballast, and material for pipe backfill.
2. For re-use in wastewater treatment, it can be a material for septic-tank absorption beds and a biological filter media.

If the slag is ground finely, it can be added to soil to make it alkaline. Slag contains calcium and magnesium alumino silicate, with minor amounts of iron, sulfur, manganese, and phosphorous.

## FOOD SOLID WASTES

Food solid wastes include: animal processing (red meat, poultry, and so on); canneries (fruit, vegetables, specialties, fish and fish products); dairies; citrus; soft drinks, breweries, and so on; and the agriculture industries (cane sugar, grain mills, and other foods).

### Animal Processing

*Meat.* For all practical purposes, the meat-packing industry may be divided into three categories: (1) slaughterhouses (killing and dressing); (2) packinghouses (killing, dressing, curing, cooking, and so on); and (3) processing plants (processing with no killing operation). Meat-packing plants carry out the slaughtering and processing of cattle, calves, hogs, and sheep for the preparation of meat products and byproducts from these animals.

For the ultimate disposal of solids, sludge screenings and skimmings are disposed of on land, with or without prior treatment. Liquid sludge can be transported for spreading on land, or dewatering processes can be employed and the filter cake hauled to a landfill. Solids in lagoons must be removed and used for landfill, or new lagoons may be built if land area is available and the cost is less than cleaning. Solid-waste generation rate for cattle amounts to 102 pounds per head per year.

*Waste Characteristics.* Total solids are composed largely of blood, meat, and fatty tissue, meat extracts, paunch, bedding, manure, hair, dirt, contaminated cooling and picking solutions, preservatives, and caustic or alkaline detergents. These wastes are generally high in proteins. In general, no parts of the meat-packing industry or red-meat industry are wasted. Prior to the water treatment, they are collected and further processed in other products. The waste recovered is treated, bagged, and marketed for many different uses including fertilizer and animal feed additives.

*Poultry.* The poultry industry is a billion-dollar annual business involving every state. Processing plants are highly mechanized, high-production facilities. Conveyors speed hundreds of turkeys or broilers per minute. The sequence of processing operation is: live bird receiving, killing, bleeding, plucking and eviscerating, chilling, and packing in ice or freezing for shipment.

*Waste Characteristics.* Following defeathering, the eviscerating and cutting operations produce solid wastes consisting of feet, heads, inedible viscera, and other parts; grit, sand, gravel, trimming, grease and blood.

*Waste Disposal.* The location of the poultry-processing plant dictates the extent and method of waste treatment. Feathers and offal, in plants not processing these materials further, are commonly removed by a truck trailer, in which the feathers and eviscerating wastes are separately compartmented. Where permitted, these materials may then be buried or incinerated. More often, poultry plants not further processing these primary solid-waste materials sell them separately to rendering companies for further processing to feather meal, and "cracking" for animal feeds and other byproducts such as grease. Innovations and improvement in rendering have resulted in a market for primary waste materials in many locations.

### Cannery Solid Wastes

Solid waste from the canning industry can be considerable. For example, tomato canneries solids may be equal to one-quarter of the total quantity of the product. Peach canneries may even be slightly higher. Peas and corn waste can easily be three-fourths of the product.

The general practice is to collect the solid waste from canneries in as dry a form as possible and haul it away with the garbage. Possible re-uses of solid wastes are: (1) ensilage from pea and corn; (2) dehydrated animal feed and molasses from citrus; (3) vinegar, powdered pectin, pectin concentrate, jelly stock, and apple cider from applewaste; (4) tomato pomace (which is used in animal and dog foods) from tomato waste; (5) apricot pits which are used as a raw material for the manufacture of sweet and bitter oils for use in the food, pharmaceutical, and cosmetic industries). Charcoal briquettes and, lately, laetrile have also been made from the pits.

*Fruit, Vegetables, and Specialties.* The fruits, vegetables, and specialties segment of the canned and preserved fruit and vegetable industry include all the subgroups of the food and kindred product industries; they are identified as major group 20 in the standard industrial classification (*SIC Manual* 1972). The industry is separated into three segments based on natural-processing activities, principal sources of wastes, and common usage. They are canned and preserved fruits, canned and preserved vegetables, canned and miscellaneous specialities. Several reports give different rates of raw-waste loads from fruits, vegetables, and specialties. The average total suspended solids (tss) in each one of the segments is fruits—4.4 pound per ton; vegetables—13.1 pound per ton, and specialties—28.5 pound per ton. Solid-waste generation rates for canned and frozen foods range from .04 to .06 tons per ton of raw product. This waste consists of culled fruits and vegetables, discarded pieces, and residues from various processing operations. The solid wastes which result after processing fruits and vegetables are used to recover valuable products for animal feeds or for the fermentation industry. Remaining wastes are incinerated or disposed of on land. The generated solid wastes are rich in sugar and are readily biodegradable. One exception of the waste utilization as animal feed occurs when excessive amounts of pesticides have been used during the growing season, and the wastes become contaminated. These contaminated wastes are then used for fertilizer or landfill. Screening devices of various designs and operating principles remove large-scale solids such as peel, pulp, cores, and seeds prior to waste water treatment. These solids are then either processed for co-products, sold for animal feed, or are landfilled (see Tables 15–6 and 15–7).

Tomato cannery waste contained 51 percent by weight as tomato seed (Geisman 1974). These seeds were found to contain about 29 percent crude protein.

*Fish and Fish Products.* Categorization of this industry is based on four broad groups of subcategories: catfish, crab, shrimp, and tuna.

1. *Catfish*—The waste generated by farm-raised, catfish-processing plants constitutes about 45 percent of the whole catfish processed. Total waste projections were 50.6 million pounds for 1975. Several methods are used for the disposal of catfish offal are: processing for petfood and catfish feed, rendering for fish meal, and burial. The catfish offal consists of heads, skin, viscera, and fat.

The proximate analysis of raw catfish is: moisture, 58 percent; crude fat, 25.5 percent; ash, 3.1 percent; and crude protein, 12.8 percent.

2. *Crab*—The solid-waste load from the conventional blue-crab processing industry for 1971 was 74 million pounds. This waste con-

Table 15-6. Estimated Production of Solid Wastes in the Processing of Each 100 Kilograms of Various Foods.

| Product | Total Waste Produced, Kg | Utilized As Byproduct, Kg | Handled As Solid Wastes, Kg |
|---|---|---|---|
| Apples | 28 | 19 | 9 |
| Beans, green | 21 | 10 | 11 |
| Beets, carrots | 41 | 21 | 20 |
| Citrus | 39 | 38 | 1 |
| Corn | 66 | 62 | 4 |
| Crab, shrimp | 57 | 11 | 46 |
| Fish | 28 | 17 | 11 |
| Olives | 14 | 12 | 2 |
| Peaches | 27 | 9 | 18 |
| Pears | 29 | 9 | 20 |
| Peas | 12 | 8 | 4 |
| Potatoes (white) | 33 | 29 | 4 |
| Tomatoes | 8 | 2 | 6 |
| Vegetables, miscellaneous | 22 | 9 | 13 |

Source: Middlebrooks (1979).

Table 15-7. Percentage of Solid Wastes from the Canning and Frozen Foods Industry Handled by Various Disposal Methods in California.

| Method of Disposal | Percent of Total Tonnage Processed |
|---|---|
| Landfill operations | 31 |
| Spread on fields | 12 |
| Animal feeds | 30 |
| Ocean disposal | 1 |
| Charcoal production | 5 |
| Other disposal methods | 10 |
| Nonfood wastes | 11 |
| Total | 100 |

Source: Middlebrooks (1979).

sists primarily of exoskeletal material with attached unrecovered flesh and viscereal material. Most of the waste is utilized in crab meal for eventual incorporation into animal feed. The mechanized blue-crab processing (184 plants) yields only about 20 to 27 percent of edible food; 80 percent is wasted. This figure is reduced to 50 percent when solids are cooked. The total amount of solid waste generated in 1972 was 25,000 tons (Agee and Cywin 1974). The composition of this waste consists of chitin, protein, calcium carbonate, flesh, and visceral materials. The protein of crab waste is low, thus reducing its value as a potential source of animal feed.

3. *Shrimp Processing*—The waste generated from canned shrimp and frozen peeled shrimp in mechanical peeling amounts from 78 to 85 percent. The composition of shrimp waste will depend on the technique applied in the process. For example, the hand-peeling process will give waste with 27 percent protein, 57 percent chitin, and 15 percent of Ca $CO_3$ (calcium carbonate). On the other hand, the mechanical-peeling process waste will consist of 22 percent protein, 42 percent chitin, and 35 percent Ca $CO_3$. Crude wastes provide sources of protein for livestock feed. Techniques are developing for utilizing all portions of a fishery resource. Wastes will be used as food additive. Protein that can be reclaimed from the shell is high quality and does not exhibit the amine odor found in fish flesh.

4. *Tuna Processing*—Four main segments characterize the entire operation: harvesting, processing for human consumption, production of pet food, and byproduct recovery.

In general, no part of the tuna which enters the processing plant is regarded as waste by the industry. Stickwaters, the non-edible portions, and the aforementioned red meat are all collected and further processed into other products.

Optimal approach to solving the waste and pollution problems in the seafood industry is to utilize the raw material fully, rather than waste most of it and subsequently treat that waste.

### Dairies

Plants handling milk are classified as receiving, bottling, condensing, dry milk manufacturing, ice cream manufacturing, cheese-making, and butter-making. Other types of milk-processing operations are of lesser importance. In addition, useful products are also made for animal consumption, specialty byproducts such as casein, lactic acid, albumin, milk, sugar, and a range of table delicacies.

The production of foodstuffs from milk is usually carried out at large centralized dairies. The handling of large volumes of fresh milk as such results in large amounts of waste. The sources of waste milk solids in dairy and milk products effluents are from spillage, leaks, milks and milk products left in churns, tankers, piping and equipment before washing out, processing losses such as discharge from bottlers, washers, and pasteurizing plants, and wastage, often deliberate, of surplus materials such as whey and skim milk. Settleable solids are not an important consideration in pollution by dairy wastes since all organic material is in a colloidial or dissolved state. Packaging areas of the plant is a major contributor to solid-waste production. Solids residue from wastewater lagoons may need final periodic disposal in sanitary landfills.

### Citrus

Citrus production, processing and marketing comprises a multibillion-dollar industry in the United States. The majority of the citrus is subject to processing within the country to produce single-strength or concentrated juice for marketing in containers. A survey in 1970 reveals a total of seventy-five citrus-processing plants in the state of Florida alone (Environmental Pollution Laboratory 1970). It is understandable that operations of such magnitude yield great quantities of waste materials to be dealt with. The industry has made substantial progress in disposing of waste through recovery of valuable byproducts. Part of these byproducts are used to feed cattle.

Citrus-processing plants are operated on a seasonal basis. The harvesting and processing season extends over eight to ten months. The main waste sources in a typical citrus-concentrated plant come from juice extractor and finisher areas. Other solid and liquid waste result from less frequent cleaning of juice storage, evaporator, blend tank, chiller and packaging areas. All the wastewaters are screened to remove pulp, peel, seeds, and other suspended solids. Total solids are composed largely of sugars which are readily biodegradable. Some plants do not process these wastes to recovery valuable products (animal feed, molasses, and so on). Instead, they prefer to spread the

208    INDUSTRIAL SOLID WASTES

solid residues on the land. In Table 15-6 Middlebrooks reports that 39 kilograms of solid wastes were produced in 1971 for each 100 kilograms of citrus processed. Of these 38 kilograms were re-used as a byproduct, and only 1 was "handled as a solid waste."

### Beverage and Soft Drink Industry

This industry can be subdivided into nonalcoholic and alcoholic beverages.

*Soft Drinks.*   Nonalcoholic beverages are manufactured from sugar, syrup, flavors, acids, water or water saturated with carbon dioxide for carbonated beverages, flavor extracts or mixtures of flavoring extracted from fruits, roots, and other plant tissues. Essential oils or aromatic chemicals in solution are also contained in the flavors. The production of canned soft drinks is a relatively simple process which is basically automatic and requires few manufacturing personnel. A flow diagram of a standard manufacturing process for bottled and canned soft drinks is shown in Figure 15-3. Waste materials are generated by bottle or can processing, and the preparation of flavoring materials and carbonated water. The characteristics of wastewaters from nonalcoholic beverage industries are summarized in Tables 15-8 and 15-8A. The solids are recovered by screening the wastewater which can then be used as animal feed or fertilizer, or it can be sent out for use as landfill.

*Brewery.*   A process flow diagram for the brewing of beer is shown in Figure 15-4. The brewing process begins with cooking the cereal grains to solubilize the starches. The cooked grains are then mixed with malt to convert the starches to sugar. The mixture of malt and grains (mash) is filter-pressed to remove the spent grain which is used as a byproduct. The filtrate (wort) is placed in a brew kettle, and hops are added for flavor. Undesirable protein (trub) is removed from the mixture by boiling. The hops are then removed, and the wort is pumped to a cooler where the trub is removed by sedimentation. Diatomaceous earth filtration is frequently used to filter the residual trub from the cooled wort. The clear wort is transported to the fermenter and yeast is added to convert the sugars to alcohol and carbon dioxide. When the fermentation is complete, the excess yeast

**Figure 15-3.** Flow Diagram of Standard Manufacturing Process for Bottled and Canned Soft Drinks.

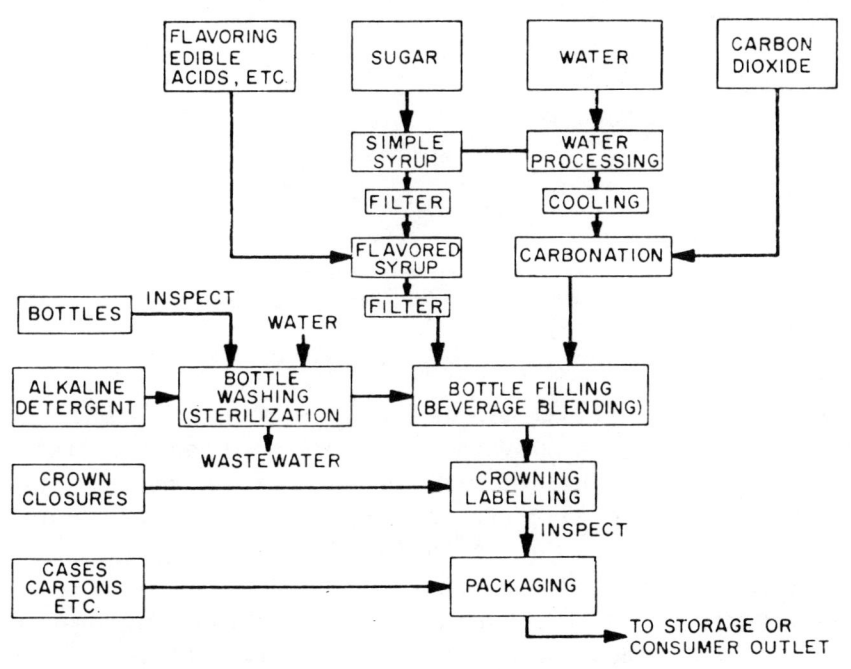

Source: Joyce and others (1977).

**Table 15-8.** Waste Analysis of the Effluents from Soft Drink Bottling Plants [*Porges and Struzeski, 1961*].

| Plant | $BOD_5$ mg/ℓ | SS mg/ℓ | Alkalinity Pheno. | Alkalinity Total | Range of pH |
|---|---|---|---|---|---|
| A | 380 | 170 | 230 | 390 | 10.1-11.4 |
| B | 660 | 160 | 100 | 250 | 10.0-11.2 |
| C | 250 | 340 | 110 | 220 | 10.4-11.2 |
| D[a] | 260 | — | — | — | 4.0-8.5 |

a. Plant D is a typical value for a medium or large-sized modern plant operating under franchise from a major national company.

Table 15-8A. Waste Loads Discharged from Soft Drink Bottling Plants per Cubic Meter Finished Product [Porges and Struzeski, 1961].

| Plant | Wastewater $m^3/m^3$ | $BOD_5$ $kg/m^3$ | SS $kg/m^3$ |
|---|---|---|---|
| A | 5.07 | 1.92 | 0.88 |
| B | 3.53 | 2.32 | 0.57 |
| C | 12.8 | 3.15 | 4.33 |
| D[a] | 2.65 | 0.58 | — |

a. Plant D is a typical value for a medium or large-sized modern plant operating under franchise from a major national company.

is removed, and the beer is cooled and transported to primary storage. A mass diagram showing the quantities of raw materials and waste products produced in the production of 1 cubic meter of beer is shown in Figure 15-5. The characteristics of brewery waste are shown in Tables 15-9 through 15-11.

Recovery of solids from process streams is a common practice in the brewery industry. Extensive recovery of the spent grains, spent hops, trub, excess yeast, and spent powder-type filter aid is practiced and should be encouraged. Most of the recovered materials are added to the spent grains and hops which are converted to animal feed.

A plant processing approximately 160 tons of corn, rye, and barley malt per day on a twenty-four-hour-day, five-day-a-week schedule illustrates product characteristics and solid-waste generation of beverage industry. The main products are beverages and industrial alcohol with Fusel oil and high-protein, livestock feed material recovered as byproduct. The amount of suspended solids in this plant wastewater varies from 200-300 milligrams per liter. This brewery discharges a predominantly carbohydrate waste. Significant variations in both the quantity and quality of this waste result because of batch processing and seasonal fluctuations in production schedules.

Most of the solid waste is combined and contained in the wastewater. These wastes are already warm when they reach the treatment plant. The solid waste recovered is treated and marketed for many different uses including the animal feed or fermentation industries.

INDUSTRIAL WASTES    211

**Figure 15-4.** Process Diagram for Malt Liquor Production.

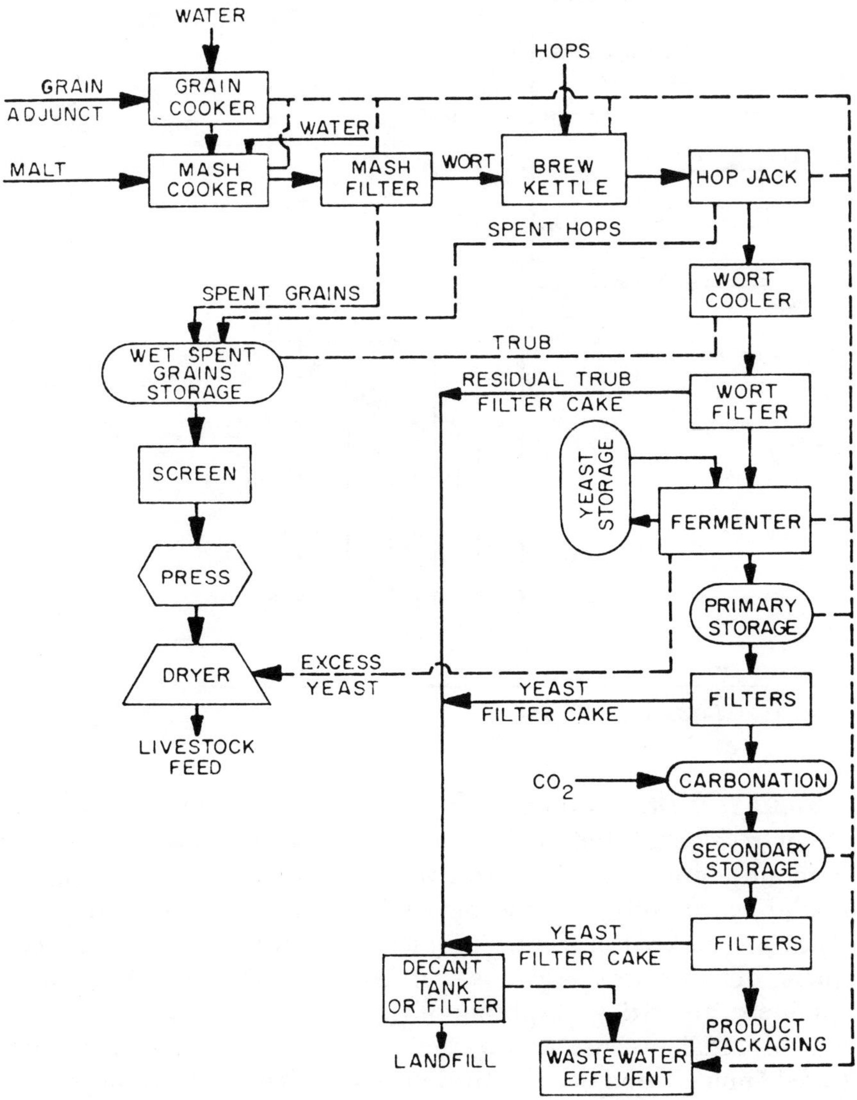

Source: Joyce and others (1977).

**Figure 15-5.** Brewery Input-Output Characteristics.

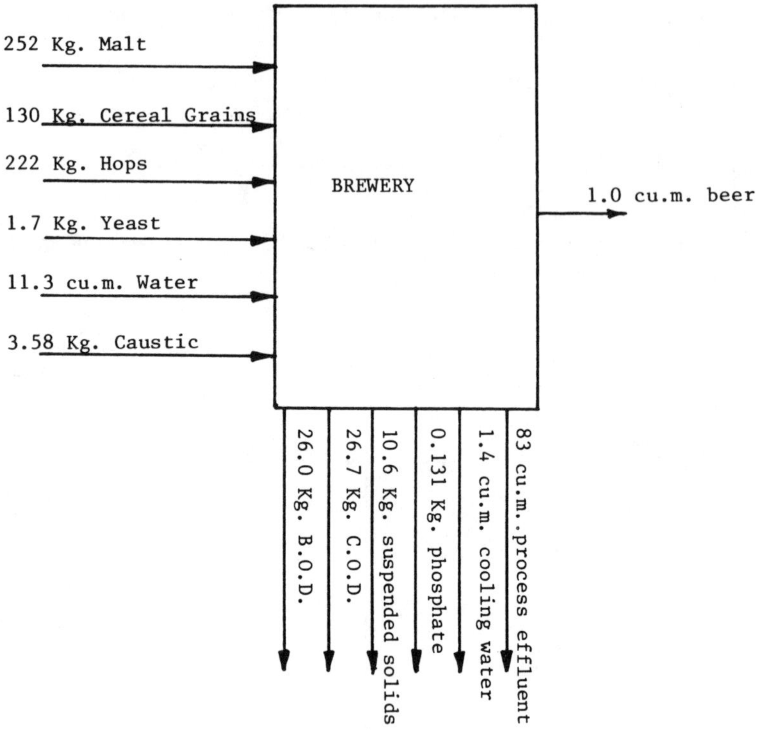

Inputs to Brewery:
- 252 Kg. Malt
- 130 Kg. Cereal Grains
- 222 Kg. Hops
- 1.7 Kg. Yeast
- 11.3 cu.m. Water
- 3.58 Kg. Caustic

Output: 1.0 cu.m. beer

Waste outputs:
- 26.0 Kg. B.O.D.
- 26.7 Kg. C.O.D.
- 10.6 Kg. suspended solids
- 0.131 Kg. phosphate
- 1.4 cu.m. cooling water
- 83 cu.m. process effluent

Source: AWARE (1971).

*Beer and Soft Drink Containers.* Beer and soft drink containers are part of residential and commercial solid waste. They contribute to the mounting need for land area for disposal and the increasing cost of solid-waste collection and disposal. Beverage and soft drink containers are an environmental problem, primarily because some consumers of beverages create social costs by littering their empty containers rather than disposing of them properly. These social costs are substantial because of the large number of beverage containers littered annually and because of these containers' high visibility.

To a larger extent, the per capita growth for any beverage of soft drink must come at the expense of other beverages. The age composition of the population plays a significant role in determining the

Table 15-9. Brewery Effluent Characteristics Obtained from a Survey of 75 Plants [AWARE, 1971c].

| Parameter | Median (50%) | Mean | 80% Value |
|---|---|---|---|
| Process Effluent Suspended Solids | 3.33 kg/m³ beer | 772 mg/ℓ<br>4.73 kg/m³ beer | 5.33 kg/m³ beer |
| Process Effluent $BOD_5$ | 9.00 kg/m³ beer | 1622 mg/ℓ<br>10.2 kg/m³ beer | 15.0 kg/m³ beer |
| Process Effluent COD | — | 2944 mg/ℓ<br>11.2 kg/m³ beer | — |
| Process Effluent pH | — | 7.6 | — |
| Process Effluent Temperature | — | 28 degrees C | — |
| Process Effluent Volume | 7.2 m³/m³ beer | 8.3 m³/m³ beer | 12.2 m³/m³ beer |
| Cooling Water Volume | — | 1.42 m³/m³ beer | — |
| Total Water Usage | — | 11.3 m³/m³ beer | — |

(0.1364 m³/barrel).

Table 15-9A. Sources of Pollutants from a Brewery [Linton, 1973].

| Source | $BOD_5$ (kg/m³ beer) | $BOD_5$ (percent) | SS (kg/m³ beer) | SS (percent) |
|---|---|---|---|---|
| Yeast | 3.71 | 30 | 2.55 | 30 |
| Trub | 3.21 | 26 | 1.24 | 14 |
| Hops | 0.39 | 3 | 0.77 | 9 |
| Pressed grain liquor | 0.85 | 7 | 0.50 | 6 |
| Drain and Rinse | 2.09 | 17 | 0.85 | 10 |
| Filter effluent | 0.50 | 4 | 1.58 | 19 |
| Bottling | 1.20 | 10 | 0.66 | 8 |
| Miscellaneous | 0.42 | 3 | 0.35 | 4 |
| Total | 12.4 | 100 | 8.50 | 100 |

Table 15-10. Typical Concentrations of Wastes Discharged from Specific Brewery Operations.

| Brewing Operation | SS (mg/ℓ) | $BOD_5$ (mg/ℓ) |
|---|---|---|
| Cereal cooker | 300 | 700 |
| Mash tun | 300 | 2,000 |
| Lauter tun | 3,000 | 10,000 |
| Spent grain tank (or press) | 10,000 | 15,000 |
| Brew kettle | 100 | 300 |
| Hot wort tank (inc. trub) | 5,000 | 10,000 |
| Wort cooler | 20 | 30 |
| Fermentation tanks | 2,000 | 5,000 |
| Ruh chiller | 30 | 700 |
| Ruh tanks (primary aging) | 20,000 | 30,000 |
| Primary filtration | 30,000 | 40,000 |
| Aging tanks | 600 | 10,000 |
| Final filtration | 500 | 100 |
| Finished beer tanks | 200 | 50 |
| *Nonreturnables* | | |
| Rinser | 3 | 20 |
| Pasteurizer | — | 50 |
| *Returnables* | | |
| Prerinse | 200 | 500 |
| Final rinse | 10 | 10 |
| Pasteurizer | 20 | 30 |
| *Kegs* | | |
| Prerinse | 100 | 1,000 |
| *Miscellaneous Wastes* | | |
| Bottle and can filler drip | — | 50,000 |
| Conveyor lube drip | 1,000 | 5,000 |
| Spray tunnel drip | 40 | 3,000 |
| Floor hosedown | — | — |

Source: Associated Water and Air Resources Engineers, Inc., Report for E.P.A. 1971.

Table 15-11. Overall Plant Raw Waste Characteristics.

| Parameter | Coors Raw Waste[a] | Brewing Industry Mean Raw Waste[b] |
|---|---|---|
| Volume | 3.5 m$^3$/m$^3$ beer | 8.3 m$^3$/m$^3$ beer BOD$_5$ |
| BOD$_5$ | 2.90 kg/m$^3$ (825 mg/ℓ) | 11.8 kg/m$^3$ beer (1622 mg/ℓ) |
| SS | 1.00 kg/m$^3$ (280 mg/ℓ) | 4.8 kg/m$^3$ beer (722 mg/ℓ) |

a. Based on average at Coors for month of June 1974.

b. Industrial Waste Survey of the Malt Liquor Industry prepared for EPA, August 1971, by Associated Water and Air Resources Engineers, Inc.

growth in the consumption of a beverage or soft drink since tastes vary by age. The beverage and soft drink container elements of solid waste is represented by the number of discarded containers. This number is directly calculated from the consumption and containerization estimates after substracting the littered containers. Re-use and recycling seems to be the one way to reduce the amount of containers of solid waste. Other mechanisms have been proposed to the government; lowered taxes and mandatory deposits appear to be the most likely to succeed.

### Agricultural Industries

According to Middlebrooks (1979), large quantities of hulls, shells, stalks, and meals are produced by processing raw materials for further preparation or storage or packaging. This may approximate 50 percent of the original raw material. Also considerable amounts of sludge are produced during the treatment of wastewater from these plants. These quantities generally vary from 2 to 15 percent of the wastewater volume.

*Sugar Cane Production*

*Industry Categorization.* Several factors are considered significant with regard to identifying potential subcategories in the cane sugar refining industry. These factors include raw material quality, refinery size, refinery age, nature of water supply, land availability, and process variation.

*Sugar Refining.* The sugar-processing industry can be divided into two major categories according to the raw material: cane and beets are the principal materials used in the production of sugar.

1. *Cane Sugar Processing.* Raw sugar production consists of the extraction of juice from sugar cane, purification of the juice, formation of sucrose crystals in the juice, and the separation of the crystals from the juice. A typical sugar cane averages 15 percent by weight of fiber and 85 percent by weight of juice, which is composed of 85 percent water, 12 percent sucrose, and 3-5 percent invert sugars and impurities.

Cane-harvesting techniques vary considerably in the United States and throughout the world. Some use mechanical harvesting; others use manual harvesting. The major disadvantage to mechanical harvesting is the introduction of much larger quantities of extraneous material. Cane is transported from the fields to the factory by truck. It is usually necessary to wash the cane before beginning the extraction process. A flow diagram of a raw cane sugar processing plant is shown in Figure 15-6. Bagasse from the last mill usually contains about 50 percent moisture and amounts to about 30 percent by weight of the cane entering the operation. Bagasse is usually used as a fuel to produce steam. Once start-up is achieved, enough bagasse is produced to supply all the steam required in the operation of the sugar factory. Steam generated by the boilers is used to operate the milling plant and the turbo-generators that burn electricity for the plant. Excess bagasse is wasted to the land environment.

Impurities in the juice from milling operation consist of fine particles of bagasse as well as fats, waxes, and gums. The coarser particles are removed from the juice by screens, and the particles are returned to the mills. The majority of the remaining impurities are removed by the clarification process. The sludge that settles in the clarifier is usually thickened by a rotary vacuum filter. The dry filter cake is difficult and expensive to handle but can very easily be spread on the fields as soil conditioner and fertilizer. The raw sugar separated by the centrifuges is shipped or stored for future use. The final molasses (black-strap) is used for production of animal feeds, rum, ethyl and butyl alcohols, and acetic and citric acids. A flow diagram showing a typical cane sugar factory is presented in Figure 15-7. The raw sugar is transported in bulk form to a sugar-refining factory. The concept of sugar refining is relatively simple, but the processes in-

INDUSTRIAL WASTES 217

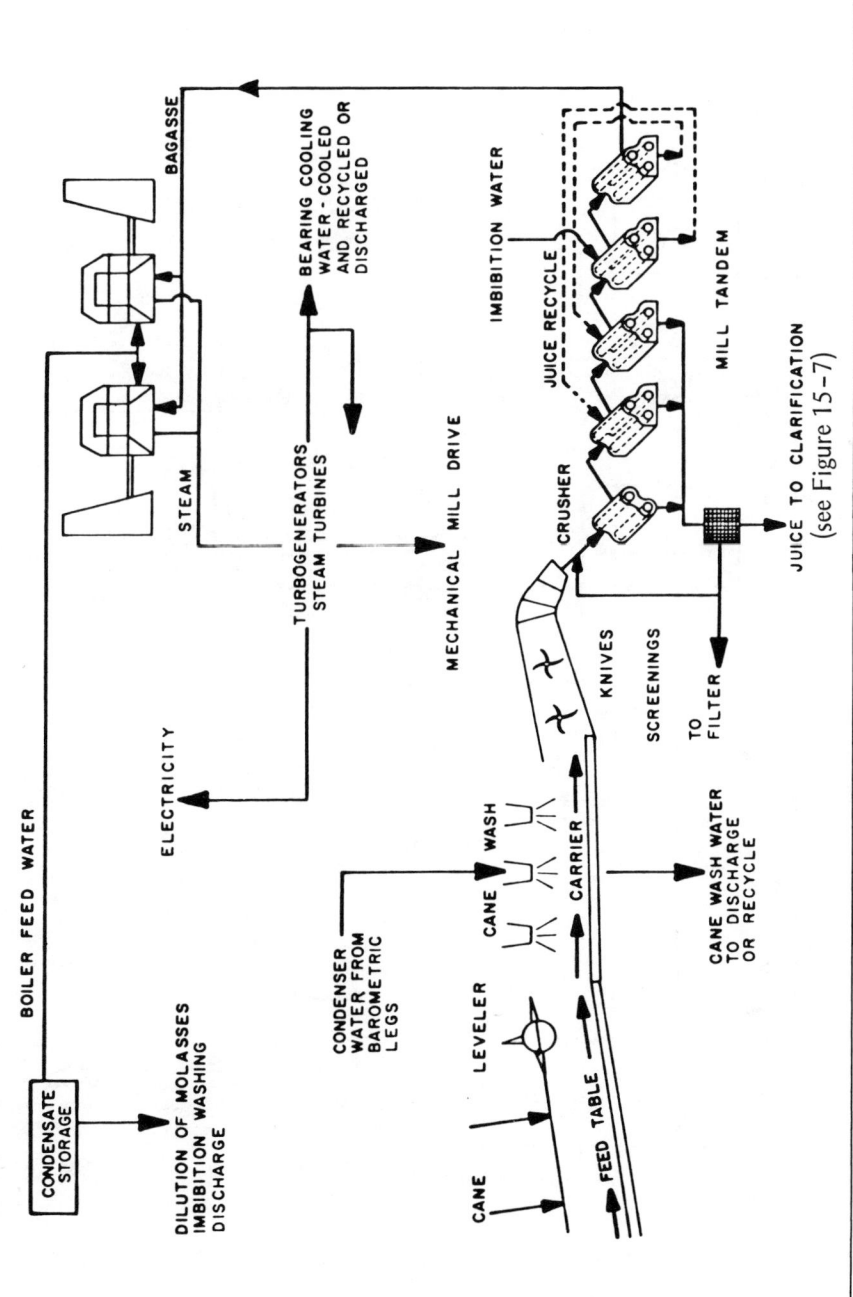

Figure 15-6. Typical Sugar Factory with Cane Wash.

Source: EPA (1975).

218    INDUSTRIAL SOLID WASTES

Figure 15-7.  Water Usage in a Typical Sugar Cane Factory.

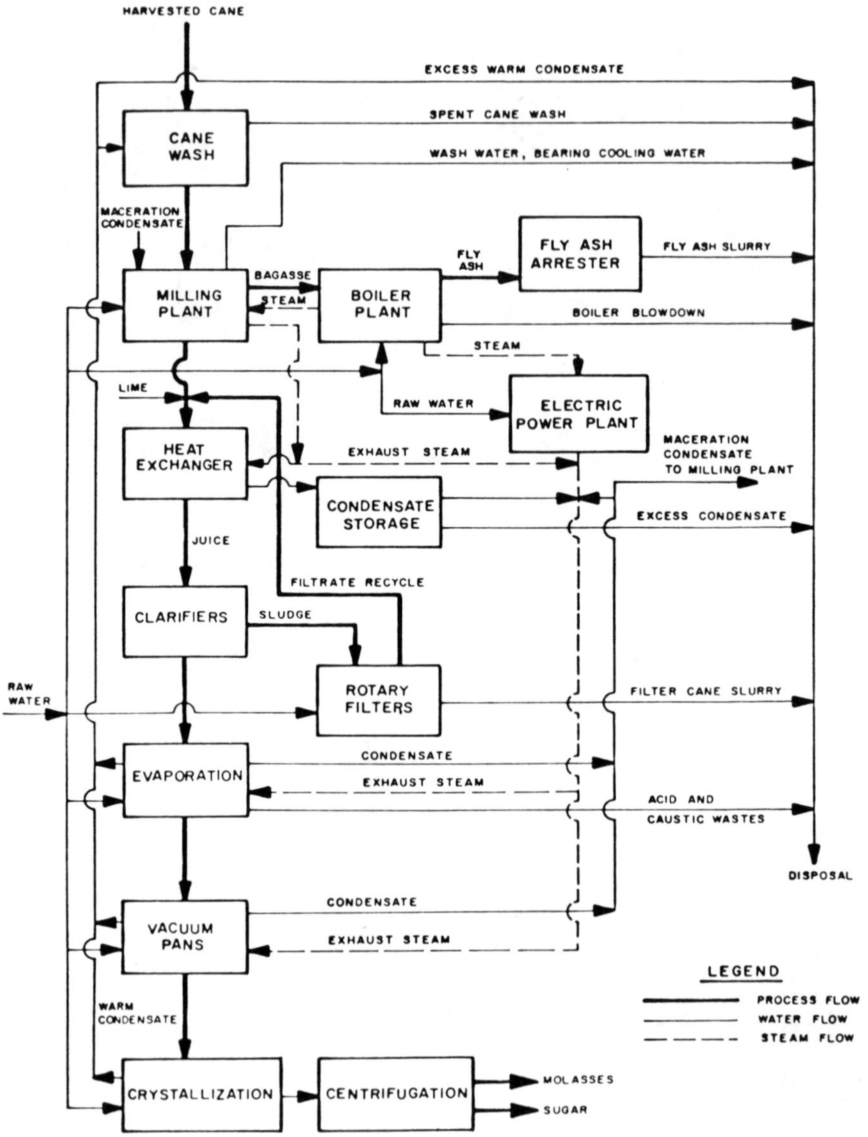

Source: EPA (1975).

volved are complex and thus vary from refinery to refinery. Figure 15-8 shows a process flow diagram for refining using carbon. Figure 15-9 shows a wastewater flow diagram for a crystalline sugar refinery.

Some of the highly concentrated wastes (baggase, filter cakes, and spent molasses) come from manufacturing operations. Fibrous residue (baggase) from grinding and screening of particle amounts to 30 percent by weight of the cane processed. The quantities of solid wastes are variable and affected by the season of the year.

*Characteristics of Wastes.* Solid wastes such as baggase, filter cakes, and spent molasses contain sucrose. Filter cake amounts to 3 to 4 percent of the weight of cane processed and has a high BOD (8,000-25,000 milligrams per liter, total solids (40,000-100,000 milligrams per liter and .21 percent sucrose (Braggi 1968). Spent molasses from the process has an exceedingly high BOD (400,000-900,000 milligrams per liter.

*Treatment and Disposal of Wastes.* Solid wastes from production of sugar cane can be reduced in different ways. These wastes can be employed for fuel, fertilizer, or byproducts, and hog and cattle feeding. Baggase is burned in the boiler plant for generation of steam and provides enough fuel for operation of the factory. An alternate use for baggase is as a raw material for production of fiberboard or paper. Filter cake can be disposed to landfill or discharged to the land as a fertilizer. Spent molasses is used for alcohol fermentation and for feeding cattle and hogs. Research by this author shows that mixtures of bagasse and cachaza can be fermented to produce valuable methane gas and a useful residue.

2. *Beet Sugar Processing.* The raw materials entering beet sugar-processing operations are sugar beets, limestone, small quantities of sulfur, fuel, and water. The products are refined sugar, dried beet pulp, and molasses. The average raw material requirements and end products produced per unit weight of clean beets processed are: for limestone, 80 pounds per ton; for dry beet pulp, 100 pounds per ton; for sugar product, 260 pounds per ton; and for molasses produced, 100 pounds per ton.

Beets are harvested mechanically and hauled by truck or rail to the plant. Sugar beets are transported to the processes by a water flume

**Figure 15-8.** Typical Carbon Refinery.

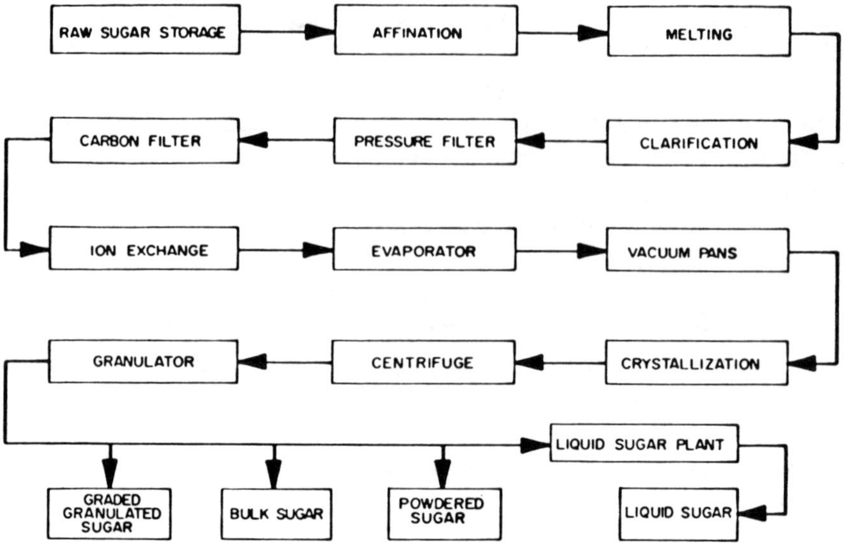

```
RAW SUGAR STORAGE → AFFINATION → MELTING
                                      ↓
CARBON FILTER ← PRESSURE FILTER ← CLARIFICATION
    ↓
ION EXCHANGE → EVAPORATOR → VACUUM PANS
                                      ↓
GRANULATOR ← CENTRIFUGE ← CRYSTALLIZATION
   ↓                          → LIQUID SUGAR PLANT
   ↓         ↓         ↓              ↓
GRADED    BULK      POWDERED      LIQUID SUGAR
GRANULATED SUGAR    SUGAR
SUGAR
```

Source: EPA (1974).

equipped with traps to remove stones and other heavy foreign material. Following washing, the beets are sliced into thin strips called "cassettes." Sugar and other soluble substances are extracted from the cassettes in the diffusers using a counter-current flow of water. The raw juice containing 10-15 percent sugar is pumped to purification stations.

The remaining beet pulp is transported to presses where the water content is reduced from approximately 95 percent to 80 percent. Following the presses, the pulp is dried to a moisture content of 5-10 percent. Water pressed from the pulp is usually returned to the diffusers, and the dried pulp is used as a livestock feed supplement. Molasses remaining after removal of crystalline sugar may be added to beet pulp and sold for animal feed, or additional sugar may be recovered by the Steffen Process. In the Steffen Process, a thick liquid called concentrated steffen filtrate may be dried with the beet pulp or used to produce monosodium glutamate or potash fertilizer salts. A process flow diagram of the beet sugar industry is shown in Figure 15-10.

INDUSTRIAL WASTES 221

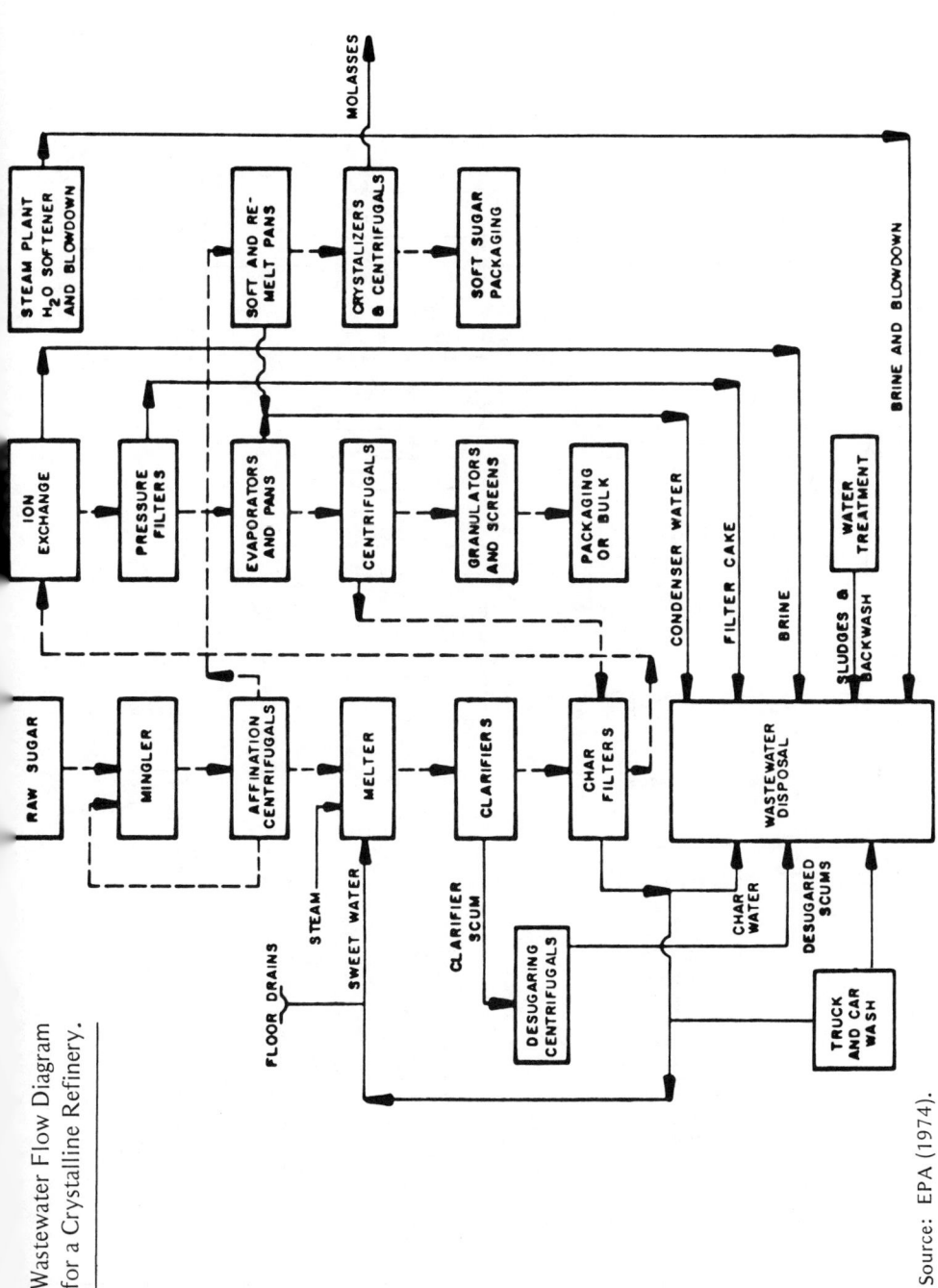

Wastewater Flow Diagram for a Crystalline Refinery.

Source: EPA (1974).

## 222 INDUSTRIAL SOLID WASTES

**Figure 15-10.** Materials Flow in Beet Sugar Processing Plant with Typical Water Utilization and Waste Disposal Pattern.

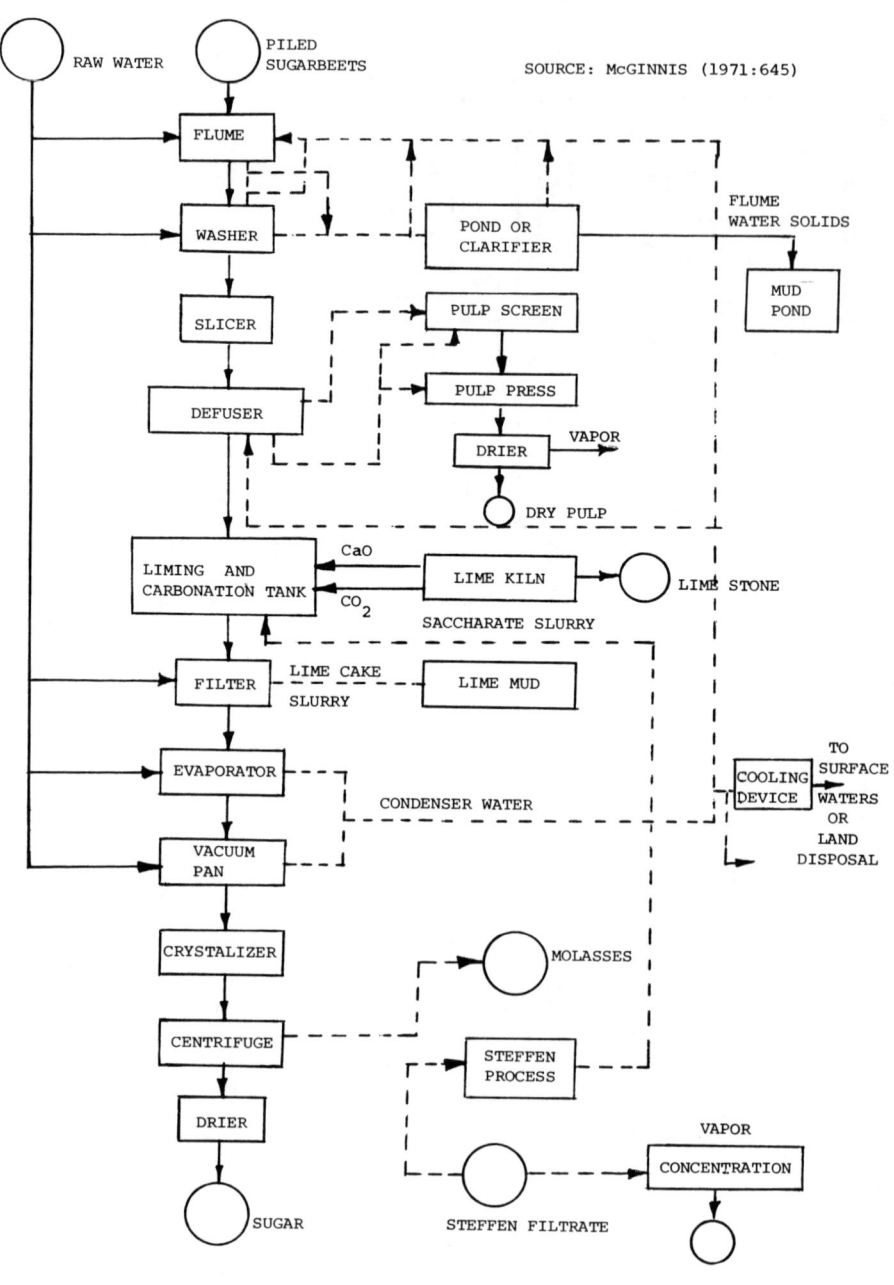

Source: McGinnis (1971: 645).

Solid waste from the sugar-refining industry is re-used as animal feed and steam generation before excess waste can be disposed in landfill, incineration, or anaerobic digestion.

*Solid-Waste Disposal.* The large volumes of dirt and solid material removed from sugar beets at the processing plant pose a perplexing problem for permanent disposal. Generally, about 100 pounds of solid waste per ton of beets sliced are contributed by a typical beet sugar-processing plant. Where holding ponds are employed, solids accumulated in the ponds are removed annually and disposed of by adding the material to pond dikes.

Sugar beets stored in large piles at the plant site or in outlying areas such as railroad sidings may be exposed to rodent activity and additional pollution from truck or railroad car unloadings.

In addition to the large volumes of soil delivered to the plant with the incoming beets, solid waste is also generated in terms of trash normally associated with municipal activities. Disposal of this material may be at the plant site, or the waste material may be collected by the local municipality for disposal by incineration or sanitary landfill. *Sanitary landfills* are generally best suited for noncombustible material and organic wastes which are not readily combustible such as decomposed beets, weeds, and peelings.

*Composting* offers a viable alternative for disposing of organic materials such as decomposed beets, weeds, and peelings. Experience with this method for the disposal of municipal wastes has proved it more costly than sanitary landfill operations. Therefore, the sanitary landfill is probably the lower cost alternative, provided that adequate land is available.

## Grain Mills

The industry is characterized into several discrete segments such as corn wet milling, corn dry milling, normal wheat-flour milling, bulgur wheat-flour milling, normal rice milling and parboiled rice milling, animal feed manufacturing, hot cereal manufacturing, ready-to-eat cereal manufacturing, and wheat starch and gluten manufacturing.

The principal raw materials used in the milling industry are the basic cereal grains: corn, wheat, oats, and rye. Vitamins and other additives are used in animal feed production, and sugar or syrup is

**Figure 15-11.** The Corn Wet Milling Process.

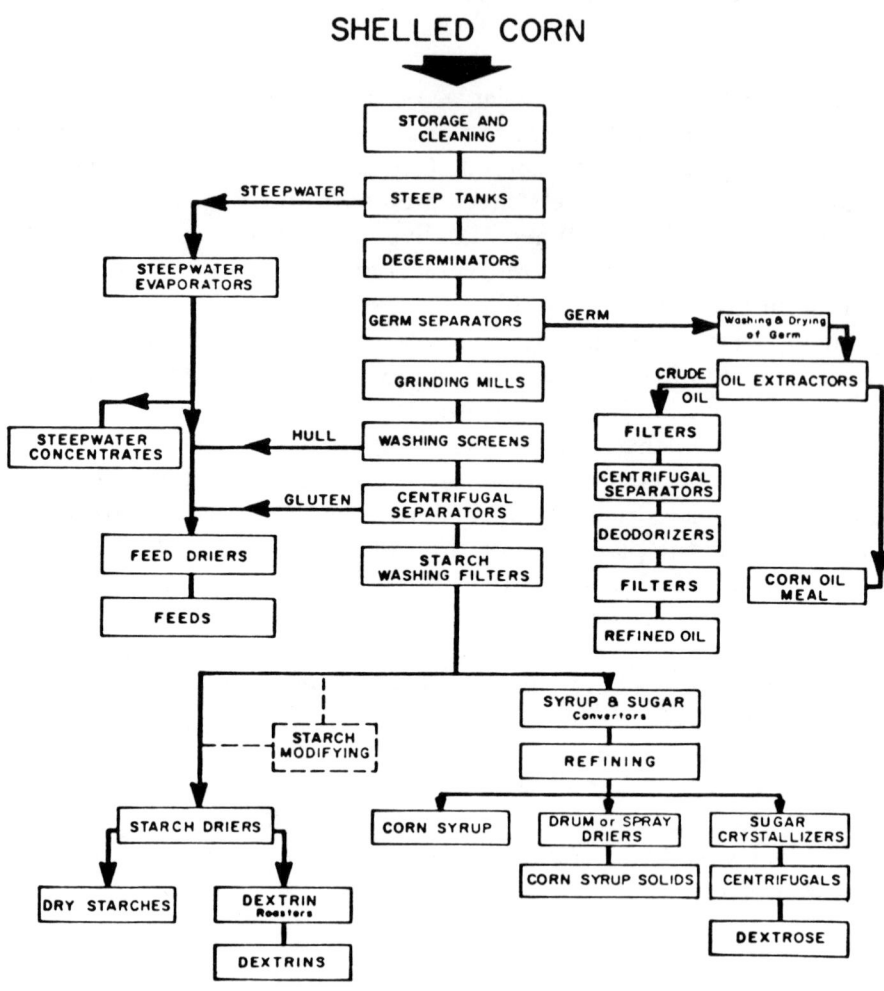

Source: EPA (1974).

frequently used in ready-to-eat breakfast cereals. Figures 15-11 to 15-19 show a flow diagram for some of these processes. The characteristics of wastewater from milling industry is summarized in Table 15-12. Solid waste consists of hulls, germ, bran and shorts, and rice polish.

Solid waste is produced in large quantities in the milling operations and the biological treatment processes. In corn wet milling the

Figure 15-12. The Corn Dry Milling Process.

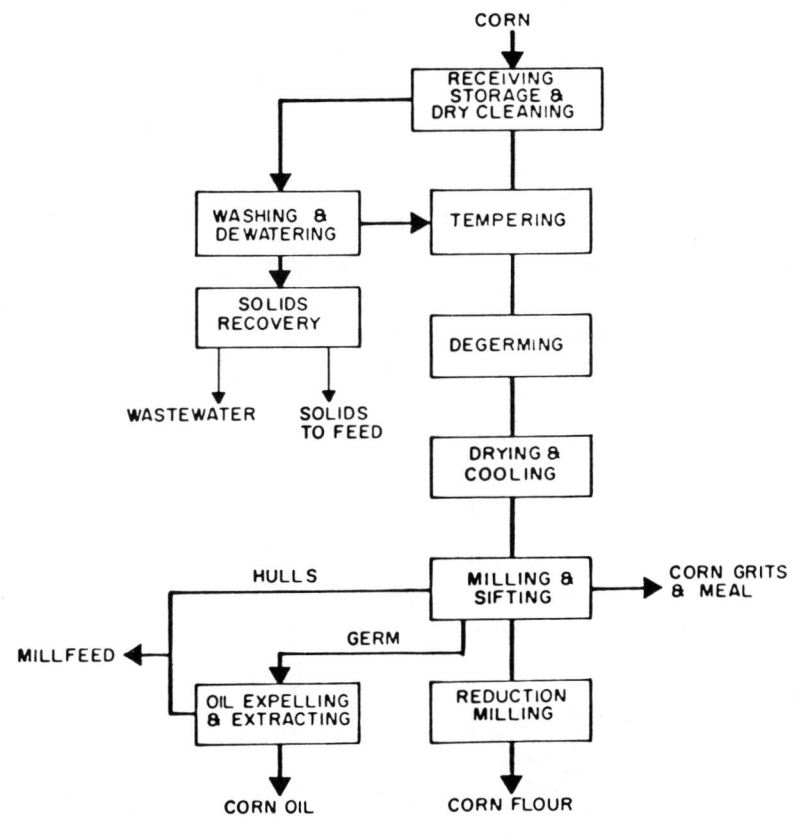

Source: EPA (1974).

amount is 526 pounds per million bushels; in rice milling, 55 pounds; corn dry milling, 2000 pounds; bulger wheat flour milling, no discharge; wheat milling, 414 pounds; and in parboiled rice milling there is no discharge. Fortunately, solids from both sources can be incorporated into animal feed. If this alternative is unavailable for some reason, solids can be handled by digestion and landfill, incineration, and other conventional methods of handling biological solids.

The treatment of wastewaters from cereal and wheat-starch plants will give rise to substantial quantities of solid wastes, particularly biological solids from activated sludge or comparable systems. Conventional methods for handling biological solids are applicable to these wastes such as digestion, dewatering, landfill, or incineration.

**Figure 15-13.** The Bulgur Process.

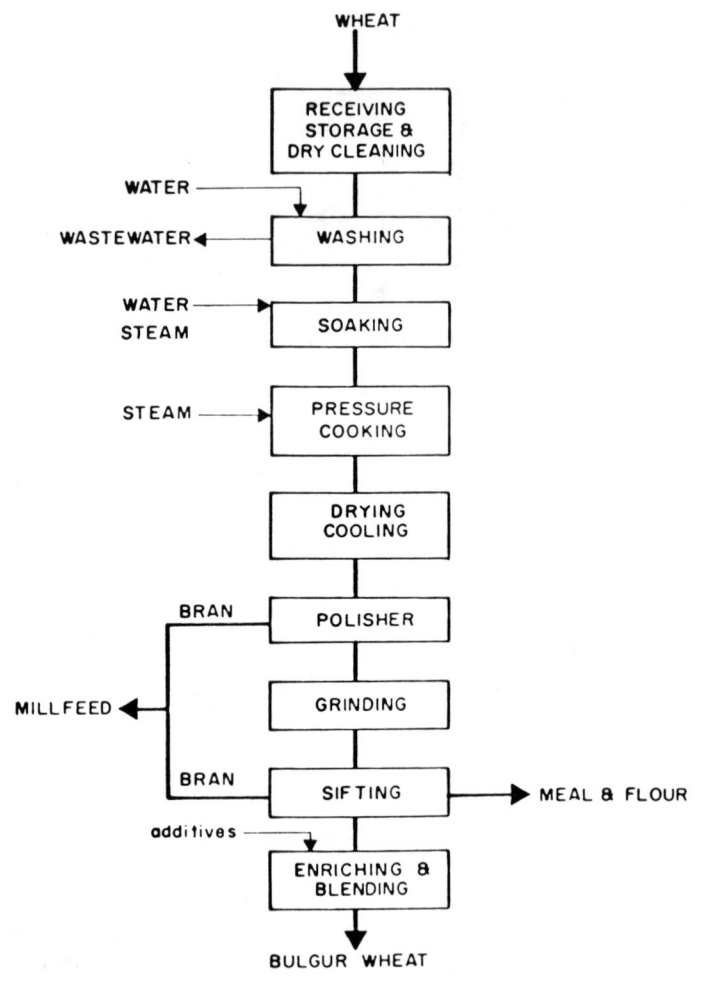

Source: EPA (1974).

Disposal of this solid material so as not to contribute to pollution of ground- or surface water is necessary.

For those waste materials considered nonhazardous, land disposal is the choice for disposal. Practices similar to proper sanitary landfill technology may be followed. For those materials considered to be hazardous, disposal will require special precautions.

**Figure 15-14.** The Wheat Milling Process.

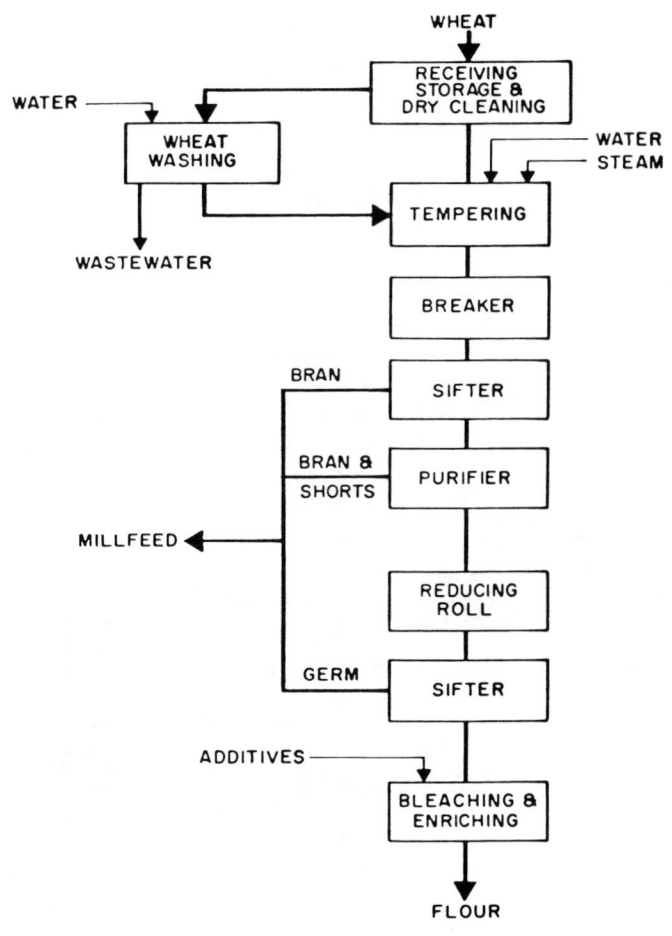

Source: EPA (1974).

### Wine and Brandy

Wines are classified into three categories: table, dessert, and sparkling burgundy. The wine-making process starts with the harvest of the grapes, followed by crushing, pressing, fermentation, and the finishing operation (racking, filtering, fining, refrigerating, aging, and bottling).

**Figure 15-15.** The Rice Milling Process.

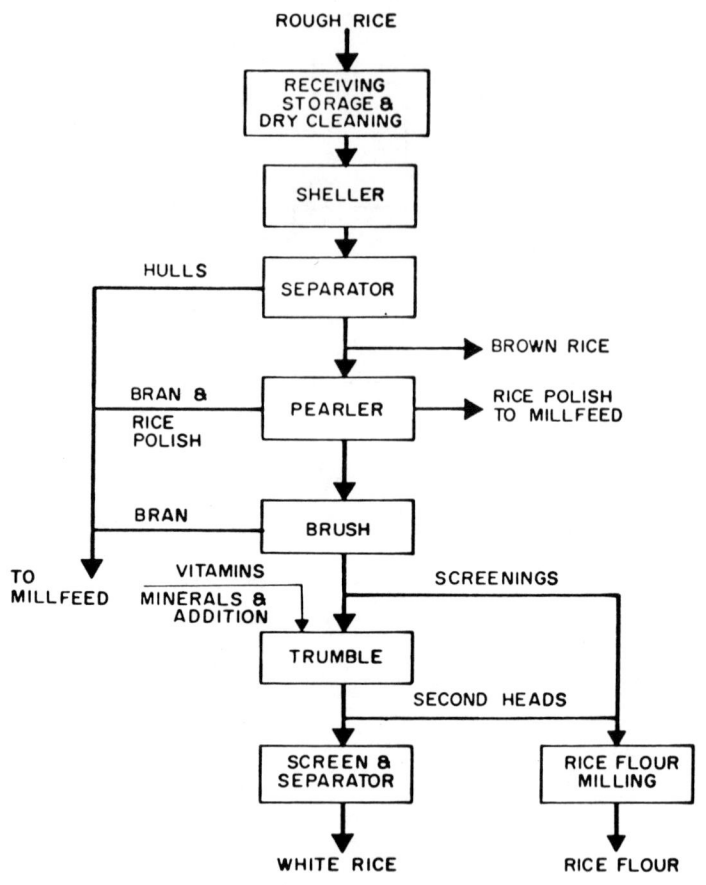

Source: EPA (1974).

Methods used to produce table wines vary with the winery, but the basic procedures are as shown in Figure 15-20 and 15-21. Grapes used to make white wine are stored in a tank to allow juice to separate from the skins, seeds, and stems. The separated juice is transferred to a fermenter. The remaining solids are pressed to remove the juice, which is used for blending or making less expensive wines. The solids, which are referred to as "pomace," constitute a major waste-disposal problem. Following fermentation, new wines

**Figure 15-16.** The Parboiled Rice Process.

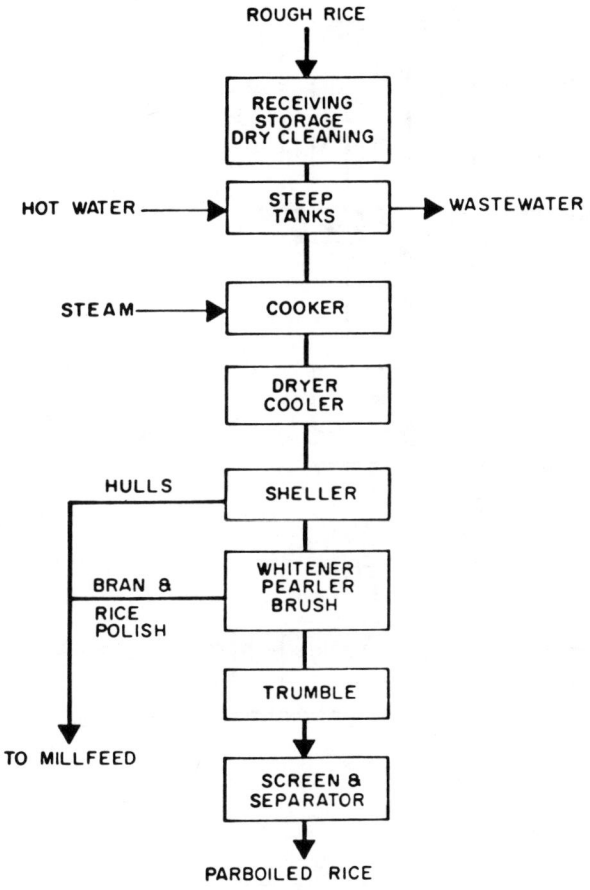

Source: EPA (1974).

are pumped to storage tanks for clarification and aging. Solids in the wines settle out, and it is necessary to transfer the clarified wine to other tanks. When the wine is transferred, a residue of lees (yeast cells and grape residue) remains behind, and it is necessary to dispose of this residue or use it as distilling material.

Wine consumption per adult per year in the United States was approaching 2.5 gallons (1960) and was expected to reach 3.4 gallons by 1980. The amount of solid waste produced by this industry as a result of expansion is expected to increase.

Figure 15-17. Animal Feed Manufacturing.

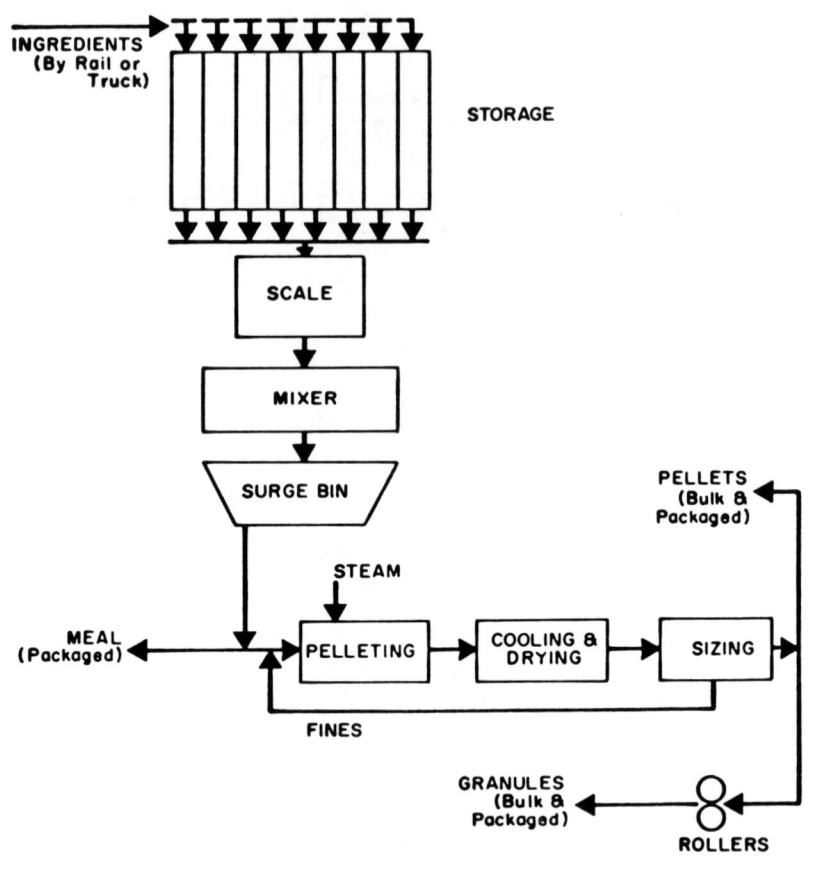

Source: EPA (1974).

*Solid-Waste Characteristics for Wine.* Part of the solid wastes, the pomace, are, as just mentioned, a major waste-disposal problem. Yeast cells and grape residue are used as distilling material. The solids from the filters are relatively dry and are usually landfilled. Total solids of wine stillage amounts to 20,100 milligrams per liter. The characteristics of winery waste are summarized in Tables 15-13 to 15-15.

Brandy is produced from wine, lees, and pomace by a distillation and aging process. The main steps in the brandy-making process include: disintegration, distilling material storage, continuous column

**Figure 15-18.** Flaked or Crisp Cereal Production.

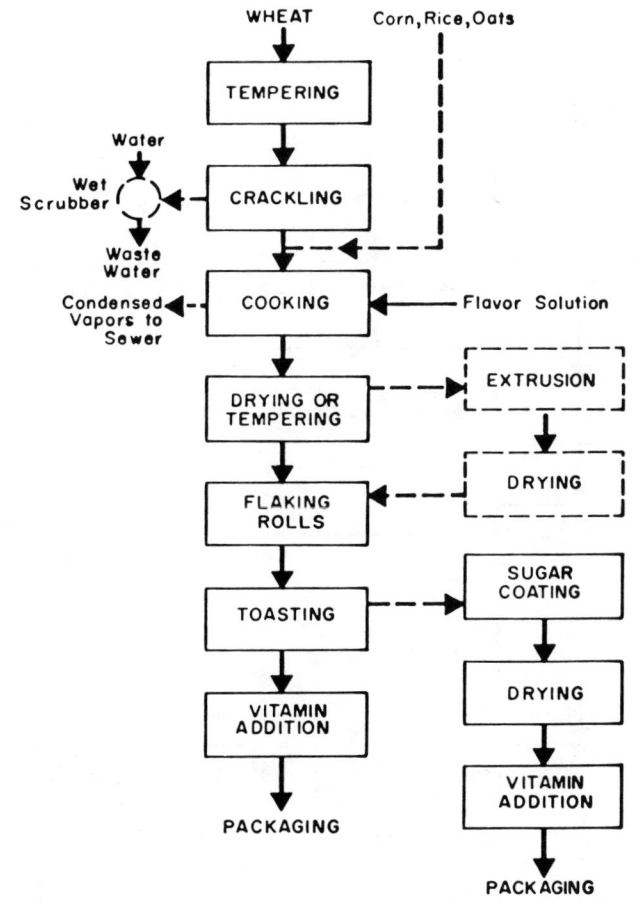

Source: EPA (1974).

still, aldehyde column dilution and aging, and bottling (see Figure 15-22).

*Solid-Waste Characteristics for Brandy.* The material remaining after removal of the alcohol (stillage) is a concentrated waste that is the most difficult waste material to be treated in the wine industry. Total solids of pomace stillage amounts to 13,180 milligrams per liter.

232   INDUSTRIAL SOLID WASTES

Figure 15-19. Wheat Starch and Gluten Manufacturing.

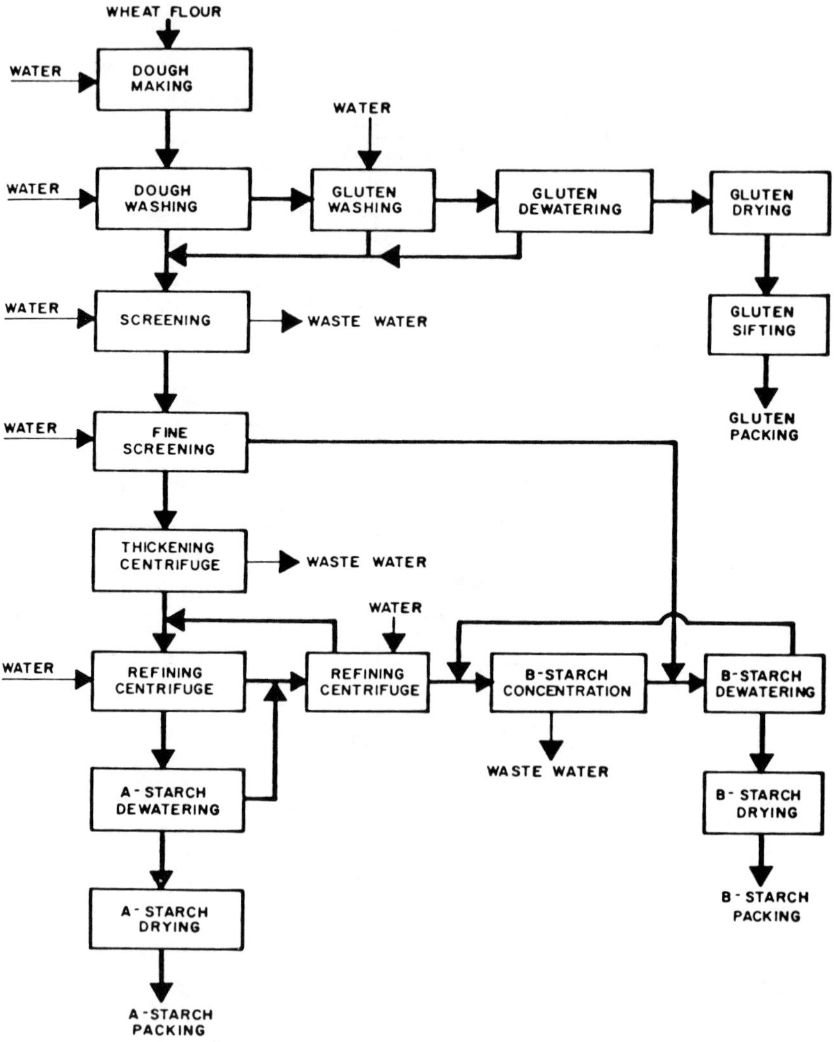

Source: EPA (1974).

INDUSTRIAL WASTES    233

Characteristics of Wastewater

| Type of Wastewater | Flow Rate | | BOD | | COD | | Suspended Solids | |
|---|---|---|---|---|---|---|---|---|
| | m³/day | m³/KKg Prod. | mg/ℓ | Kg/KKg Prod. | mg/ℓ | Kg/KKg Prod. | mg/ℓ | Kg/KKg Prod. |
| Milling | | | | | | | | |
| Corn wet milling | 3,785 to 189,000 | 3.1 to 41.7 | 225 to 4,450 | 2.1 to 12.5 | 473 to 1,560 | 6.8 to 22.3 | 81 to 2,451 | 0.5 to 9.8 |
| Corn dry milling | 227 to 900 | 0.48 to 0.90 | 603 to 2,748 | 1.14 | 1,795 to 1,901 | | 1,038 to 3,485 | 1.62 |
| Wheat milling normal | No wastewater discharge | | | | | | | |
| Wheat bulgur milling | 38 to 115 | | 238 to 521 | 0.11 | 800 | | 294 to 414 | 0.10 |
| Rice milling normal | No wastewater discharge | | | | | | | |
| Rice milling parboiled | | | 1,280 to 1,305 | 1.8 | 2,810 to 3,271 | | 33 to 77 | 0.07 |
| Animal feed manufacturing | No wastewater discharge | | | | | | | |
| Hot cereal manufacturing | No wastewater discharge | | | | | | | |
| Ready-to-eat cereal manufacturing | 2.50 to 9.59 | | 420 to 2,500 | 2.20 to 18.21 | 804 to 4,434 | 5.71 to 42.40 | 80 to 1,073 | 0.97 to 2.70 |
| Wheat starch gluten manufacturing | | 7.42 to 12.42 | 6,200 to 14,633 | 80.5 to 103.4 | 9,300 to 35,057 | 115.6 to 259.6 | 4,176 to 14,824 | 51.9 to 109.8 |

Source: Development Document for Effluent Limitations Guidelines and New Source Performance Standards for the Beet Sugar Processing Subcategory of the Sugar Processing Point Source Category EPA/1-74-002-b, Washington, D.C. 1974.

Figure 15-20. White Wine Production Diagram.

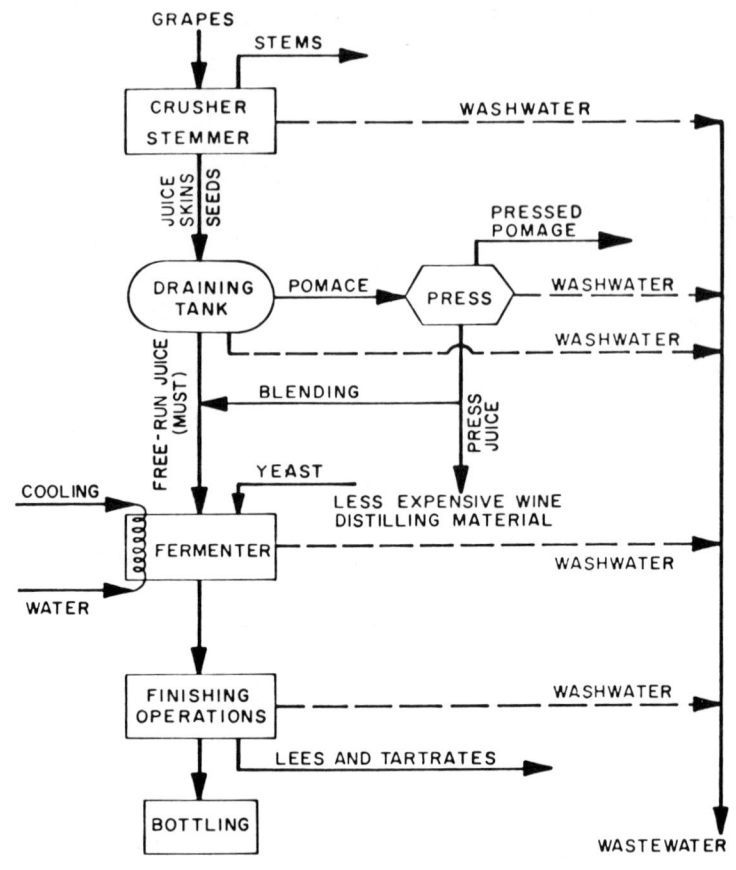

Source: Joyce and others (1977).

Both wine and brandy stillages are high in BOD, $CaCO_3$, ash, $NH_3$-N, and phosphorous and have a low pH (3-4).

*Treatment Process of Solid Wastes.* Pomace has been used for hundreds of years as a soil conditioner and fertilizer, which is a convenient disposal technique and also provides a useful product. Pomace can also be used as a livestock feed supplement. Grape-seed oil is another byproduct that has good potential. Oil recovery from grape seed was practiced in Europe and in one plant in the United States for several years. Waste stems are usually spread on the fields.

**Figure 15-21.** Red Wine Production Diagram.

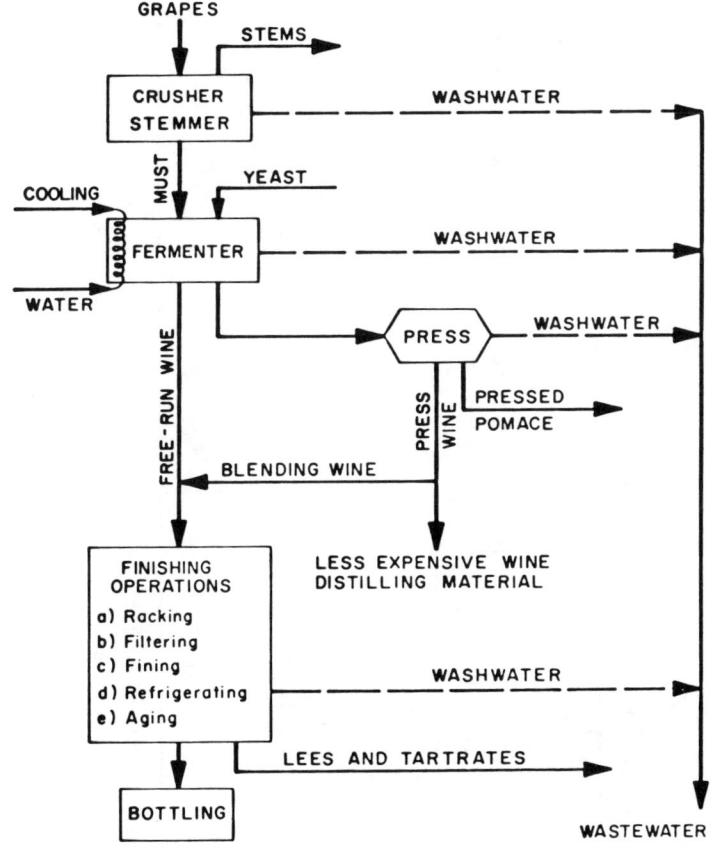

Source: Joyce and others (1977).

Land application of stillage is the most successful method for treating this waste. To be effective the stillage must seep into the soil rapidly; and adequate drying periods between applications must be allowed.

As with most wastewater treatment systems, there are drawbacks to the land application of stillage. The major disadvantage of this method of treatment is the large land area required, ranging from three-quarters to one and a half hectares per 100,000 liters of stillage produced per day. The other serious problem is the pollution of groundwater by the leachate from the stillage.

Table 15-13. Characteristics of Winery Wastewater Sources.

| Characteristic[a] | Units | Crusher Wash[b] | Pomace Conveyor Wash | Fermentation Tank Wash | Press and Area Wash | Storage and Bottle Wash | Storage Tank Floor Wash | Cooling and Refrigeration Blowdown and Miscellaneous |
|---|---|---|---|---|---|---|---|---|
| pH | — | 3.85 | 4.20 | 4.08 | 3.80 | 6.6 | 7.13 | 6.65 |
| Suspended Solids | mg/ℓ | 3,220 | 3,050 | 2,440 | 1,046 | 290 | 108 | 4 |
| $BOD_5$ | mg/ℓ | 27,300 | 4,650 | 8,300 | 1,540 | 1,130 | 2,800 | 373 |
| Portion of Daily Flow | % | 2.5 | 5 | 10 | 7.5 | 50 | 10 | 15 |

a. Data from Christian Brothers, South St. Helena and Greystone Wineries—1965-66.
b. Washdown from Crush, Pomace Conveyor and Press Area occurs during the one hour cleanup period from 4 p.m.-5 p.m.

Source: Ryder (1973).

Table 15-14. Wastewater Characteristics for Nondistilling California Wineries.

| Characteristics | Crushing Season | | Noncrushing Season | |
|---|---|---|---|---|
| | Range | Mean | Range | Mean |
| pH | 3.5-5.5 | 4.1 | | 4.8 |
| Dissolved oxygen | 0 | 0 | 0 | 0 |
| $BOD_5$ | 2000-5000 | 2500 | 2000-5000 | 2400 |
| COD | 4000-10,000 | 5000 | 4000-10,000 | 4000 |
| Grease | 5-30 | 15 | 5-50 | 40 |
| Settleable solids | 25-100 | 80 | 2-10 | 2.5 |
| Suspended solids | 200-800 | 500 | 100-400 | 400 |
| Volatile suspended solids | 150-700 | 450 | 80-350 | 300 |
| Dissolved solids | 300-600 | 800 | 400-800 | 700 |
| Nitrogen | 5-40 | 20 | 10-50 | 40 |
| Phosphorus | 5-10 | 10 | 10-25 | 25 |
| Sodium | 100-200 | 150 | 100 | 140 |
| Alkalinity ($CaCO_3$) | 40-120 | 115 | 10-100 | 50 |
| Chloride | 100-250 | 150 | 100-250 | 150 |
| Sulfate | 20-75 | 50 | 20-75 | 50 |
| Boron | 0-0.2 | 0.1 | 0.2 | 0.1 |

Source: Ryder (1973).
Note: All units are mg/ℓ except pH.

### Livestock Wastes

There are radical variations in waste characteristics among different types of livestock operations. In its simple form, livestock production can be based upon allowing animals to obtain their food supply from fields and discharging their waste directly on the land. With relatively low livestock population densities, little or no waste escapes the land to enter receiving streams and the water pollution potential is very low.

The types of livestock production activities, their organization, and magnitude vary radically among the many countries of the world, but some general types of production are common to many regions.

Table 15-15. Wine Stillage Characteristics.

| Component (mg/ℓ) Except pH or Otherwise Specified | Wine Stillage (44 Wineries) | Average Values Wine Stillage Detartrated (7 Wineries) | Pomace Stillage (8 Wineries) |
|---|---|---|---|
| Alcohol content (% by volume before distilling) | 6.37 | 5.82 | 5.05 |
| pH | 3.74 | 4.34 | 3.72 |
| Acidity (as $CaCO_3$) | 3,700 | 2,300 | 3,800 |
| Total Solids | 16,700 | 13,950 | 29,780 |
| SS | 4,470 | 2,940 | 18,660 |
| Soluble Solids | 12,410 | 10,900 | 13,410 |
| Volatile Solids (Total) | 13,120 | 10,420 | 27,140 |
| Total Ash | 2,900 | 3,490 | 3,440 |
| Soluble Volatile Solids | 8,870 | — | 9,380 |
| Soluble Ash | 2,400 | — | 2,610 |
| Total $BOD_5$ | 12,300 | 9,825 | 17,840 |
| Soluble $BOD_5$ | 9,660 | 7,745 | 11,330 |
| | Wine Stillage | Average Values Lee Stillage | Pomace Stillage |
| pH | 4.7 | 3.8 | 6.8 |
| Acidity (as $CaCO_3$) | 3,170 | 9,860 | 1,220 |
| Total Solids | 20,100 | 68,000 | 13,180 |
| Volatile Solids (% of TS) | 87.4 | 86.5 | 77.0 |
| SS | 3,120 | 59,000 | — |
| Extractable Acids (as acetic) | 1,900 | 2,480 | 380 |
| Total Nitrogen (as N) | 271 | 1,532 | 330 |
| $NH_3$-N (as N) | 2.8 | 45.1 | 4.0 |
| Total Phosphorus (as P) | 11,150 | 4,284 | 1,310 |
| $BOD_5$ | 11,000 | 20,000 | 2,400 |

Sources: Coast Laboratories (1946); Pearson (1955).

INDUSTRIAL WASTES 239

**Figure 15-22.** Brandy and Wine Spirits Production Diagram.

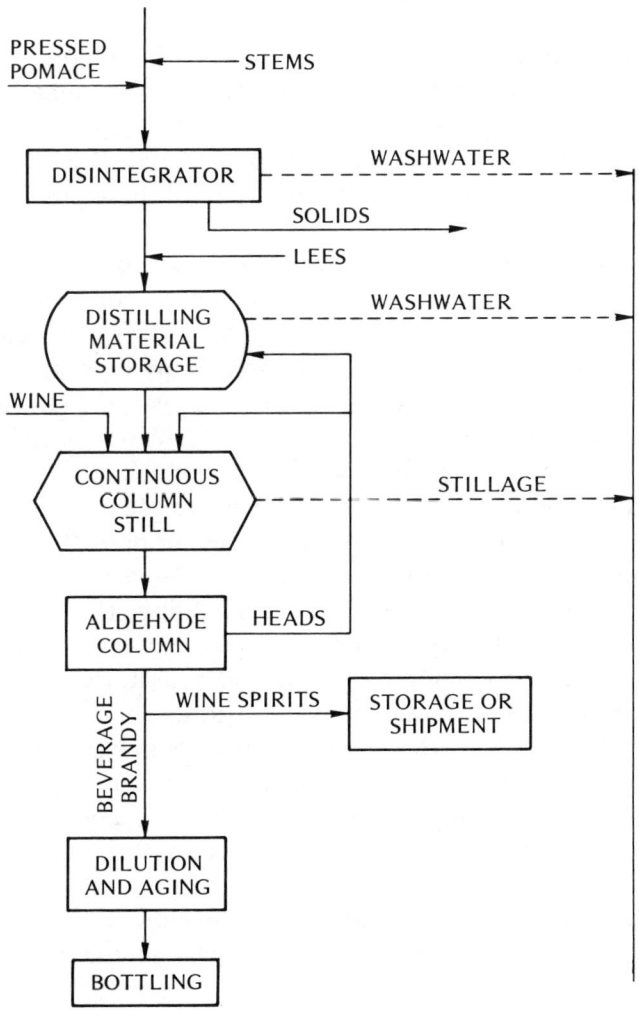

Source: Joyce and others (1977).

The livestock industry in the United States is considered to consist principally of production of four major types of animals: cattle, poultry, swine, and sheep. These animals for human consumption are fattened by grass feeding or by feed concentrates, such as corn and other products with mineral and various components added to the animal's ration to produce a high rate of weight gain. These operations are numerous. They are commonly carried out with animals confined to feed lots. Most direct methods for waste collection include: building with slotted floors, units with concrete slab floors, and open dirt or pasture lots. Rate of manure from major domestic animals are given by Whetstone: beef cattle (1,000 pounds live weight) total solids, pounds per day per animal, 7.2; dairy cattle, 11.2; poultry, 1.12; swine and sheep, 1.12 (all these numbers are for a peak production).

The characteristics of the waste consist of liquid and solid elements. The dairy cattle industry, of course, is concerned with the same general type of animal, but the waste differs in its characteristics from those of the fattening beef cows. Many cow-milking operations employ water flush systems, used for washing down animals before milking and to clean up after milking. Waste production and handling problems arising from swine, sheep, chickens grown as broilers or layers are similar. Most of these facilities use water carriage systems to remove waste from feed lots, cages, or pens. They minimize labor required to clean these areas. Mechanical sweepers are also used to remove the solid waste. Prior to stockpiling, the solid waste is spread out to dry. Runoff water is trapped in retention ponds or dikes to reduce the chance of groundwater pollution, then is sent into shallow lagoons or ponds for evaporation.

A clear picture about the character of livestock waste is difficult to obtain. Differences in waste among the various types of animal may be attributed to ranges in size, diet, and metabolism. Thus one must not assume that data collected for one type of livestock will apply to other species.

The Environmental Protection Agency has divided the animal industry into the following subcategories for characterization:

1. Beef cattle, open lot
2. Beef cattle, house lot
3. Dairy, stall barn
4. Dairy, free stalled barn
5. Dairy, cow yard
6. Swine, solid concrete floor
7. Swine, slotted floor house
8. Swine, open dirt or pasture
9. Chickens, broilers

10. Chickens, layer, breed and replacement
11. Chickens, layers
12. Sheep, housed lot
13. Sheep, open lot
14. Turkeys, housed lot
15. Turkeys, open lot
16. Ducks, dry lot
17. Ducks, wet lot
18. Horses, stables

Characteristic open and housed beef feedlot facilities are given in several references used. Feeding and watering arrangements, manure collection, and cleaning may vary from one location to the other. Figures 15-23 to 15-29 show the animal production, materials utilized, and waste production. Table 15-16 and 15-17 as prepared by Middlebrooks (1979) show waste characteristics from livestock industry.

*Treatment and Disposal of Waste.* The predominant method of utilization or disposal of animal wastes is by land-spreading. The other methods of utilization or disposal of manure have been classified as microbiological, macrobiological, and thermochemical. These methods are described as follows:

1. *Microbiological* consists of anaerobic treatment, anaerobic lagoons, anaerobic digestion, aerobic treatment, aerobic lagoon, oxidation ditches, and composting.
2. *Macrobiological process* includes refeeding of manure to animals.
3. *Thermochemical* is pyrolysis.

All the solid waste can be used as a fertilizer.

### Other Foods

Gouin and Shanks (1981) found that in the production of gelatin, made from animal hides, the resulting sludge and tankage is an oily-gray, smelly material containing 75 percent water after centrifugation with a pH of 9.0. Sulfur is added to this solid waste prior to composting at a ratio of 1:2.5 by volume with mixed wood chips

Figure 15-23. Beef Cattle Industry Structure and Mass Flow Diagram.

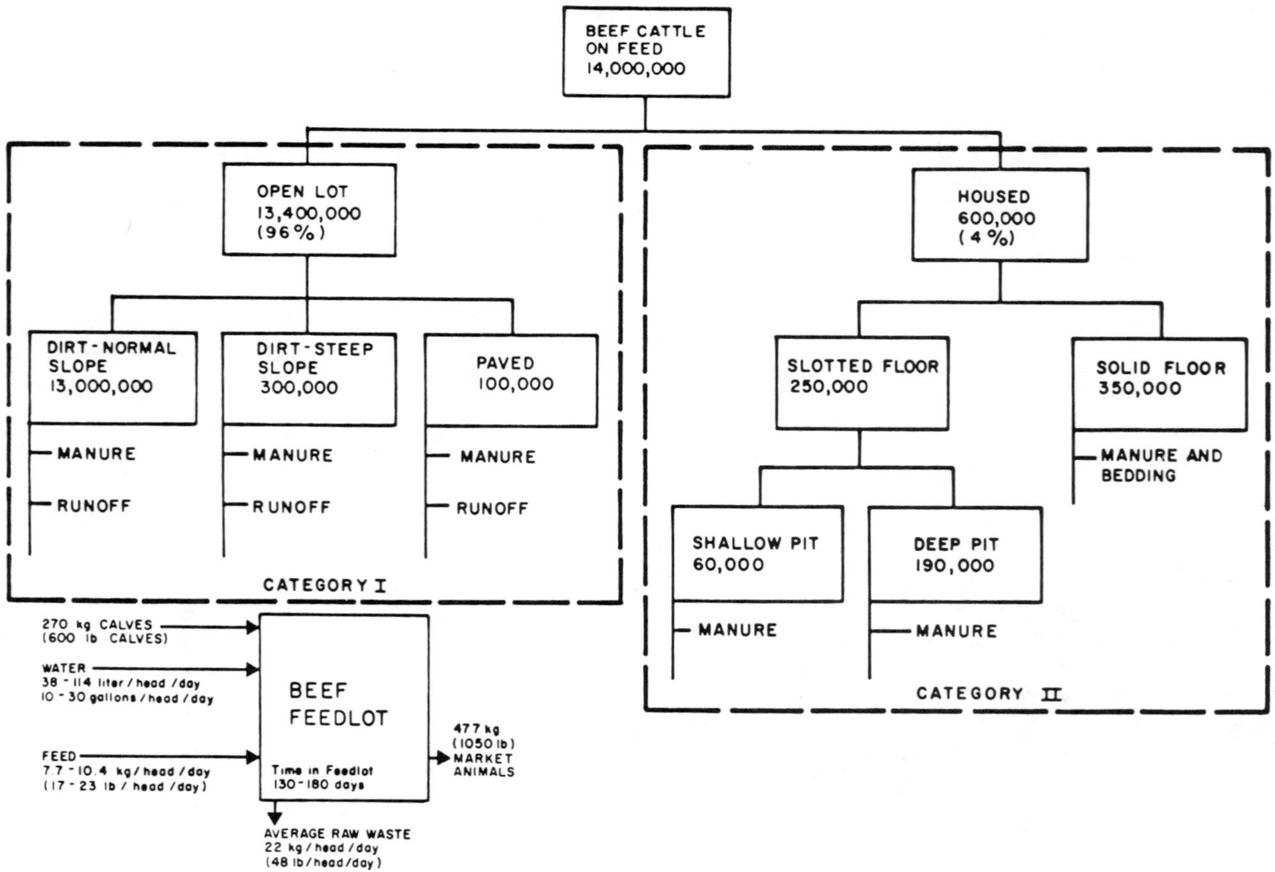

Figure 15-24. Swine Industry Structure—Mass Flow Diagram.

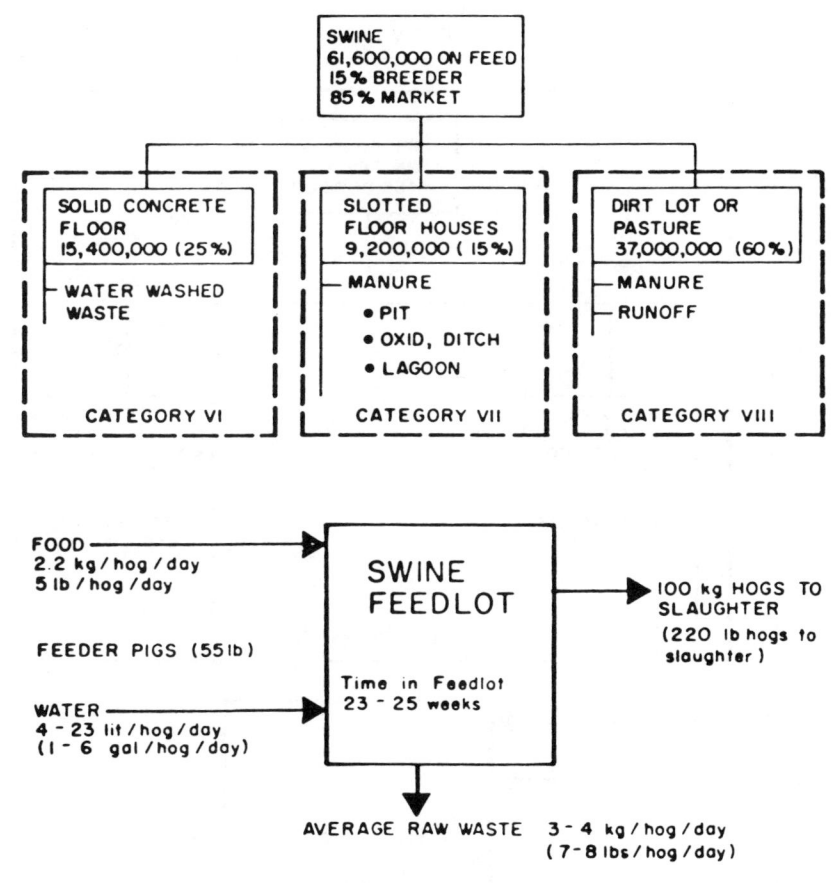

Source: EPA (1974).

from roadside tree chippers. After composting and curing by the Beltsville Aerated Pile Method (Wilson et al. 1980), the resulting compost has an earthy color and is relatively odor-free. The coarse material is removed by passing through a No. 3 screen. The final product contains: nitrogen, 1.2 percent; phosphorous, 155 parts per million (ppm); potassium, 397 ppm; calcium, 11,888 ppm; magnesium, 4,400 ppm; iron, 11,390 ppm; manganese, 210 ppm; zinc, 100 ppm; soluble salts, 5,714 ppm; and a pH of 6.9.

**Figure 15-25.** Broiler Industry Structure — Mass Flow Diagram.

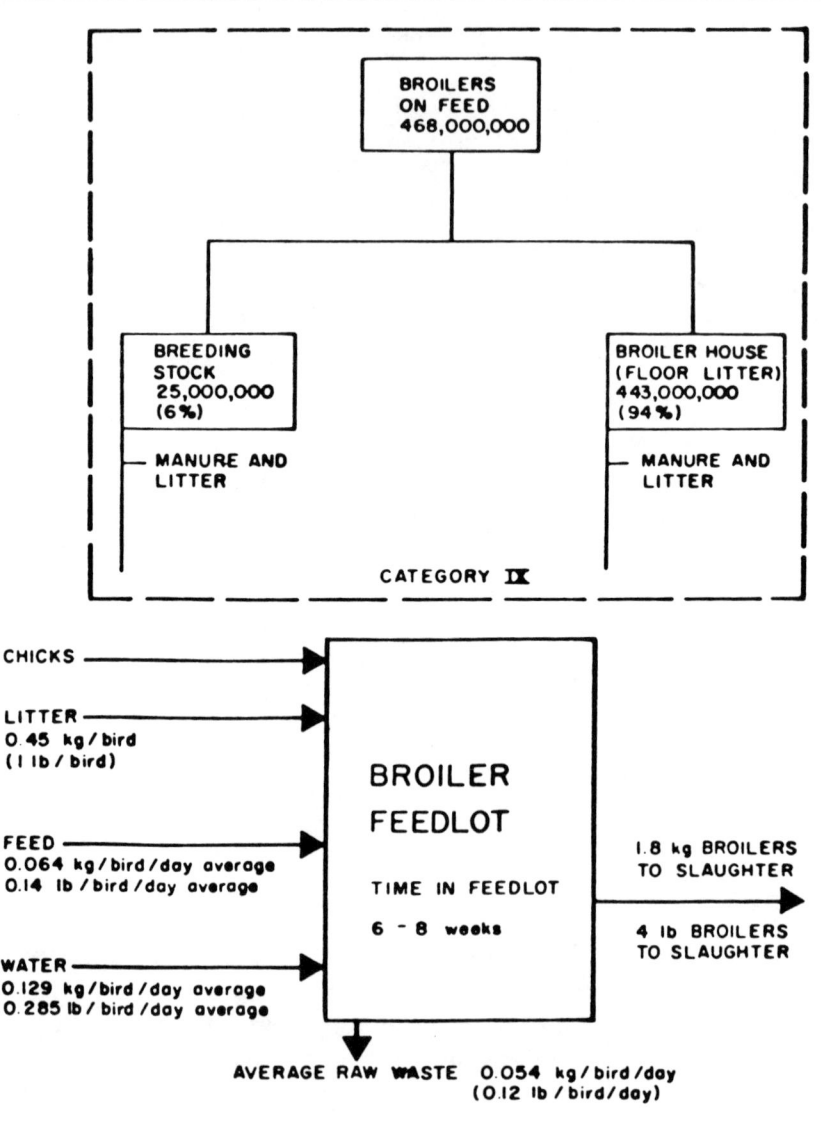

Source: EPA (1974).

INDUSTRIAL WASTES 245

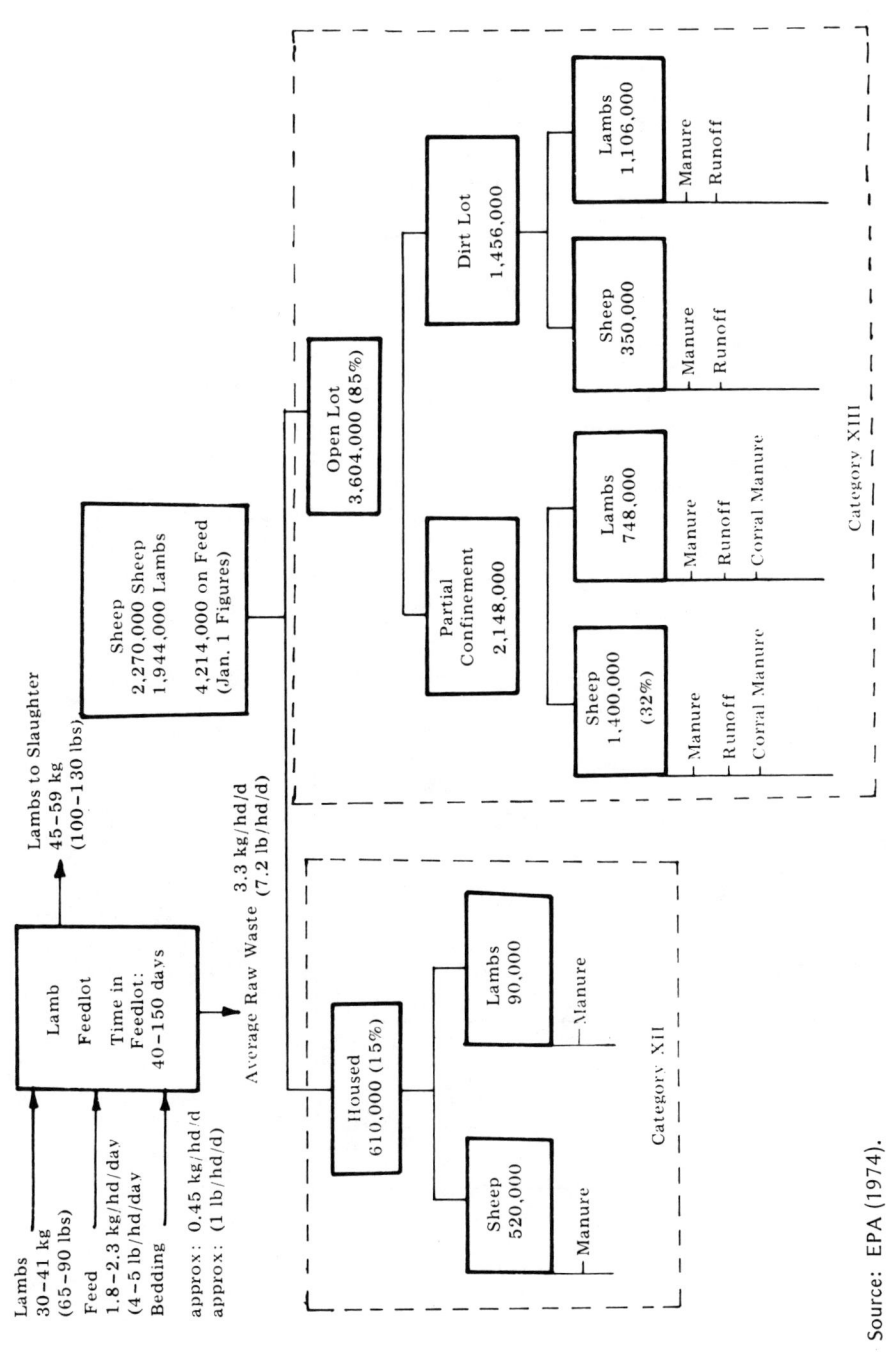

Figure 13-20. Sheep Industry Structure and Mass Flow Diagram.

Source: EPA (1974).

**Figure 15-27.** Turkey Industry Structure and Mass Flow Diagram.

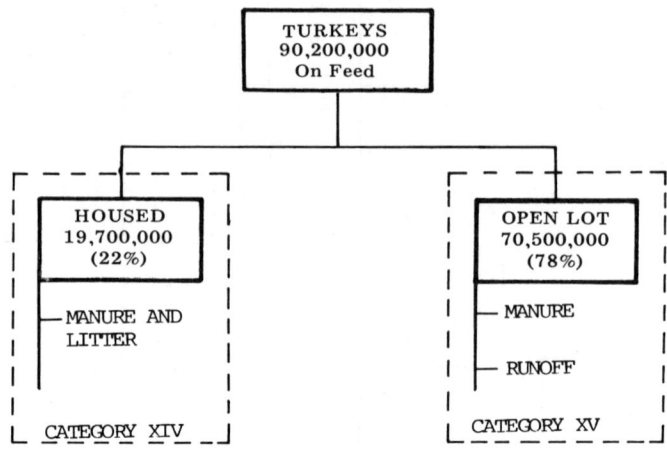

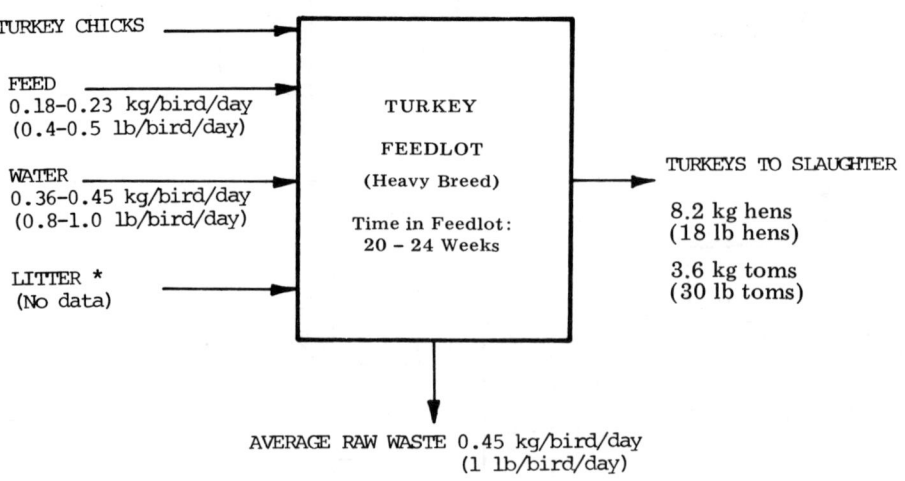

*For housed feedlots only.
Source: EPA (1974).

**Figure 15-28.** Duck Industry Structure and Mass Flow Diagram.

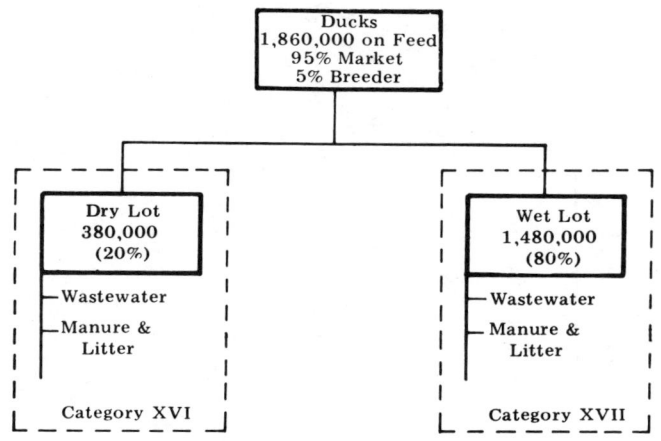

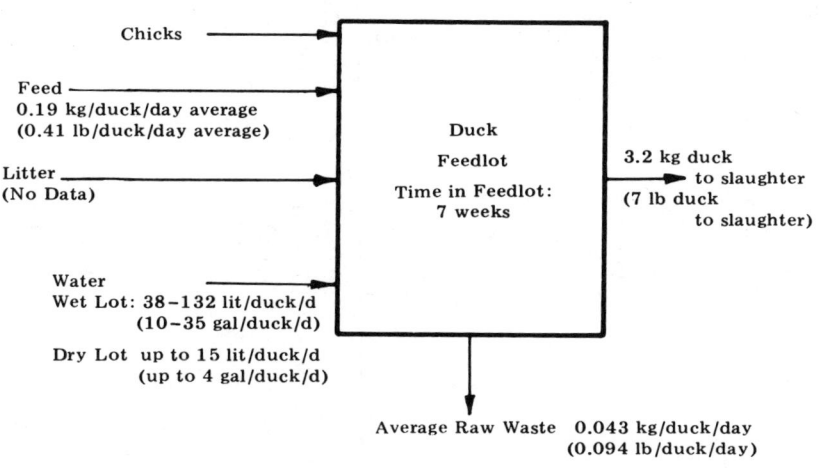

Source: EPA (1974).

Table 15-16. Physical Characteristics of Livestock Manure (from Middlebrooks, 1979).

| Animal | Wet Manure ($gg^{-1}$ of Animal-day) | Total Solids ($gg^{-1}$ of Animal-day) | Volatile Solids ($gg^{-1}$ of Animal-day) | References |
|---|---|---|---|---|
| Poultry | 0.0234 | 0.011 | — | Moore, 1969 |
| | 0.027–0.087 | 0.011–0.022 | 0.0084–0.017 | Taiganides and Hazen, 1966 |
| | 0.074 | 0.021 | — | Hart, 1960 |
| | — | 0.014 | 0.0098 | Dornbush and Anderson, 1964 |
| | — | 0.013 | 0.0101 | Hart and Turner, 1965 |
| | — | 0.013–0.019 | — | Department of Scientific and Industrial Research, 1964 |
| | 0.072 | 0.0086 | 0.0054 | Townshend, et al., 1969 |
| | 0.083 | — | — | Kearl, 1965 |
| Average | 0.062 | 0.014 | 0.0096 | |
| Swine | 0.084 | 0.011 | — | Moore, 1969 |
| | 0.028–0.095 | 0.008–0.016 | 0.0068–0.0136 | Taiganides and Hazen, 1966 |
| | 0.087 | 0.016 | — | Hart, 1960 |
| | — | 0.0080 | 0.0063 | Hart and Turner, 1965 |
| | — | 0.0097 | 0.0080 | Taiganides, et al., 1964 |
| | — | 0.0050 | 0.0035 | Clark, 1965 |
| | — | 0.0071 | — | Department of Scientific and Industrial Research, 1964 |
| | — | 0.0048 | 0.0033 | Humenik, 1972 |
| | 0.074 | 0.0059 | 0.0047 | Schmid and Lipper, 1969 |
| | — | 0.0099 | 0.0070 | Townshend, et al., 1969 |
| Recommended Value | 0.074 | 0.0089 | 0.0054 | |

| | | | | |
|---|---|---|---|---|
| Cattle | (Dairy) | 0.071 | 0.0114 | — | Moore, 1969 |
| | (Dairy) | 0.058 | 0.0087 | — | Hart, 1960 |
| | (Dairy) | — | 0.0104 | 0.0083 | Hart and Turner, 1965 |
| | (Dairy) | — | 0.0068 | 0.0057 | Witzel, et al., 1966 |
| | (Dairy) | — | 0.0075 | — | Department of Scientific and Industrial Research, 1964 |
| | (Dairy) | 0.124 | 0.0025 | 0.0018 | Townshend, et al., 1969 |
| | (Beef) | 0.082 | 0.0197 | — | Moore, 1969 |
| | (Beef) | 0.039–0.074 | 0.0095–0.0114 | — | Taiganides and Hazen, 1966; Taiganides, 1964 |
| | (Beef) | 0.063 | 0.0095 | — | Hart, 1960 |
| | (Beef) | 0.067 | 0.0090 | 0.0069 | Loehr and Agnew, 1967 |
| | (Beef) | — | 0.0036 | 0.0032 | Witzel, et al., 1966 |
| | (Beef) | 0.063 | 0.0050 | 0.0040 | Townshend, et al., 1969 |
| | (Beef) | — | 0.0091 | — | Dale and Day, 1967 |
| Average | (D) | 0.084 | 0.0079 | 0.0053 | |
| Average | (B) | 0.066 | 0.0095 | 0.0047 | |
| Sheep | | 0.072 | 0.016 | — | Hart, 1960 |
| Ducks | | — | 0.016 | — | FWPCA, 1966 |

**Figure 15-29.** Horse Industry Structure and Mass Flow Diagram.

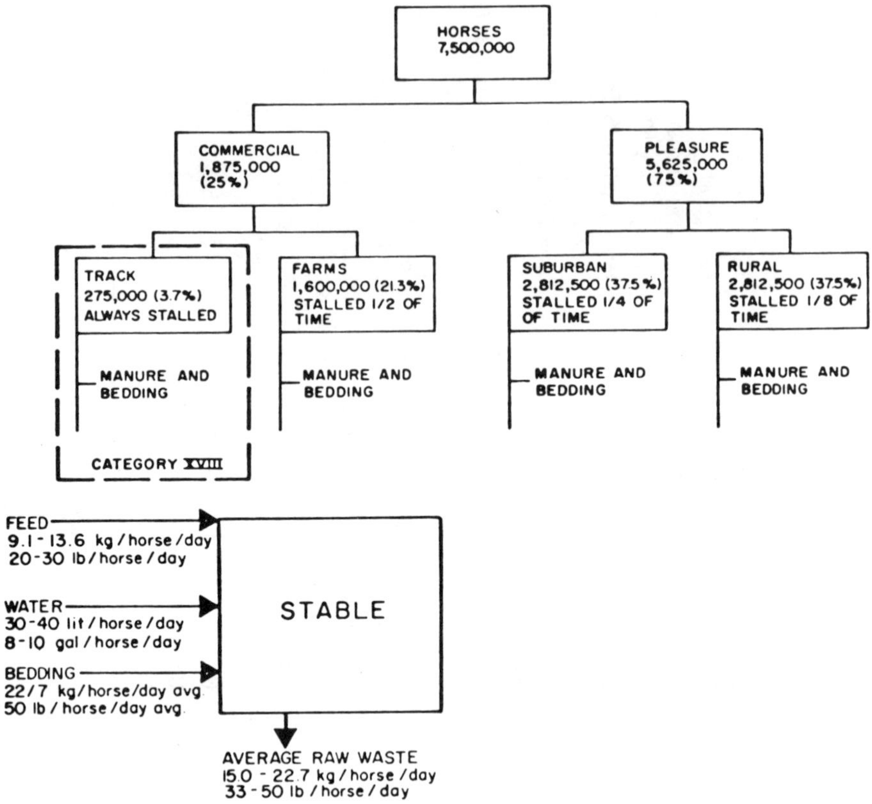

```
                          ┌─────────────────┐
                          │     HORSES      │
                          │    7,500,000    │
                          └────────┬────────┘
                   ┌───────────────┴───────────────┐
           ┌───────┴────────┐              ┌───────┴────────┐
           │   COMMERCIAL   │              │    PLEASURE    │
           │   1,875,000    │              │   5,625,000    │
           │     (25%)      │              │     (75%)      │
           └───────┬────────┘              └───────┬────────┘
          ┌────────┴────────┐             ┌────────┴────────┐
   ┌──────┴──────┐   ┌──────┴──────┐ ┌────┴─────┐   ┌───────┴──────┐
   │    TRACK    │   │    FARMS    │ │ SUBURBAN │   │    RURAL     │
   │275,000(3.7%)│   │1,600,000    │ │2,812,500 │   │2,812,500     │
   │ALWAYS       │   │(21.3%)      │ │(37.5%)   │   │(37.5%)       │
   │STALLED      │   │STALLED 1/2  │ │STALLED   │   │STALLED 1/8   │
   │             │   │OF TIME      │ │1/4 OF    │   │OF TIME       │
   │             │   │             │ │TIME      │   │              │
   └──────┬──────┘   └──────┬──────┘ └────┬─────┘   └───────┬──────┘
   │ MANURE AND  │   │ MANURE AND  │ │MANURE AND│   │ MANURE AND   │
   │ BEDDING     │   │ BEDDING     │ │BEDDING   │   │ BEDDING      │
   │             │
   │ CATEGORY XVIII│
```

FEED ────▶
9.1 - 13.6 kg/horse/day
20 - 30 lb/horse/day

WATER ────▶    ┌──────────┐
30 - 40 lit/horse/day   │  STABLE  │
8 - 10 gal/horse/day    │          │
                        └────┬─────┘
BEDDING ────▶                │
22/7 kg/horse/day avg        ▼
50 lb/horse/day avg   AVERAGE RAW WASTE
                      15.0 - 22.7 kg/horse/day
                      33 - 50 lb/horse/day

Source: EPA (1974).

The authors recommended the use of this gelatin compost in blending media for growing poinsettias and chrysanthamums and most annual bedding plants.

## POULTRY

A general flow diagram for the poultry industry is shown in Figure 15-30. Live poultry are held for only a short period (a few hours) prior to slaughter and processing. Very little feeding or fattening of the poultry is practiced at processing plants; however, a certain

Table 15-17. Characteristics of Runoff from Cattle Feedlots.

| | Range of Values for Constituents ($mg\ \ell^{-1}$) | | | | | | | |
|---|---|---|---|---|---|---|---|---|
| Suspended Solids | Orthophosphate ($PO_4$) | Organic Nitrogen | Ammonia Nitrogen | Nitrate Nitrogen | BOD | COD | References |
| 3400–13,400 | — | — | — | — | 500–3300 | — | Owens and Griffin, 1968 |
| — | — | 6–800 | 2–770 | 0–1270 | 1000–12,000 | 2400–38,000 | Wells, et al., 1970 |
| 1000–7000[a] | | | | | 300–6000 | | Norton and Hansen, 1969 |
| — | 15–80 | — | — | — | 1500–9000 | 4000–15,000 | Loehr, 1969 |
| 1400–12,000 | | — | 1–139 | 0.1–11 | — | 2500–15,000 | Miner, et al., 1966 |
| — | 20–30 | 600–630 | 270–410 | — | 5000–11,000 | 16,000–40,000 | Loehr, 1969 |
| 1500–12,000 | — | — | 16–140 | — | — | 3000–11,000 | Miner, 1967 |
| 1400–12,000 | 62–1460[b] | 265–3400 | — | — | 800–7500 | — | Townshend, et al., 1970 |

a. Volatile solids.
b. Total phosphorus as $PO_4$.
Source: Middlebrooks (1979).

**Figure 15-30.** Flow Chart of Poultry Processing Plant.

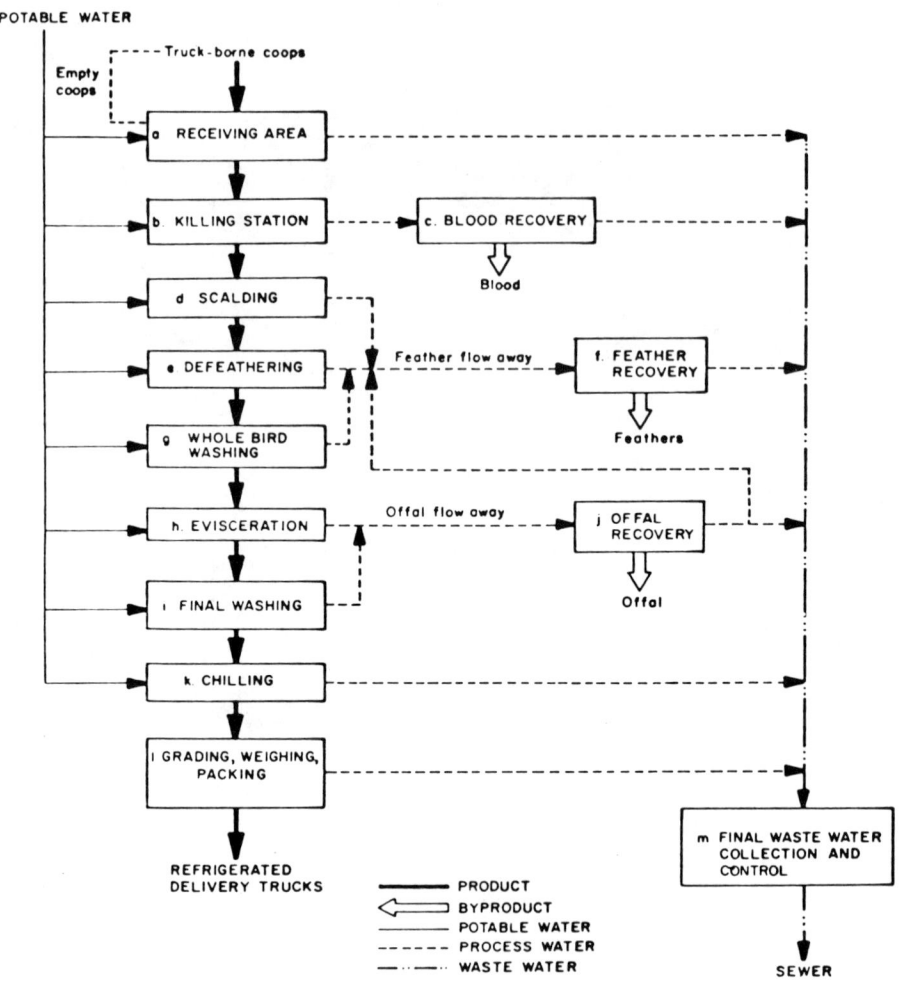

Source: EPA (1973).

amount of waste material is deposited in the receiving area of the poultry-processing plant. Birds are usually fastened by their feet to a moving conveyor which transports them to a room where they are slaughtered by cutting the jugular vein. Blood is usually trapped and recovered. After the blood is drained from the carcass, the feathers are removed by scalding with hot water. Feathers are removed from the carcass by a mechanical process, and the feathers are washed from the bird and transferred to a collection site. Pin feathers are removed by manual labor or by dipping in wax. A gas flame is also used to singe hair and odd pin feathers. The poultry is eviscerated in an area separated from the remainder of the plant to prevent contamination. Poultry to be prepared for distribution as parts are processed immediately following evisceration and washing.

Poultry waste contain blood, grease, feet, heads, sand, inedible viscera, feathers, and offal. The characteristics and concentration of poultry processing is listed in Table 15-18. The defeathering process has a pollution potential of great magnitude because of the volume of feathers involved. One thousand birds will yield approximately seventy pounds of feathers.

The utilization of byproducts has proven economical to processors. Dry or wet feathers may be rendered into proteinaceous animal feed; collected offal may be converted to animal feed; and recovered blood may be used to produce animal feed. Therefore, all the solid waste from the processing of poultry can be used as animal feed. If there is some waste, like sand and nonreusables, they can be incinerated or buried.

The manure and spilled feed and feathers from the receiving area should be handled in a dry fashion by sweeping and trucking away. This material can be disposed of as fertilizer either directly to the field or to fertilizer manufacturers.

## PLASTIC PLANTS

The U.S. production of plastic materials (resins) during 1970 has been estimated at 9-35 million tons. This quantity is increasing year by year and is expected to increase to 6 percent in a decade, that is, in the year 1990. The main plastic products are polyethylene, polyvinyl chloride (P.V.C.), polystyrene and polyurethane. In the industrial production, plastic products commonly represents a potential fire hazard. When heated, they burn and emit acid fumes. They may react with oxidizing materials. Polyethylene is odorless. The high molecular weight materials are tough, white, leathery, resinous material. Densities and heat values of prevalent plastics are given in Table 15-19.

While plastic industrial solid wastes were estimated to be less than 12 percent of the total of all plastic solid waste generated in 1969,

Table 15-18. Coefficients Used in Estimating Byproducts, Water Use, and Waste Loads of Poultry-Slaughtering Plants.

| Variable | Unit | Value per 1,000 Pounds[a] |
|---|---|---|
| Byproducts | | |
| Blood: | | |
|   Young chickens | Pounds | 70 |
|   Mature chickens | Pounds | 70 |
|   Turkeys | Pounds | 70 |
|   Other poultry | Pounds | 70 |
| Offal: | | |
|   Young chickens | Pounds | 175 |
|   Mature chickens | Pounds | 170 |
|   Turkeys | Pounds | 125 |
|   Other poultry | Pounds | 140 |
| Feathers: | | |
|   Young chickens | Pounds | 70 |
|   Mature chickens | Pounds | 70 |
|   Turkeys | Pounds | 70 |
| Water Use | | |
|   Young chickens | Gallons | 2,198 |
|   Mature chickens | Gallons | 2,173 |
|   Turkeys | Gallons | 1,700 |
|   Other poultry | Gallons | 2,100 |
|   Cut-up | Gallons | 500 |
|   Further processing | Gallons | 500 |
| Wasteloads | | |
| BOD: | | |
|   Young chickens | Pounds | 8.2 |
|   Mature chickens | Pounds | 8.7 |
|   Turkeys | Pounds | 8.0 |
|   Other poultry | Pounds | 8.0 |
| Suspended solids: | | |
|   Young chickens | Pounds | 6.3 |
|   Mature chickens | Pounds | 5.4 |
|   Turkeys | Pounds | 5.0 |
|   Other poultry | Pounds | 5.0 |
| Timespan of Operation[b] | | |
|   Young chicken, mature chicken, and other poultry plants | Days | 234 |
|   Turkey plants | Days | 130 |

## Notes to Table 15-18

a. Live weight except for cut-up and further processed coefficients which are ready-to-cook weight.

b. These coefficients are based on a maximum of 260 operating days per year. It was assumed that the chicken and other poultry plants operated at 90 percent capacity—0.90 × 260 = 234. Turkey plants were assumed to operate at 50 percent capacity—0.50 × 260 = 130.

Sources: EPA (1971, 1973); U.S. Department of Agriculture (1957); and industry contacts.

Table 15-19. Densities and Calorific Values of Typical Plastics.

| Type of Plastic | Density (grams/cc) | Calorific Value | |
|---|---|---|---|
| | | (Btu/#) | (K Cal/kg) |
| Polyvinylchloride | 1.313 | 9500 | 5250 |
| Polystyrene | 1.055 | 19,500 | 10,800 |
| High-density Polyethylene | 0.958 | 18,000 | 10,000 |
| Low-density | 0.916 | | |
| Polypropylene | 0.901 | | |
| Polymethylmethacrylate | — | 14,000 | 7,750 |
| Polyamides | — | 13-15,000 | 7200-8200 |
| Polyesters | — | 13,000 | 7200 |

estimated at less than 500,000 tons—some plastic-producing companies may want to dispose of their own wastes. Incinerators built to consume these relatively small quantities may be difficult to justify economically unless there are other types of waste to be disposed of concurrently. When some plastics are incinerated, they emit potentially toxic gases in the form of smoke. When these same plastics are buried in landfills, they resist degradation and hence persist in the soil for extensive periods of time. An environmentally sound method of ultimate disposal of plastics defies a satisfactory solution at the present time.

The nonwaste quality aspects of the treatment technologies found in the synthesis and plastic industries are related to: (1) the disposal of solids or slurries resulting from waste, water treatment, and in process-plant control methods; (2) the generation of a byproduct of commercial value; (3) disposal of scrap products; and (4) the creation of problems of air pollution and land utilization. For complete treatment of small volumes of wastes, these plants use liquid waste incineration for removing phenol compounds from epoxy, acrylics, and phenolic wastes.

### Solids and Slurries

Biological sludges are the principal disposal problems resulting from the end of treatment of waste waters. Of occasional concern is chemical sludge, such as from neutralization and precipitation of an inorganic chemical. Biological sludges are most frequently subjected to degradation. When lagoons are operated in the extended aeration mode, the solids accumulate in these lagoons or in polishing lagoons. The long-term consequence of these operations is a gradual filling of the lagoons, at which point they must be degraded or abandoned.

### Plastic Recycling

Homogeneous plastic scrap is recycled by plastic processors; however, hetergenous plastic mixtures cannot be recycled in the same way. Because plastics may deteriorate during recycling, a stabilizer is usually added. Table 15-20 reports the quantity of plastic wastes in the United States.

Table 15-20. The Quantity of Plastic Wastes Generated in the United States.

| | Amount of Plastic Wastes $(10^6 \, T/Year)$ | | |
|---|---|---|---|
| *Type of Plastic Waste* | *1971* | *1975* | *1980* |
| Waste containing a single type of plastic | 0.2 | 0.3 | 0.6 |
| Waste containing several types of plastic | 0.3 | 0.7 | 1.4 |
| Waste containing plastic and nonplastic refuse | 1.5 | 3.0 | 6.0 |
| Total | 2.0 | 4.0 | 8.0 |

A New York City research report (1971) recommended incineration of its refuse with 5 percent or more of plastics. The following conclusions were reached:

1. During the burning of the plastics, there was no problem as the temperature was controlled.
2. No smoke increase occurred.
3. No plastics melted or dripped through the grates.
4. Hydrogen chloride was noted during the burning of polyvinyl chloride.
5. The chloride ion concentration was 511 parts per million in the flue gas.
6. Polyethylene and polystyrene had no effect on chlorine emission because they contain no chlorine.
7. Anti-air-pollution devices were not provided.
8. Obviously, if all plastics are removed from incinerator refuse, materials remaining would contain a substantial amount of chlorides and other corrosive and polluting agents.
9. It is also anticipated that the 6 percent level of plastics waste may be reached in a decade (1990).

Despite this study, some difficulties with plastic material incineration can develop. They can pre-ignite outside the combustion chamber or can upset the combustion conditions within the chamber. Conversely, flame-resistant varieties burn with difficulty, and large quantities of black smoke can be liberated. Finally, combustion of PVC results in the liberation of hydrogen chloride gas, an air pollutant which is also corrosive within the incinerators. The even distribution of plastics in municipal refuse is not easily achieved.

Thus recycling and re-use of plastics is preferred, and pyrolysis of residual waste plastics may be selected rather than incineration.

The author has found that plastics in municipal refuse depresses gas production during anaerobic digestion.

## TEXTILE PLANTS

### Introduction

A study by EPA's Office of Solid Waste Management (1976) estimated that the textile industry generated 1,760,000 metric tons of solid wastes (wet basis) in 1974. These wastes were comprised of: innocuous process-related materials such as dirt, vegetable matter, fiber, flock, yarn, fabric, and so on (8.5 percent); potentially hazardous dye and chemical containers with residual dyestuff and chemicals (0.5 percent); and potentially hazardous waste water treatment sludges (91 percent). The study projected that in 1977 the industry would generate 1,940,000 metric tons of solid wastes with essentially the same percent distribution. The sludges contain heavy metals, including arsenic, cadmium, chromium, cobalt, copper, lead, mercury, nickel, and zinc, and chlorinated organics such as trichlorobenzene, polyvinyl chloride, and perchloroethylene. Current sludge management practices in the industry include: storing or retaining sludge in disposal ponds or in the bottom of ponds or lagoons that are used for aeration; dumping sludge on land generally off the plant site; spreading sludge on land for fertilizer value, again generally off the plant site; and sending sludge to general-purpose landfills.

The textile industry is comprised of over 5,000 textile plants in the United States. In 1974 it was estimated that the textile industry produced 310,175 and 2,098,575 thousands of kilograms per year of dry and wet waste, respectively. The corresponding figures for 1977 and the projections for 1983 are: 336,274 and 2,221,399, and 533,602 and 1,120,759 kkg per year.

A synoptic breakdown in Table 15-21 per category of textile industry process is given in Table 15-22, and a geographical distribution across the United States is given in Figure 15-31 ("Assessment of Industrial Hazardous Waste Practices, Textiles Industry 1979).

The textile industry produced approximately $35 billion worth of fabrics (as measured by plant shipments) for various uses in 1974.

Table 15-21. Potentially Hazardous Waste (*in 1000's of kilograms per year*).

| Year | Total Potentially Hazardous | Total Hazardous Constituents (Dry) |
|---|---|---|
| 1974 | 48,400/1,770,000 | 1,020 |
| 1977 | 49,700/1,870,000 | 1,080 |
| 1985 | 179,000/716,800 | 9,360 |

(1000's kilograms per year).

| Industry Category | Total Wastes (Dry/Wet) | | | Total Potentially Hazardous Wastes (Dry/Wet) | | | Total Hazardous Constituents (Dry) | | | Specific Hazardous Constituents |
|---|---|---|---|---|---|---|---|---|---|---|
| | 1974 | 1977 | 1983 | 1974 | 1977 | 1983 | 1974 | 1977 | 1983 | |
| A—Wool Scouring | 32,000/ 261,600 | 32,000/ 261,600 | 20,900/ 63,800 | 25,500/ 255,000 | 25,500/ 255,000 | 14,300/ 57,200 | 134 | 134 | 76 | heavy metals,[a] chlorinated organics[b] |
| B—Wool Fabric Dyeing and Finishing | 19,438/ 43,533 | 19,438/ 43,588 | 46,488/ 150,958 | 895/ 1,720 | 895/ 1,720 | 27,900/ 111,600 | 7.6 | 7.6 | 2,040 | heavy metals, dyestuff[c] and chemicals[c] |
| C—Greige Goods | 159,000/ 159,000 | 174,000/ 174,000 | 207,000/ 207,000 | 0 | 0 | 0 | 0 | 0 | 0 | none |
| D—Woven Fabric Dyeing and Finishing | 35,616/ 1,522,477 | 37,702/ 1,618,203 | 77,224/ 227,070 | 15,300/ 1,500,000 | 16,200/ 1,600,000 | 51,400/ 205,600 | 842 | 892 | 2,980 | heavy metals, chlorinated organics, dyestuff and chemicals |
| E—Knit Fabric Dyeing and Finishing | 10,448/ 13,239 | 11,073/ 14,065 | 50,002/ 162,272 | 1,400/ 2,590 | 1,490/ 2,760 | 38,500/ 154,000 | 3.4 | 3.7 | 2,020 | heavy metals, chlorinated organics, dyestuff and chemicals |
| F—Carpet Dyeing and Finishing | 23,539/ 27,359 | 30,061/ 34,344 | 67,849/ 116,522 | 210/ 1,170 | 263/ 1,470 | 14,600/ 58,400 | 1.0 | 1.3 | 817 | heavy metals, chlorinated organics, dyestuff and chemicals |
| G—Yarn and Stock Dyeing and Finishing | 30,132/ 71,367 | 32,000/ 75,599 | 64,139/ 193,137 | 5,080/ 6,340 | 5,400/ 6,740 | 32,500/ 130,000 | 36.5 | 38.7 | 1,430 | heavy metals, chlorinated organics, dyestuff and chemicals |
| Total textiles industry | 310,173/ 2,098,575 | 336,274/ 1,221,399 | 533,602/ 1,120,759 | 48,400/ 1,770,000 | 49,700/ 1,870,000 | 179,000/ 716,800 | 1,020 | 1,080 | 9,360 | see above |

a. Includes arsenic, barium, cadmium, chromium, cobalt, copper, iron, lead, manganese, mercury, nickel and zinc.
b. Individual chlorinated organic compounds were not identified in the laboratory, only total quantities.
c. See Section 3.2 of this report for explanation of types of dyestuff and chemicals.

Source: EPA/PB-258-953 June 1976.

Figure 15-31. Estimated Quantities of Total Waste to Land Disposal, 1974 (*Dry/Wet Weight*).

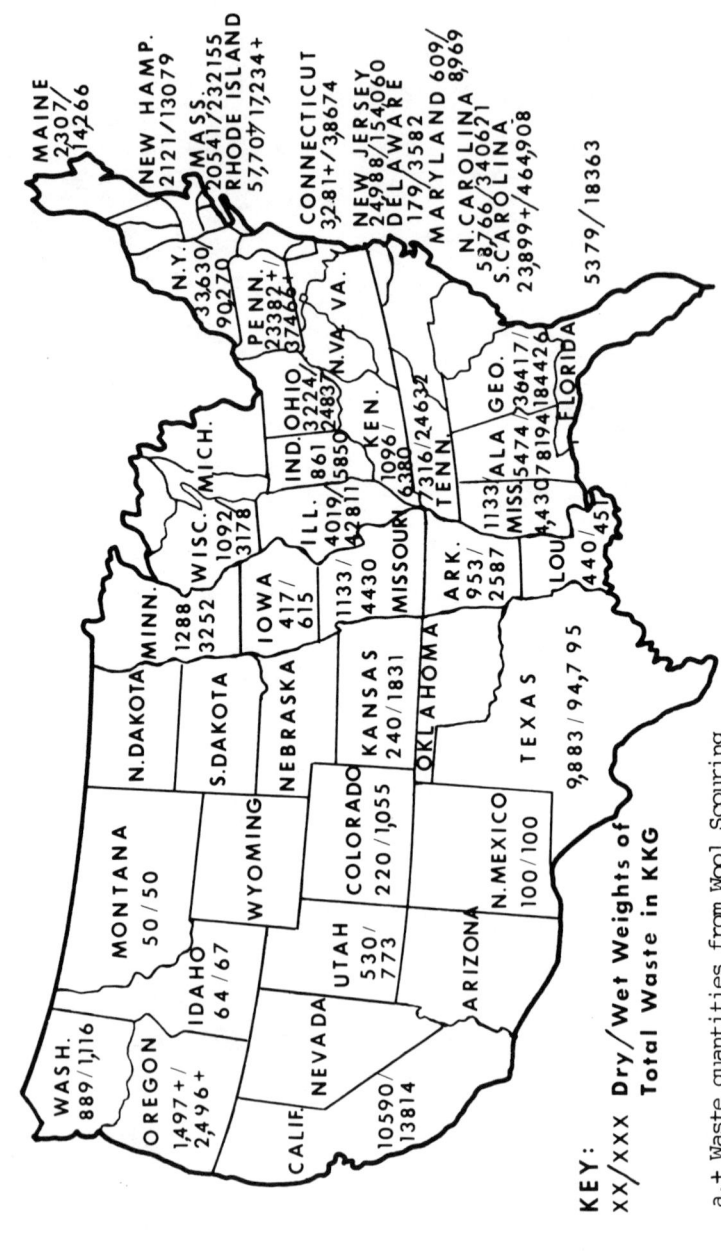

The major uses were: apparel (38 percent); home furnishings (31 percent); other consumer products (11 percent); industrial products (17 percent); and exports (3 percent).

The industry has been categorized, by process, as follows:

1. Wool scouring (Figure 15-32), sludge character (Table 15-23).
2. Wool fabric dyeing and finishing (Figure 15-33), sludge character (Table 15-24).
3. Greige goods (Figure 15-34).
4. Woven fabric dyeing and finishing (Figure 15-35), sludge character (Table 15-25).
5. Knit fabric dyeing and finishing (Figure 15-36), sludge character (Table 15-26).
6. Carpet mills (Figure 15-37), sludge character (Table 15-27).
7. Yarn and stock dyeing and finishing (Figure 15-38), (Table 15-28).

### Origin

The textile industry comprises of a variety of plants that each manufacture different types of textile products. In general, the process of manufacturing involves the addition of various kinds of dyes and chemicals, and one of the major components of solid wastes in the textile industry arises out of the empty containers for these dyes and chemicals. The other type of solid waste that is often found in a textile industry is the waste cotton, wool, rayon, and other fibers produced in the initial operations of spinning and weaving. In addition to these solid wastes which are obtained from the manufacturing processes, solid waste in the form of sludge produced at the different effluent treatment units in the textile mill also contribute to the total volume of solid waste. In the finishing operations of a textile mill, a large volume of lint cuttings of processed textile fiber and cloth pieces are produced which add to the volume of solid waste,

Potentially hazardous waste streams include wastewater treatment sludge, dye containers, and chemical containers.

A discarded dye container carries approximately 28 to 56 grams of residual dyestuff to the disposal site, which is often a county or municipal landfill. The residual dyestuff which ended up in a landfill for 1974 amounted to 11.7 kkg and was projected to increase to

262  INDUSTRIAL SOLID WASTES

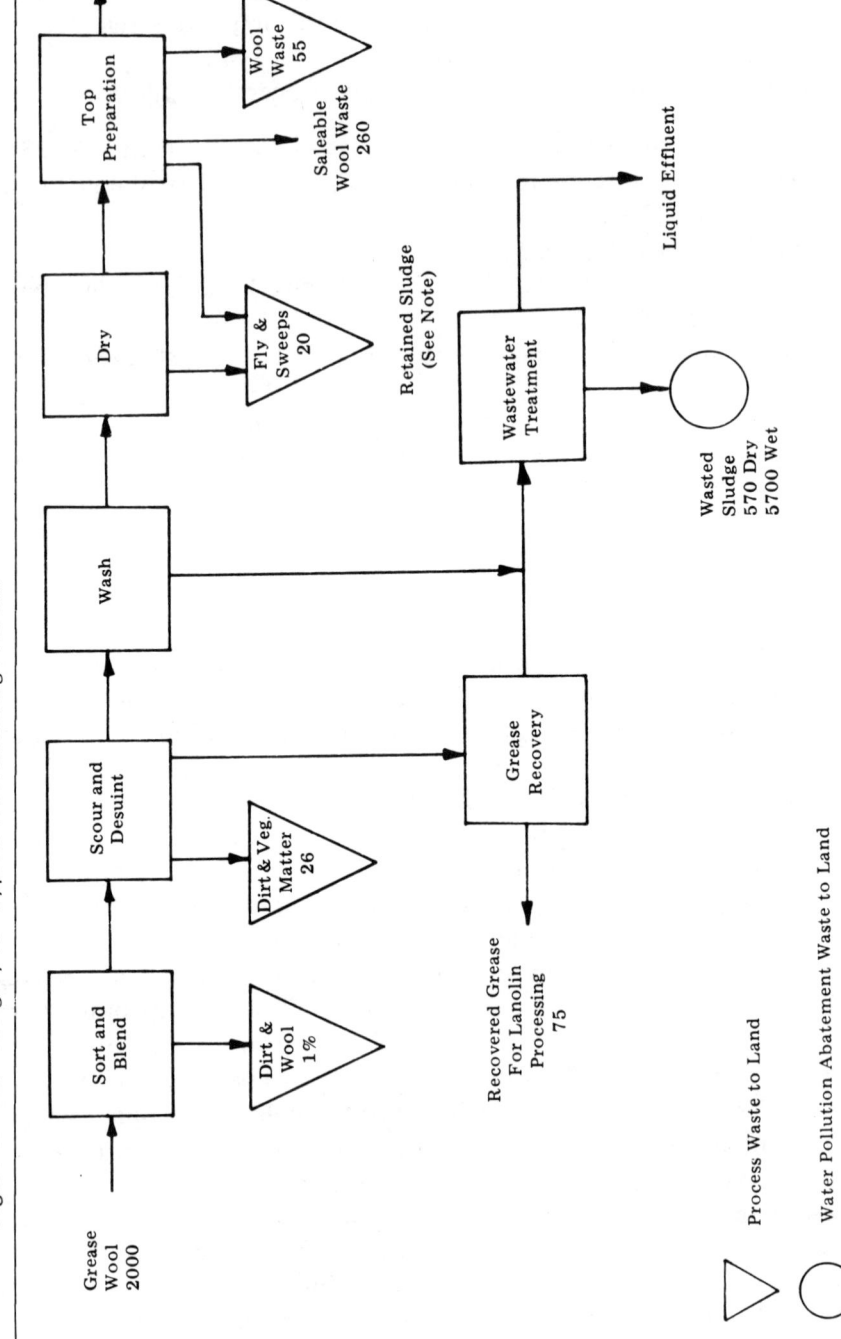

Figure 15-32. Category A—Typical Wool Scouring Process.

| Waste | Source | Quantity (Kg of waste per kKg of product) |
|---|---|---|
| Dirt and wool | Sorting and blending | 12 |
| Dirt and Veg. Matter | Scouring | 26 |
| Fly and sweeps | Drying, top preparation | 2 |
| Wool Waste | Top preparation | 55 |
| Wasted Sludge | Wastewater treatment | 570 (Dry) and 5700 (Wet) |
| Retained Sludge | Wastewater treatment | 780 (Dry) and 7800 (Wet) |

Potential Hazardous constituents: Total Heavy Metals—3.0
Total Chlorinated Organics—730 × 10⁶

Note 1. An average of 780 Kg (Dry) and 7800 Kg (Wet) sludge is retained containing 64.1 Kg total heavy metals, and $1.0 \times 10^3$ total chlorinated organics (Not included in mass balance).

Note 2.

a. Potential hazardous wastestream to land.
b. The retained sludge quantity is an accumulation over the life of the pond.

Source: EPA (1979).

Table 15-23. Category A—Wool Scouring Sludge Analyses (*milligrams per kilogram of dry sludge*).

| Parameter | Drinking Water Limit[a] (ppm) | Average[b] |
|---|---|---|
| Arsenic | 0.05 | < 0.1[c] |
| Barium | 1.0 | 59 |
| Cadmium | 0.01 | 1.2 |
| Chromium | 0.05 | 19 |
| Cobalt | d | 4.2 |
| Copper | 1.0 | 18 |
| Iron | 0.3 | 4,820 |
| Lead | 0.05 | 28 |
| Manganese | 0.05 | 205 |
| Mercury | 0.002 | < 0.01 |
| Molybdenum | d | < 2 |
| Nickel | d | 12.5 |
| Zinc | 5.0 | 106 |
| Total heavy metals | | |
| Aluminum | d | 4,860 |
| Magnesium | 60.0 | 5,560 |
| Potassium | d | 9,240 |
| Sodium | d | 675 |
| Strontium | d | 21.6 |
| Total chlorinated organics | 0.7 | 1.28 |
| Suspended Solids (%) | d | 9.8 |
| Total Solids (%) | d | 10.1 |

a. Average of four measurements from one plant.
b. Less-than values were considered to be at the maximum in computing the totals.
c. U.S. Public Health Service's Drinking Water Standards (1962).
d. No drinking water standards have been set for these metals.
Source: EPA/PB-258-953 June 1976.

Table 15-24. Category B—Wool Fabric Dyeing and Finishing Sludge Analyses (*milligrams per kilogram of dry sludge*).

| Parameter | Drinking Water Limit[a] (ppm) | Average[b] |
|---|---|---|
| Arsenic | 0.05 | < 17[c] |
| Barium | 1.0 | < 170 |
| Cadmium | 0.01 | < 17 |
| Chromium | 0.05 | 267 |
| Cobalt | d | < 67 |
| Copper | 1.0 | 117 |
| Iron | 0.3 | 1,100 |
| Lead | 0.05 | < 170 |
| Manganese | 0.05 | 8,000 |
| Mercury | 0.002 | <1.7 |
| Molybdenum | d | < 333 |
| Nickel | d | < 33 |
| Zinc | 5.0 | 1,130 |
| Total heavy metals | | 11,423 |
| Aluminum | d | 11,500 |
| Magnesium | 60.0 | 12,000 |
| Potassium | d | 14,000 |
| Sodium | d | 137,000 |
| Strontium | d | 170 |
| Total chlorinated organics | 0.7 | 0.11 |
| Suspended Solids (%) | d | 0.008 |
| Total Solids (%) | d | 0.06 |

a. Average of four measurements from one plant.
b. Less-than values were considered to be at the maximum in computing the totals.
c. U.S. Public Health Service's Drinking Water Standards (1962).
d. No drinking water standards have been set for these metals.
Source: EPA/PB-258-953 June 1976.

266  INDUSTRIAL SOLID WASTES

Figure 15-33. Category B—Typical Wool or Wool Blend Fabric Dyeing and Finishing Process.

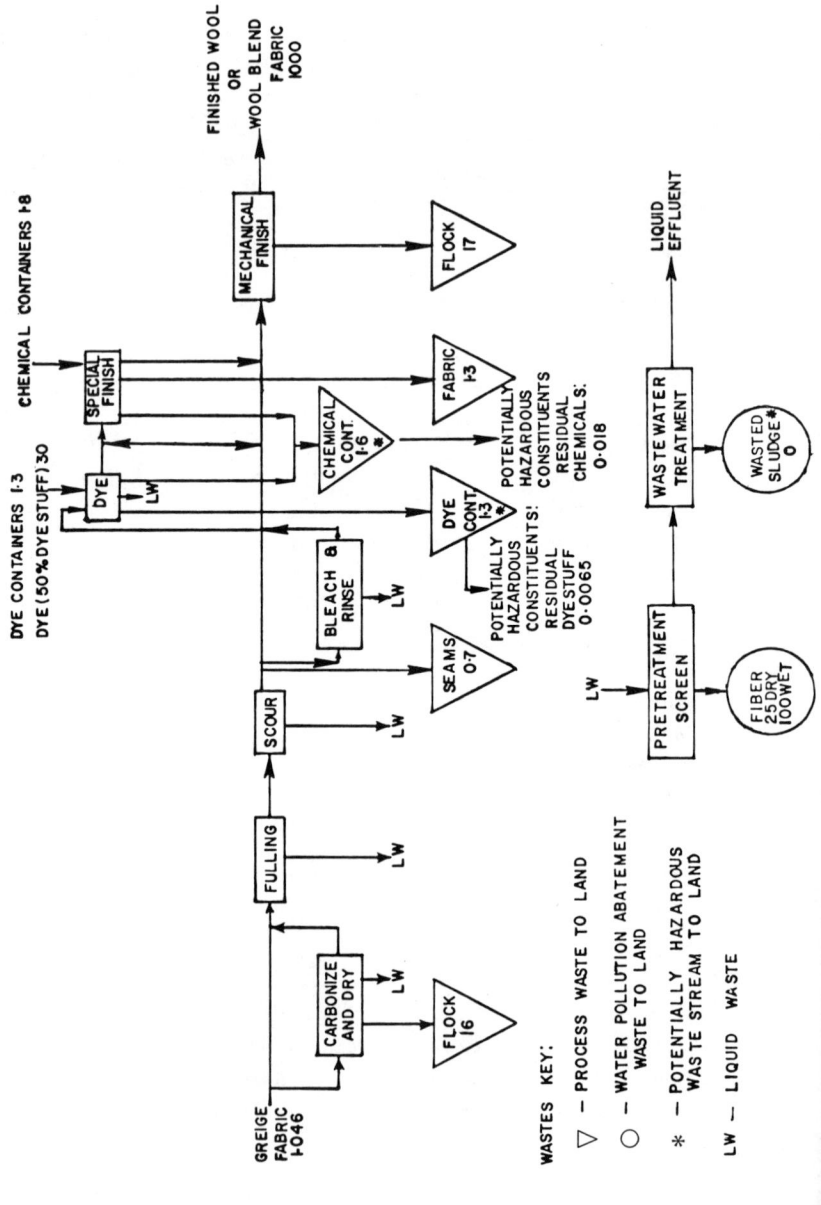

Note 1. An average of 1.6 Kg (Dry), 20,000 Kg (Wet) sludge is retained containing $34 \times 10^{-3}$ Kg total heavy metals. $108 \times 10^{-9}$ Kg total chlorinated organics and 0.08 Kg dye stuff (Not included in mass balance).

Note 2. This category's typical plant land-destined waste streams are:

| Waste | Source | Quantity (kg of Waste/ kkg of product) |
|---|---|---|
| Flock | Carbonizing and drying | 16 |
| Seams | Scouring | 0.7 |
| Dye containers | Dyeing | 1.3 |
| Chemical containers | Dyeing, special finishing | 1.6 |
| Fabric | Special finishing | 1.3 |
| Flock | Mechanical finishing | 17 |
| Fiber | Wastewater pretreatment screening | 25 (dry) 100 (wet) |
| Wasted sludge | Wastewater treatment | none |
| Retained sludge[a] | Wastewater treatment | 1.6 Kg (dry) 20,000 Kg (wet) |

a. The retained sludge quantity is an accumulation over the life of the pond.

Source: EPA (1979).

268  INDUSTRIAL SOLID WASTES

Figure 15-34. Category C—Typical Greige Goods Process.

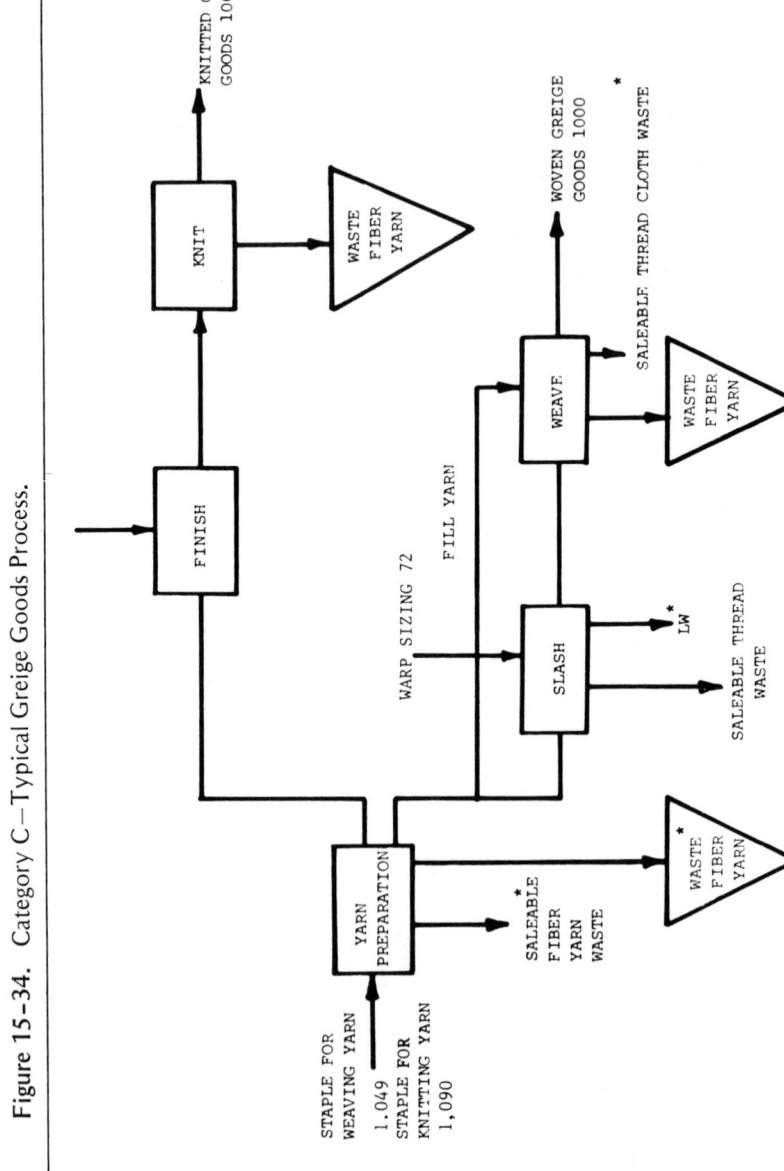

*Liquid waste contains mostly BOD. There are no potentially hazardous wastes destined for land disposal.

Note: This category's land-destined waste streams are:

| Waste | Source | Quantity (kg of waste/ kkg of product) |
|---|---|---|
| Fiber and Yarn | Yarn Preparation | 32 |
| Fiber, Yarn and Cloth | Knitting | 10 |
| Fiber, Yarn and Cloth | Weaving | 11 |

Much of the waste fibers and yarns in this category can be sold (for garnetting) or reprocessed within the yarn preparation operation (especially in wool yarn manufacture).

Source: EPA (1979).

270　INDUSTRIAL SOLID WASTES

Figure 15-35. Category D—Typical Woven Fabric Dyeing and Finishing Process.

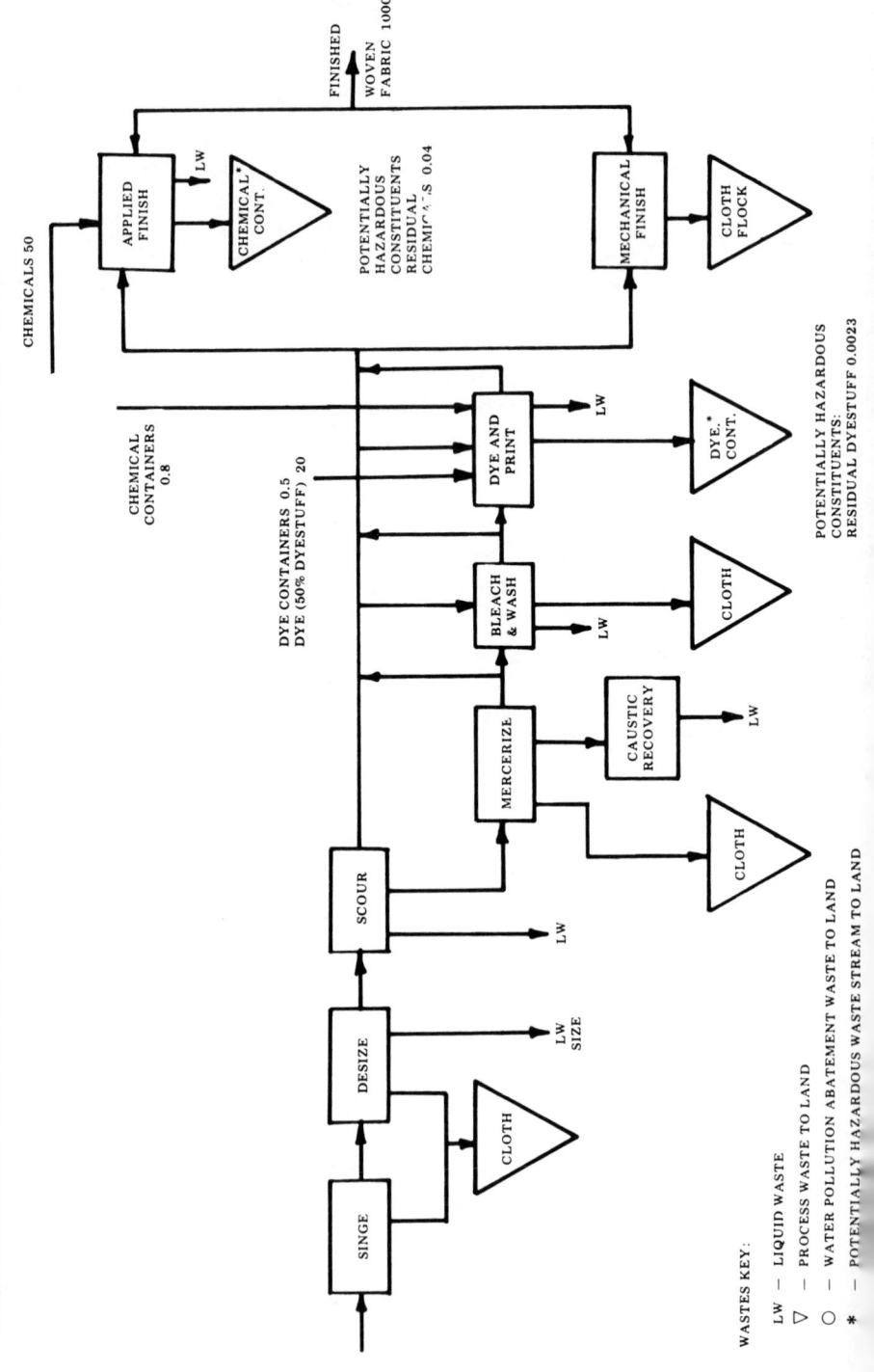

INDUSTRIAL WASTES 271

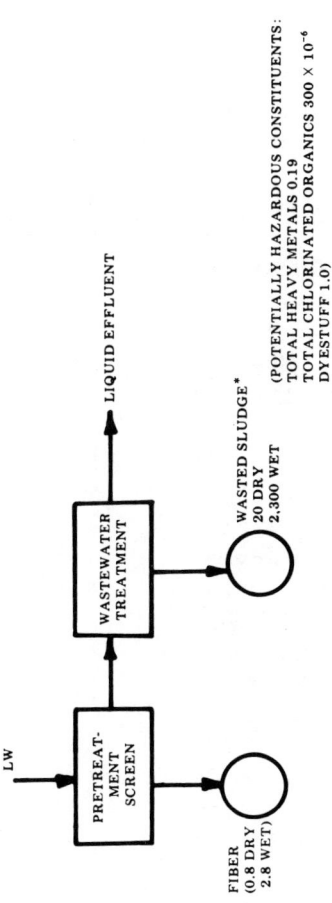

WASTED SLUDGE*
20 DRY
2,300 WET

(POTENTIALLY HAZARDOUS CONSTITUENTS:
TOTAL HEAVY METALS 0.19
TOTAL CHLORINATED ORGANICS 300 × 10⁻⁶
DYESTUFF 1.0)

FIBER
(0.8 DRY
2.8 WET)

Note 1. An average of 67 Kg (Dry) and 7,300 Kg (Wet) sludge is retained containing 0.83 Kg total heavy metals, $1.0 \times 10^{-3}$ Kg total chlorinated organics and 3.4 Kg dyestuff (Not included in mass balance).

Note 2. This category's typical plant land-destined waste streams are:

| Waste | Source | Quantity (Kg of waste/kkg of product) |
|---|---|---|
| Cloth | Singe and Desize | 0.2 |
| Cloth | Mercerize | 0.1 |
| Cloth | Bleach and Wash | 0.2 |
| Cloth | Mechanical Finish | 6 |
| Cloth | Mechanical Finish | 4 |
| Dye Containers | Dye and/or Print | 0.5 |
| Chemical Containers | Dye and/or Print, Applied Finish | 0.8 |
| Fiber | Wastewater Pretreatment and Screening | 0.8 (Dry)<br>2.6 (Wet) |
| Wasted Sludge | Wastewater Treatment | 20 (Dry), 2,300 (Wet) |
| Retained Sludge[a] | Wastewater Treatment | 67 (Dry), 7,300 (Wet) |

a. The retained sludge quantity is an accumulation over the life of the pond.

Table 15-25. Category D—Woven Fabric Dyeing and Finishing Sludge Analyses (*milligrams per kilogram of dry sludge*).

| Parameter | Drinking Water Limit[a] (ppm) | Range[b] | Average[c] |
|---|---|---|---|
| Arsenic | 0.05 | <0.6-<1.4 | <1[d] |
| Barium | 1.0 | 12-85 | 39 |
| Cadmium | 0.01 | <1.4-10.8 | 4.4 |
| Chromium | 0.05 | 89-3,969 | 1,196 |
| Cobalt | e | <2.8-109 | 26 |
| Copper | 1.0 | 193-1,130 | 652 |
| Iron | 0.3 | 917-13,600 | 4,910 |
| Lead | 0.05 | <16-68 | 36 |
| Manganese | 0.05 | 42-318 | 128 |
| Mercury | 0.002 | 0.1-0.7 | 0.35 |
| Molybdenum | e | <0.2-<28 | <17 |
| Nickel | e | 12-88 | 32 |
| Zinc | 5.0 | 318-7,791 | 2,370 |
| Total heavy metals | | | 9,412 |
| Aluminum | e | 1,420-12,800 | 4,640 |
| Magnesium | 60.0 | 1,340-5,730 | 2,820 |
| Potassium | e | 1,420-6,350 | 3,580 |
| Sodium | e | 19,400-94,700 | 51,300 |
| Strontium | e | 2.4-21 | 16 |
| Total chlorinated organics | 0.7 | 4.3-27.8 | 15.2 |
| Suspended Solids (%) | e | 0.42-1.34 | 0.88 |
| Total Solids (%) | e | 0.72-2.04 | 1.26 |

a. Range of the individual plant averages.
b. Grand average of twenty measurements from five plants.
c. Less-than values were considered to be at the maximum in computing totals.
d. U.S. Public Health Service's Drinking Water Standards (1962).
e. No drinking water standards have been set for these metals.

Table 15-26. Category E—Knit Fabric Dyeing and Finishing Sludge Analyses (*milligrams per kilogram of dry sludge*).

| Parameter | Drinking Water Limit[a] (ppm) | Range[b] | Average[c] |
|---|---|---|---|
| Arsenic | 0.55 | <0.85-<12 | <4.8[d] |
| Barium | 1.0 | <15-<125 | <53 |
| Cadmium | 0.01 | <0.7-<12 | <4.5 |
| Chromium | 0.05 | | 33 |
| Cobalt | e | <3.7-62 | <23 |
| Copper | 1.0 | 89-1,030 | 410 |
| Iron | 0.3 | 1,557-8,260 | 3,840 |
| Lead | 0.05 | <7-<125 | <52 |
| Manganese | 0.05 | 12.6-112 | 51 |
| Mercury | 0.002 | 0.7-1.9 | 1.4 |
| Molybdenum | e | <15-<250 | <94 |
| Nickel | e | <3.7-<62 | <25 |
| Zinc | 5.0 | 120-1,250 | 550 |
| Total heavy metals | | | 5,117 |
| Aluminum | e | 1,293-6,625 | 3,180 |
| Magnesium | 60.0 | 963-1,625 | 1,210 |
| Potassium | e | 1,560-4,040 | 2,850 |
| Sodium | e | 12,800-87,500 | 54,200 |
| Strontium | e | 3.7-<38 | 15 |
| Total chlorinated organics | 0.7 | 2.24-181 | 64.7 |
| Suspended Solids (%) | e | 0.02-1.1 | 0.69 |
| Total Solids (%) | e | 0.08-1.35 | 0.87 |

a. Range of the individual plant averages.
b. Grand average of twelve measurements from three plants.
c. Less-than values were considered to be at the maximum in computing totals.
d. U.S. Public Health Service's Drinking Water Standards (1962)
e. No drinking water standards have been set for these metals.

274  INDUSTRIAL SOLID WASTES

Figure 15-36. Category E—Typical Knit Fabric Dyeing and Finish Process.

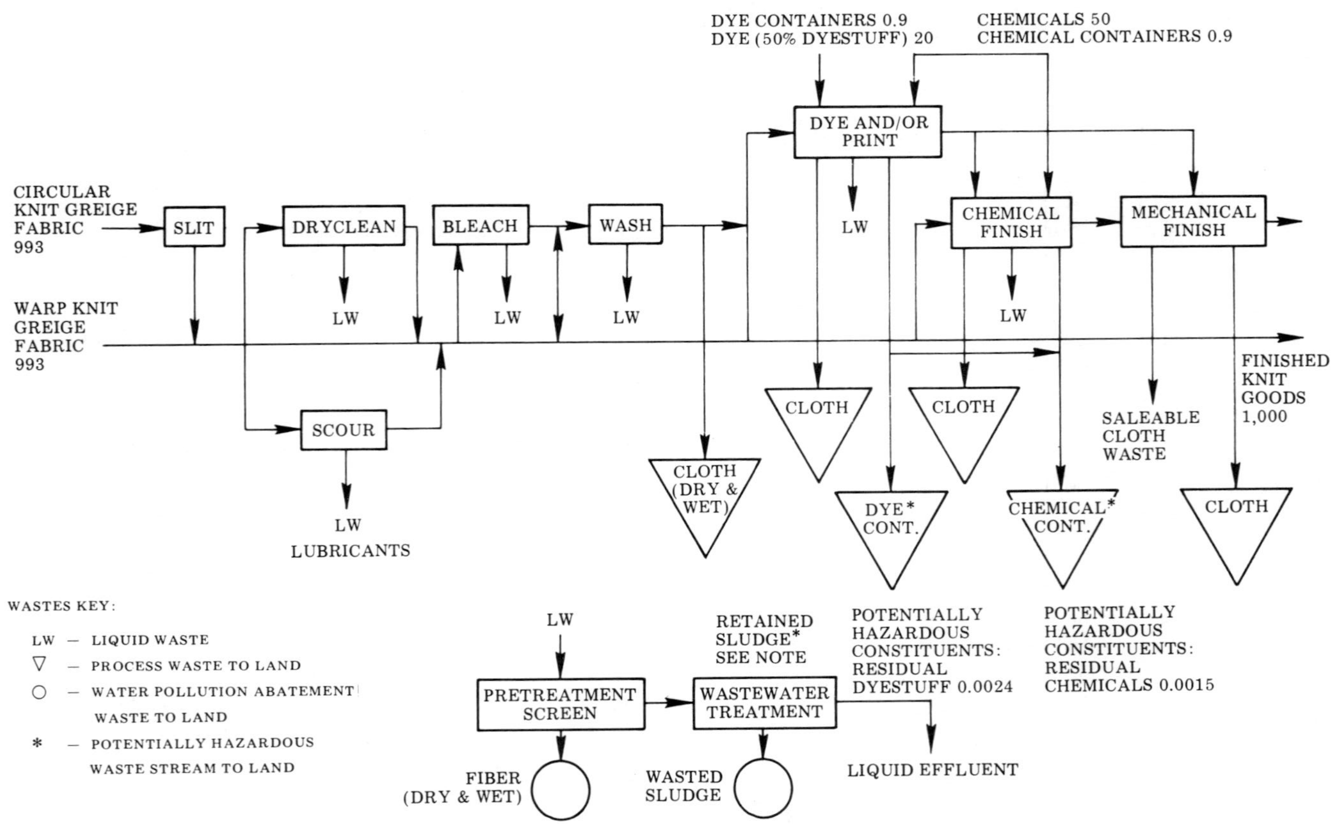

WASTES KEY:

LW — LIQUID WASTE
▽ — PROCESS WASTE TO LAND
○ — WATER POLLUTION ABATEMENT
     WASTE TO LAND
* — POTENTIALLY HAZARDOUS
     WASTE STREAM TO LAND

Note 1. An average of 64 Kg (Dry), 9600 Kg (Wet) sludge is retained containing 0.32 Kg heavy metals, 4.1 × 10⁻³ Kg total chlorinated organics and 3.2 Kg of dyestuff (Not included in mass balance).

Note 2. This category's typical plant land-destined waste streams are:

| Waste | Source | Quantity (kg of waste/ kkg of product) |
|---|---|---|
| cloth | dye and/or print | 2 |
| cloth | chemical finish | 4 |
| cloth | mechanical finish | 3 |
| dye containers | dye and/or print | 0.9 |
| chemical containers | dye and/or print and chemical finish | 0.9 |
| cloth | wash | 2 (dry) 4 (wet) |
| fiber | wastewater pretreatment screening | 0.8 (dry) 2.8 (wet) |
| wasted sludge | wastewater treatment | typically none |
| retained sludge[a] | wastewater treatment | 64 kg (dry) 9,600 kg (wet) |

a. The retained sludge quantity is an accumulation over the life of the pond.

**Figure 15-37.** Category F—Typical Tufted Carpet Dyeing and Finishing Process.

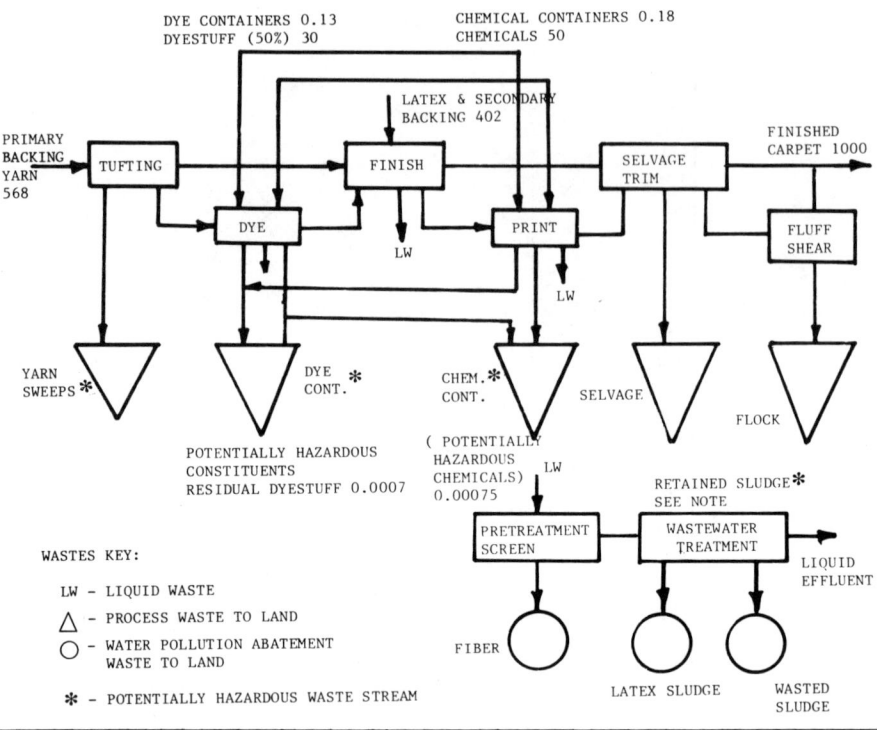

Note 1. An average of 5.2 Kg (Dry) and 22,000 Kg (Wet) sludge is retained containing 0.41 Kg heavy metals, $140 \times 10^{-6}$ Kg total chlorinated organics and 0.28 Kg of dyestuff (Not included in mass balance).

Note 2. This category's typical plant land destined waste streams are:

| Waste | Source | Quantity (kg of waste/ kkg of product) |
| --- | --- | --- |
| yarn and sweeps | tufting | 1.0 |
| selvage | selvage trim | 26 |
| flock | fluff and shear | 4 |
| dye containers | dyeing and printing | 0.13 |
| chemical containers | dyeing and printing | 0.18 |
| fiber | wastewater pretreatment | 1.2 (dry) |
|  | screen | 2.0 (wet) |
| latex sludge | wastewater treatment | 2.3 (dry) 4.9 (wet) |
| wasted sludge | wastewater treatment | typically none |
| retained sludge[a] | wastewater treatment | 5.2 kg (dry) 22,000 kg (wet) |

a. The retained sludge quantity is an accumulation over the life of the pond.

Table 15-27. Category F—Tufted Carpet Dyeing and Finishing Sludge Analyses (*milligrams per kilogram of dry sludge*).

| Parameter | Drinking Water Limit[a] (ppm) | Range[b] | Average[c] |
|---|---|---|---|
| Arsenic | 0.05 | < 7-< 12 | < 10 |
| Barium | 1.0 | < 70-< 120 | < 95 |
| Cadmium | 0.01 | < 7-< 12 | < 10 |
| Chromium | 0.05 | 100-123 | 112 |
| Cobalt | d | < 36-212 | 124 |
| Copper | 1.0 | 22-400 | 211 |
| Iron | 0.3 | 660-9,750 | 5,200 |
| Lead | 0.05 | < 70-150 | 110 |
| Manganese | 0.05 | 101-412 | 256 |
| Molybdenum | d | < 145-< 250 | < 198 |
| Nickel | d | < 36-< 62 | < 49 |
| Zinc | 5.0 | 254-3,325 | 1,790 |
| Total heavy metals |  |  | 8,117 |
| Aluminum | d | 1,740-7,120 | 4,430 |
| Magnesium | 60.0 | 1,580-2,060 | 1,820 |
| Potassium | d | 1,490-6,540 | 4,020 |
| Sodium | d | 41,000-91,250 | 66,100 |
| Strontium | d | 29-< 38 | 33 |
| Total chlorinated organics | 0.7 | 1.03-51.4 | 26.2 |
| Suspended Solids (%) | d | 0.016-0.03 | 0.024 |
| Total Solids (%) | d | 0.08-0.14 | 0.11 |

a. Range of the individual plant averages.
b. Grand average of twelve measurements from three plants.
c. U.S. Public Health Service's Drinking Water Standards (1962).
d. No drinking water standards have been set for these metals.

## 278　INDUSTRIAL SOLID WASTES

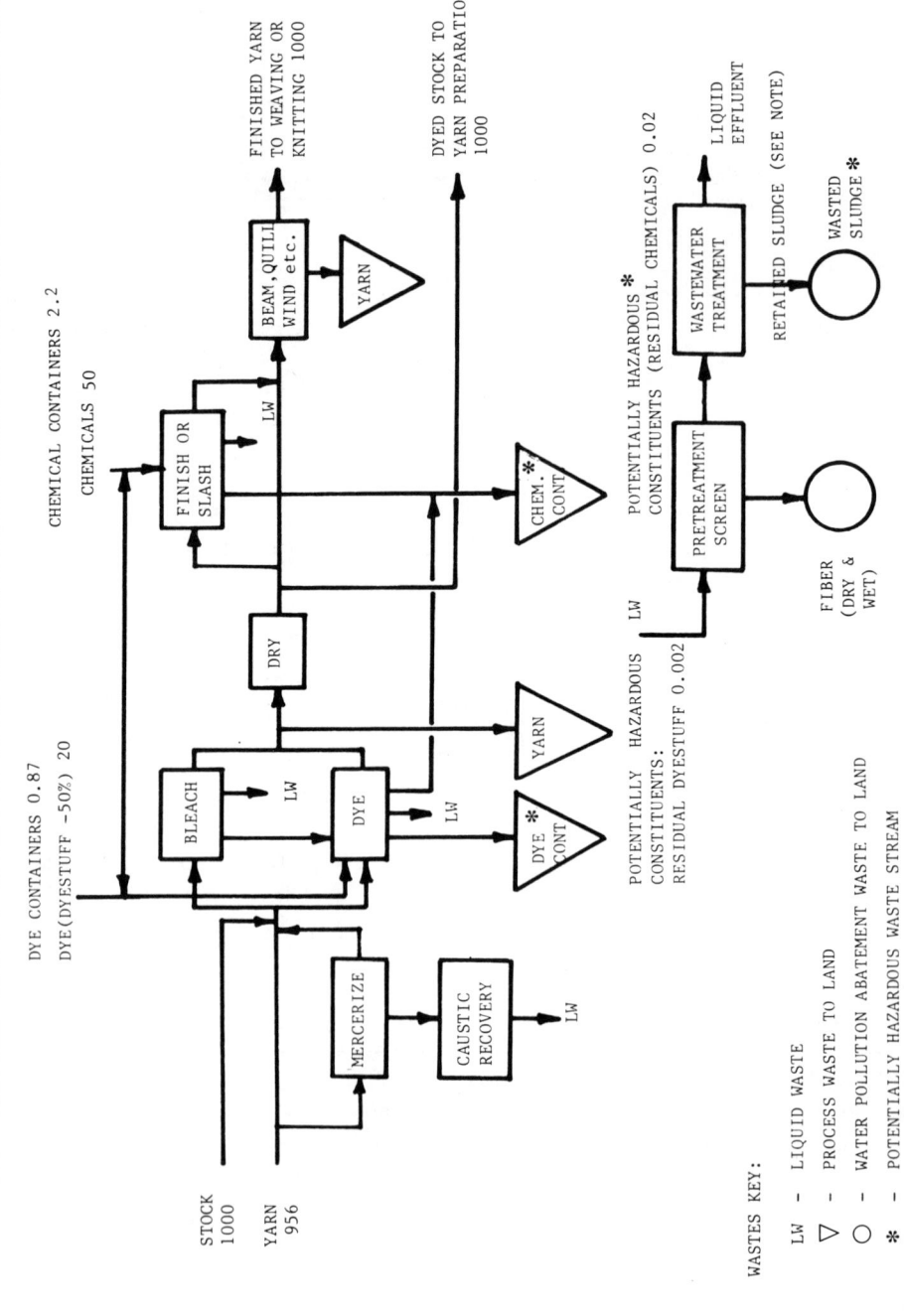

Figure 15-38. Category G—Typical Yarn and Stock Dyeing and Finishing Process.

Note: An average of 2.9 Kg (Dry) and 20,000 Kg (Wet) sludge is retained containing 0.01 Kg heavy metals, 120 × 10⁻⁶ Kg total chlorinated organics, and 0.14 Kg of dyestuff (Not included in mass balance).

The land-destined waste streams from a typical plant in this category are:

| Waste | Source | Quantity, (kg of waste/ kkg of product) |
|---|---|---|
| yarn | bleaching/dyeing | 0.7 |
| yarn | beaming/quilling/ winding, etc. | 5.4 |
| dye containers | dyeing | 0.87 |
| chemical containers | dyeing and finishing | 2.2 |
| fiber | wastewater pretreatment screen | 9.0 (dry) 33 (wet) |
| wasted sludge | wastewater treatment | typically none |
| retained sludge [a] | wastewater treatment | 2.9 kg (dry) 20,000 kg (wet) |

a. The retained sludge quantity is an accumulation over the life of the pond.

Table 15-28. Category G—Yarn and Stock Dyeing and Finishing Sludge Analyses (*milligrams per kilogram of dry sludge*).

| Parameter | Drinking Water Limit[a] (ppm) | Range[b] | Average[c] |
|---|---|---|---|
| Arsenic | 0.05 | < 0.01-< 5 | < 2.5[d] |
| Barium | 1.0 | < 0.1-< 50 | < 25 |
| Cadmium | 0.01 | < 0.01-< 5 | < 2.5 |
| Chromium | 0.05 | 24.4-38 | 31 |
| Cobalt | e | < 0.05-< 24 | 12 |
| Copper | 1.0 | 105-423 | 264 |
| Iron | 0.3 | 605-2,715 | 1,660 |
| Lead | 0.05 | < 0.1-< 50 | < 25 |
| Manganese | 0.05 | 10-122 | 66 |
| Mercury | 0.002 | < 0.5-0.81 | 0.66 |
| Molybdenum | e | < 0.2-< 100 | 50 |
| Nickel | e | < 0.05-< 24 | 12 |
| Zinc | 5.0 | 571-2439 | 1,505 |
| Total heavy metals | | | 3,656 |
| Aluminum | e | 357-2,276 | 1,320 |
| Magnesium | 60.0 | 405-6,772 | 3,590 |
| Potassium | e | 2,100-7,431 | 4,770 |
| Sodium | e | 221,000-497,000 | 359,000 |
| Strontium | e | 14-65 | 40 |
| Total chlorinated organics | 0.7 | 3.3-76.8 | 40.1 |
| Suspended Solids (%) | e | 0.013-0.018 | 0.015 |
| Total Solids (%) | e | 0.12-0.21 | 0.165 |

a. Range of the individual plant averages.
b. Grand average of eight measurements from two plants.
c. Less-than values were considered to be at the maximum in computing totals.
d. U.S. Public Health Service's Drinking Water Standards (1962).
e. No drinking water standards have been set for these metals.

12.5 kkg for 1977 and 14.8 kkg for 1983. This waste stream presents environmental hazards such as toxicity and biodegradability hazards and the possibility of anaerobic degradation of dyes in a landfill to toxic or carcinogenic metabolites.

In regard to chemical containers, in chemical bags or drums one can find, among other things, dichromate salts (oxidizing agents), sodium hydrosulfite (reducing and stripping agent), zinc nitrate and magnesium chloride (catalyst), polyvinyl chloride, tetrakis (hydroxymethyl), and phosphonium chlorite. The amount of residual hazardous chemicals which reach landfill sites in discarded packaging amounted to 11.7 kkg in 1974 and was projected to increase to 117.4 kkg in 1977 and 139 kkg in 1983. Because of the multitude and variety of heavy metal-containing chemicals and chlorinated organic chemicals used in the textile industry and the known persistency and possible toxicities of some of them, this stream is considered potentially hazardous.

Nonhazardous waste streams include fly, flock, and cotton dust, which are usually landfilled in containers such as plastic bags or cardboard boxes, and dried fabric, stock, and yarn, and also rags.

### Characteristics

The general characteristic of the solid waste consists of empty containers and drums of dyes and chemicals. However, the contents of these containers are mostly synthetic dyes and chemicals which are classified as potential hazardous wastes and are generally resistant to biodegradation.

### Disposal

In general, most textile mills practice disposal by land dumping, that is, landfills or lagooning of both wastewater sludges and process wastes. This is not desirable since it may cause leaching of hazardous dyes and chemicals from the landfill. In some cases, the plants send the returnable drums of dyes, chemicals, and solvents back to the suppliers, thereby reducing the generation of solid waste. However, there are plants which practice proper treatment of the solid waste from the wastewater treatment sludge and process wastes. One of the suitable methods adopted for treatment of the sludge is wet oxidation. The process uses liquid-phase oxidation of wastes at high temperatures and pressures. It reduces the amount of sludge and makes the remaining sludge easier to dewater. It also converts much of the nonbiodegradable organic matter to either oxidized innocuous components or biologically degradable material which can be recycled to the plant's wastewater treatment units for destruction. An interesting study has been conducted to ascertain the possibility of utilizing the solid waste generated in a woolen textile industry for a useful purpose. One such study indicated that waste wool samples, mainly dyed wool-synthetic blends discarded from wool-manufacturing processes, have the ability to absorb mercury from aqueous solutions of mercuric chloride. The amount of mercury absorbed parallels the keratin protein (wool) content, reaching 90 percent absorption from 100 parts per million mercury solutions. Waste wool samples, which are available in large quantities and at virtually little cost, therefore offer a real potential for the removal of mercury to acceptable limits from industrial effluents, laboratories and particularly the effluents from the chlor-alkali industry. But then the question remains as to what should be done for the disposal of waste wool samples containing the absorbed mercury. One study being conducted is aimed at developing an economically viable method of recovering the absorbed mercury using strong acidic or alkaline solutions. The use of sodium chloride solutions to recover the absorbed mercury may be of use in the chlor-alkali industry in a cyclic recovery process. One common method of disposal of the lint-cutting of processed cloth is by selling these to manufacturing industries where they are often used as wiping cloths, for machines in the lathe shop, tool room, and so on.

## CEMENT, CONSTRUCTION, AND DEMOLITION

### Origin

The cement, construction, and demolition work produce a considerable amount of solid wastes. The construction of buildings, bridges, dams, and other structures generate solid wastes during construction activity. The demolotion of structures also produce a large volume of solid wastes which are described below.

### Characteristics

The solid wastes generated by construction and demolition activities mainly consist of construction material and rubbish like broken bricks, plaster, stone aggregates, wooden planks, short pieces of reinforcing bars, used building paper, insulation materials, and so on. These materials are inert in nature, and the waste volume produced is quite high.

*Cement.* The waste generated (kiln dust) amounts to 10 to 20 percent of kiln feed. The smallest plant would generate more than 1 million killogram of waste per month (Thomas, Ramo, Wallridge 1979).

Thirty-four percent of plants (out of 166 plants producing from $0.18 \times 10^6$ to $27 \times 10^6$ tons annually) return kiln dust to kiln. Some plants leach the dust with water to reduce alkali content. Leached dust is returned to kiln (Thomas, Ramo, Wallridge 1979).

*Construction and Demolition.* Except for portions composed of asphalt, paper, wood, plastics, textiles, and other organic materials, building rubble consists of concrete, bricks, rock, masonry, plaster, clay, glass, nonferrous metals, and steel. These materials occur when buildings and other structures are torn down, roads are taken out, sewer systems are changed, and other construction is altered.

Combustion Engineering, Inc. estimated that in 1966, 19.1 million tons of demolition wastes were generated and that virtually all this material went to disposal without salvage.

Salvage is practiced, but it is losing its importance except in those instances where large quantities of a valuable material can be recovered (steel from steel-frame buildings and nonferrous metal fixtures and installations). The nonmetalic inorganic fraction of demolition wastes is becoming less and less desirable to salvage; the salvable items include brick and stone; concrete (much like thermosetting plastics) is useless except perhaps as fill material. The scarcity of semiskilled labor necessary to justify the considerable hand labor needed for salvage and the rising costs of labor are probably responsible for the demise of demolition salvage.

Inorganic building rubble is used for landfill in preference to organic wastes. It provides a good substructure for subsequent construction when used in combination with other fill materials.

## Disposal

In general, the common method of disposal for this kind of solid waste is to landfill. Sometimes filling under the construction of structures is carried out with this material. However, before landfilling is done, some of the usable materials like wooden planks, doors and window panels, steel bars, or insulation materials are manually salvaged for re-use. The remaining rubbish material consisting of broken bricks, plaster, aggregates, and so on, are suitable material for landfilling or as an underlayer for highway construction. Construction and demolition solid wastes for cities of certain populations were also surveyed by Golueke and McGauhey (1970) and given in rates in Table 15-29.

In south Florida contractors have been burning construction debris for years under construction sites. As reported by the *Miami Herald* during 1981, "Inspectors found a ravine approximately seven feet deep, 20 feet wide, and more than 600 feet long." The main complaint of the Metro Building and Zoning Enforcement Division Chief was that "no materials that would decompose underground can be buried according to the State Department of Environmental Regulations Manual." Also builders are required to request a public hearing to excavate any hole deeper than eighteen inches. If the excavation is approved at the public hearing, regulations specify the type of materials that can be buried. Besides land settling, officials

Table 15-29. Estimation of Industrial Wastes Generation Rates.

| Population of City | Construction and Demolition Waste Multiplier, Pound/Capita Year[a] | Commercial Organization and Services Waste Multiplier, Pound/Capita Year[a] |
|---|---|---|
| > 100,000 | 500 | 3.5 |
| 10,001-100,000 | 250 | 2.5 |
| 1,001-10,000 | 100 | 2.0 |
| Alternate waste multiplier | 3.8101 ton/ employee year | 41.25205 ton/ employee year |

a. *Status of Solid Waste Management in California*, (1968).
Source: Golueke (1970: 51).

worry that buried debris could threaten groundwater and breed termites.

## POWER PLANTS

In 1977, total installed generating capacity amounted to 576,366 megawatts (EPA, 1979) in the United States.

Coal is the most extensively used fuel source in the electric utility industry, and it is expected to remain so into the future. Electric generation by each of the principal fuel types is shown in Table 15-30 (EPA 1979). For 1977, coal-burning utilities generated slightly more than half of the country's electricity. The remaining fuel use was divided among oil (18.5 percent), gas (16.1 percent), and nuclear (13.2 percent). The percentage of total electrical energy attributed to coal use has fallen since 1965 when coal accounted for two-thirds of the nation's total. In the same period, the use of oil and nuclear fuels increased substantially. However, current projections indicate that coal use will not decline further. One such projection, made by Edison Electric Institute, shows the percentage of total power generated from coal remaining very stable through 1986. The importance of coal, and therefore the importance of environmental regulations on coal-burning, varies significantly among regions.

A different view of the industry can be formulated when electrical generation is classified according to the type of generator used. Most of the country's electricity is derived from steam-driven generators. In 1977, 78 percent of the total electricity supply was provided by steam plants. The remaining supply was divided among hydroelectric generators (10 percent), nuclear-driven generators (12 percent), and internal combustion engines (less than half of 1 percent).

The principal wolid-waste disposal issue for electric utilities is the potential environmental problems associated with the disposal of coal ash and Flue Gas Desulfurization scrubber sludge.

The aggregate quantity of coal ash generated in a year by the electric utility industry is estimated annually in a survey of all coal-fired utility plants by the National Ash Association. In 1977, the survey reported the following aggregate quantities of ash by type: fly ash—44.0 million metric tons (48.5 million tons); bottom ash—12.8 million metric tons (14.1 million tons); and boiler slag—4.7 million metric tons (5.2 million tons). The total ash generation in the survey report is 61 million metric tons.

Table 15-30. Generation, by Fuel Type.[a]

| Year | Total Thermal Generation[b] | Coal MM kWh | Coal Percent of Total | Oil MM kWh | Oil Percent of Total | Gas MM kWh | Gas Percent of Total | Nuclear MM kWh | Nuclear Percent of Total |
|---|---|---|---|---|---|---|---|---|---|
| 1977[c] | 1,899,527 | 985,443 | 51.9 | 357,849 | 18.8 | 305,353 | 16.1 | 250,882 | 13.2 |
| 1976[c] | 1,748,914 | 943,877 | 54.0 | 319,499 | 18.3 | 294,427 | 16.8 | 191,111 | 10.9 |
| 1975 | 1,614,117 | 852,972 | 52.8 | 288,873 | 17.9 | 299,766 | 18.6 | 172,506 | 10.7 |
| 1970 | 1,283,271 | 706,102 | 55.0 | 182,488 | 14.2 | 372,884 | 29.1 | 21,797 | 1.7 |
| 1965 | 860,940 | 570,926 | 66.3 | 64,801 | 7.5 | 221,559 | 25.8 | 3,657 | 0.4 |

a. Edison Electric Institute (1976: 22) and "Advance Release of Data for the Statistical Year Book of the Electric Utility Industry—Year 1977."
b. Excludes generation by geothermal wood and waste.
c. Preliminary figure.

Air-pollution control has caused these plants to install precipitators or baghouse filters to remove fly ash from the gaseous wastes. The recovered fly ash is mainly used as an additive (pozzolanic) to Portland cement concrete (Brackett 1973: 13; Theis 1975: 219). Other uses of fly ash waste solids include (1) filler in asphalt mixes for highways, (2) additive for lightweight concrete aggregate, (3) grouting agent for oil wells, (4) stabilizer for highway roadbeds and railway roadbeds. Fly ash is reported (EPA 1978) to have excellent anti-skid and fire-resistant properties. For this reason it may be re-used by public works departments in snow and ice climates to be mixed with sand or salt for highway care. It is also used to control fires in working mines or alkalyze and neutralize agricultural soils which might otherwise be acid from degradation products or toxic from heavy metals (Plank, Martens, and Hallock 1975).

### Waste Characteristics

*Coal Ash.* The residual ash which results from the combustion of coal is primarily derived from the inorganic mineral matter in the coal. The amount of the residual is directly related to ash content of the coal, which can vary widely depending upon the circumstances in which the coal deposit was formed. The range of observed ash contents among the major types of coal is given in Table 15-31. In each case the possible range is very large. On the average, Eastern coals have a higher ash content than Western coals. However, most Eastern coal (bituminous coal) has a higher heat value than the subbituminous coal and lignite found in the West. Therefore, the difference in ash generation rates is less than the difference between ash contents.

Table 15-31. Range of Ash Content in Coal.

| Type of Coal | Ash (percent) |
|---|---|
| Anthracite | 4-19 |
| Bituminous | 3-32 |
| Subbituminous | 3-16 |
| Lignite | 4-19 |

Source: O'Gorman and Walker (1972).

The ash residual is divided into fly ash and bottom ash or boiler slag. Fly ash is that part of the ash which is entrained in the combustion gas leaving the boiler. It is collected downstream with mechanical collectors or electrostatic precipitators or in flue-gas-desulfurization systems. The bottom ash is created from fused particles which are sufficiently heavy to drop out of the furnace gas stream. These particles are collected in the bottom of the furnace. Boilers are characterized in terms of their ash-handling equipment as being either wet-bottom or dry-bottom. The former require high temperatures in order to cause ash to soften and form slag (thus, wet-bottom), while the latter handle bottom ash as a solid material or as cinders.

The division of the ash residual is significant in that potentially hazardous ash constituents may be found in varying amounts in either component.

*Physical Properties of Ash.* The fly ash and bottom ash are removed separately from the power plant although they are generally combined for ultimate disposal. In this section, the characteristics of the ash are discussed.

Fly ash consists largely of small glassy spheroids which are largely siliceous. The spherical particulates range in diameter from 0.5 to 100 microns. The fly ash spans a color range of light tan to gray and to black. The darker colors are evident in ash with an increased carbon content. The lighter, tan-colored ash indicates high iron content. The pH of the fly ash is generally from 8 to 12. Fly ash is a siliceous material that has little or no cementitious value itself, but in finely divided form and in the presence of moisture is able to react chemically with alkaline earth hydroxides to form a cementitious compound.

Bottom ash is formed as either solid materials or cinders (in dry-bottom boilers) or slag (in wet-bottom boilers). In the first case, a gray or black material in angular shaped particles is formed. The dry-bottom boiler ash has porous, dull, reflective surfaces. Boiler slag consists of angular black particles with a smoother, glossy appearance. Bottom ash particles are larger than fly ash particles. Samples of bottom ash from a dry-bottom boiler have been measured to be 0.7 millimeters to 40 millimeters in diameter.

*Chemical Constituents.* The inorganic constituents of ash are the same as those typically found in rocks and soil. The major compounds are oxides of silicon, aluminum, iron, and calcium. These compounds comprise over 90 percent of the composition of ash.

The distribution of these elements is evenly spread between fly ash and bottom ash. For none of six samples analyzed are there order-of-magnitude differences in the amount of a given chemical found in the fly ash and bottom ash portions.

*Trace Element Contents.* The trace elements found in coal ash and the average amounts of each are displayed in Table 15-32. A number of these and other substances are referenced in the U.S. EPA Safe

Table 15-32. Average Amounts of Trace Elements in Coal Ashes $(ppm)$.[a]

| Element | Anthracites | High-Volatile Bituminous | Medium-Volatile Bituminous | Low-Volatile Bituminous | Lignites and Sub-Bituminous |
|---|---|---|---|---|---|
| Ag | 1[b] | 3[b] | 1.4[b] | 1[b] | 50[b] |
| B | 90 | 770 | 218 | 123 | 1,020 |
| Ba | 866 | 1,253 | 896 | 740 | 5,027 |
| Be | 9 | 17 | 13 | 16 | 6 |
| Co | 81 | 64 | 105 | 172 | 45 |
| Cr | 304 | 193 | 169 | 221 | 54 |
| Cu | 405 | 293 | 313 | 379 | 655 |
| Ga | 42 | 40 | 52[b] | 41 | 23 |
| Ge | 20[b] | 285[b] | 20[b] | 20[b] | 100[b] |
| La | 142 | 111 | 83 | 110 | 62 |
| Mn | 270 | 170 | 1,432 | 280 | 688 |
| Ni | 220 | 154 | 263 | 141 | 129 |
| Pb | 81 | 183 | 96 | 89 | 60 |
| Sc | 61 | 32 | 56 | 50 | 18 |
| Sn | 962 | 171 | 75 | 92 | 156 |
| Sr | 177 | 1,987 | 668 | 818 | 4,660 |
| V | 248 | 249 | 390 | 278 | 125 |
| Y | 106 | 102 | 151 | 152 | 51 |
| Yb | 8 | 10 | 9 | 10 | 4 |
| Zn | 350[b] | 310 | 195 | 231 | 320[b] |
| Zr | 688 | 411 | 326 | 458 | 245 |

a. O'Gorman and Walker (1972).
b. Insufficient data to compute an average value; maximum value is shown in table.

Drinking Water Standards as being potentially toxic if they migrate to water sources. In general, the concentrations of the trace metals are high and exhibit a wide variance. Leachate quality is dependent upon the solubility of the metals in ponds.

*Sludge from Flue-Gas-Desulfurization Systems.* A number of electric utilities have been fitted with flue-gas-desulfurization systems for the control of air emissions. The inception of New Source Performance Standards for coal-fired utility boilers will result in a rapid increase in the percentage of utility boilers which are fitted with these systems. A major concern with these systems is the disposal or re-use of the FGD system byproducts.

The use of throwaway flue-gas-desulfurization systems or scrubbers usually results in a waste stream of a thin water slurry of from 5 to 15 percent solids. Sulfur dioxide gas in the air flow is contacted with the system reagent, lime (CaO), or limestone ($CaCO_3$), and the resulting solid mixture includes calcium hydroxide, calcium carbonate, calcium sulfate, and calcium sulfite. Fly ash may also be removed by the FGD system and removed simultaneously with the scrubber bleed stream.

*Chemical Characteristics of FGD Sludge.* The chemical pollutants in FGD scrubber sludge will be dominated by the characteristics of the coal burned if fly ash is collected with the scrubbing operation. However, it is appropriate to consider FGD sludge separately since in some cases the two waste streams are produced separately and could be handled separately. The concentration levels for the FGD sludge as measured at five plants are shown in Table 15-33.

When oil instead of coal is burned by electric utilities, small quantities of fly ash and bottom ash are produced. The amount of ash is less than 1 percent of that produced by equivalent coal-burning installations. Oil fly ash has higher metal concentrations than coal ash. The makeup of a sample of oil ash residual is shown in Table 15-34. High concentrations of various metals represent the entire mass of the ash sample. The sample includes nearly a 20 percent concentration of vanadium, a metal which is sufficiently valuable to encourage recovery. As a result, a number of utilities sell their oil ash. Organic material may also represent 50 percent of the ash volume.

*Ash Disposal.* Ash disposal is done by either wet (ponding) or dry (landfilling) methods. A survey by the National Ash Association

Table 15-33. Trace Element Analyses of Lime and Limestone Samples (*ppm*).[a,b]

| Elements | Station Number | | | | |
|---|---|---|---|---|---|
| | *1* | *2* | *3* | *4* | *5* |
| Sb | 5.3 | 3.1 | < 0.80 | 1.3 | 3.2 |
| As | 3.0 | 2.9 | 0.83 | 0.66 | 2.7 |
| Ba | < 30.0 | < 30.0 | < 30.0 | < 30.0 | < 30.0 |
| Be | 3.0 | 0.14 | < 0.27 | 0.37 | 0.17 |
| B | 6.45 | 45.1 | 11.2 | 10.8 | 17.4 |
| Cd | 0.28 | 0.24 | 0.92 | 0.90 | 0.65 |
| Cr | 1.2 | 3.5 | 0.61 | 0.57 | < 0.80 |
| F | 105.0 | 134.0 | 307.0 | 103.0 | 117.0 |
| Ge | < 1.0 | < 0.1 | < 0.1 | < 0.1 | < 0.11 |
| Hg | < 0.01 | 0.012 | < 0.01 | < 0.01 | 0.02 |
| Pb | 1.3 | 10.0 | 26.0 | 14.0 | 13.0 |
| Mn | 29.8 | 83.5 | 43.6 | 20.3 | 290.0 |
| Mo | 150.0 | 0.04 | 15.0 | 8.9 | 12.0 |
| Ni | 4.3 | 4.62 | < 12.0 | < 6.0 | 6.2 |
| Se | 0.08 | 0.17 | 0.086 | 0.3 | 0.22 |
| V | < 50.0 | < 24.0 | < 160.0 | < 24.0 | 160.0 |
| Zn | 9.6 | – | 71.4 | 28.0 | 48.0 |
| Cu | 5.8 | 14.0 | 9.3 | 15.6 | 2.4 |

a. Holland and others (1975).

b. All results are reported as the average of duplicate analyses. Analyses are in ppm on a dry sample basis. Original source does not report which samples are lime or limestone.

(1978) reported that 50 percent of the industry used each method. The Association also reports that the current trend appears to be toward dry disposal. Ash re-uses and disposal systems are reported in Table 15-35.

For ponding disposal, a wet slurry is generally used to convey ash to the disposal pond. The water may then be recycled to the plant or discharged, in which case a National Pollutant Discharge Environmental Standard (NPDES) permit would be required. Slurry systems work most easily if they can use gravitational forces. As a result, many utilities have chosen low-lying areas for ash ponds.

Ash settles to a high solids content of from 60 to 90 percent. In some cases a utility will build a pond with a sloped basin. At the

Table 15-34. Typical Residual Oil Ash Analysis.[a]

| Constituent | Weight (percent) |
|---|---|
| Iron | 22.99 |
| Aluminum | 21.90 |
| Vanadium | 19.60 |
| Silicon | 16.42 |
| Nickel | 11.86 |
| Magnesium | 1.78 |
| Chromium | 1.37 |
| Calcium | 1.14 |
| Sodium | 1.00 |
| Cobalt | 0.91 |
| Titanium | 0.55 |
| Molybdenum | 0.23 |
| Lead | 0.17 |
| Copper | 0.05 |
| Silver | 0.03 |
| Total | 100.00 |

a. Danielson (1973).

low end of the basin, facilities are constructed for recycling water to the plant. Ash is then dumped at the high end, and the water content migrates downhill. This practice provides for more rapid dewatering of the ash.

Landfill techniques vary among utilities. A number of firms, particularly those operating in states with RCRA-type legislation, have actively upgraded their landfill practices. Some firms report they have been able to reduce leachate production to a minimum. Other firms, however, appear to be using simple landfilling or landplacing (no cover). At such sites, the only compaction of the ash may be due to the truck traffic over the land which is generated by the dumping operation.

The Hong Kong Electric (*Hongkong Standard*, June 29, 1982) will produce Pulverized Fly Ash (PFA) and bottom ash and plans to recover and re-use both. The PFA will be sold to an international company for use as a cement additive. Some residual PFA and the bottom ash will be sold to another company for the manufacture of cement blocks. All transfers of PFA ash to and from enclosed silos would be made through a pneumatic system. Bottom ash will be

Table 15-35. Ash Utilization.

| Method of Utilization | Fly Ash (percent) | Bottom Ash (percent) | Bottom Slag (percent) |
|---|---|---|---|
| Commercial utilization | | | |
|   Cement uses | 37 | 2 | 3 |
|   Lightweight aggregate | 2 | 3 | — |
|   Fill material for roads, dikes, construction | 20 | 20 | 8 |
|   Stabilizer for road bases, parking areas, and so on | 3 | 5 | 2 |
|   Filler in asphalt mix | 2 | — | — |
|   Ice control | — | 22 | 13 |
|   Blast grit and roofing granules | — | — | 48 |
|   Miscellaneous[a] | 3 | 9 | 22 |
| Other ash removed from plant site | 7 | 17 | 4 |
| Ash utilized after initial disposal | 26 | 22 | 1 |
| Total | 100 | 100 | 100 |
| Total ash utilized[b] (MM metric tons) | 5.7 | 4.2 | 2.8 |
| Percent of total ash collected | 13 | 32.6 | 60 |

a. Ultimate use not specified.
b. National Ash Association (1978).

transferred in large pots under a vacuum to prevent dust leakage. Remaining PFA and bottom ash would be kept moist to eliminate dust problems and later transported by barge to its final re-use destination.

## LEATHER TANNING AND FINISHING

Two major solid wastes originate from tanning and finishing animal hides; blue trim and shavings, leather trimmings. The first are re-used for fertilizer (SCS Engineers, 1976), hog feed supplement, and glue. The second are largely used to manufacture glue and sometimes certain artistic leather goods. Other sources include buffing dust and wastewater sludge.

### Origin

In the leather tanning and finishing industry, solid wastes of various kinds are generated from the unit operations used for processing the rawhide or skin into finished leather. The salted hides which come in as raw material are either "green" or dried hides and are trimmed to remove irregular matter, and thus produce raw trimmings. In the next step, dusting of the hides generates dried blood and dung. The hides are then treated with lime and sulfide to remove the hair. Depending on the dose of lime and sulfide used, the solid waste generated in this operation would be either as hair (hair saving) itself or as a pulp (hair burning). At this stage, it is normal to flesh the hide, producing a solid waste of fleshings and fat together with some hide pieces.

Lime sludge produced in these liming pits also contributes to the total solid waste generated, but is usually allowed to flow into the effluent stream rather than as a separate solid waste. Following the beamhouse operations, the dehaired pelt is passed through the deliming, bating, pickling, and tanning operations. If the chrome tanning method is used, less solid waste is produced. In the case of the vegetable tanning method, tan bark and tan liquor sludge is produced as solid waste. In the finishing operations of machining the leather by shaving, splitting, and buffing, solid wastes are generated in addition to trimmings.

### Characteristics

Solid waste from a tannery with a wastewater pretreatment or treatment system would generally include most of the following components: fleshings; hair; hide trimmings; tanned hide, trim, and shavings; leather trimmings, buffing dust, leather finishing residues, and wastewater treatment sludges (Table 15-36).

In general, the typical character of solid waste from a tannery would be high in organic content, high suspended solids, and putre-

Table 15-36. Origins and Concentrations of the Six Major Processing Areas.

| Source | Constituent | Average Concentration Wet (mg) | Average Concentration Dry (kg) | Concentration Range Wet (mg) | Concentration Range Dry (kg) |
|---|---|---|---|---|---|
| Tanning shavings | Chromium | 15,000 | 17,000 | 3,600 | 42,000 |
|  | Lead | 110 | 130 | 3 | 530 |
| Dry buffing dust | Chromium | 20,000 | 22,000 | 1,200 | 60,000 |
|  | Lead | 71 | 77 | 44 | 120 |
| Finishing area | Chromium | 525 | 1,700 | — | — |
|  | Lead | 1,100 | 3,600 | — | — |
|  | Zinc | 105 | 340 | — | — |
| From trimming | Chromium | 19,100 | 21,200 | 7,600 | 45,000 |
|  | Lead | 250 | 280 | 120 | 460 |
| Waste water screens | Chromium | 965 | 4,200 | 5 | 14,000 |
|  | Lead | 40 | 176 | 43 | 190 |
| Sewer sumps | Chromium | 2,700 | 24,000 | 280 | 75,000 |
|  | Copper | 190 | 1,700 | 20 | 13,300 |
|  | Lead | 25 | 230 | 40 | 380 |
| Waste water sludge | Chromium | 7,700 | 38,800 | 15,500 | 75,000 |
|  | Copper | 420 | 2,000 | 50 | 5,800 |
|  | Lead | 25 | 310 | < 10 | 800 |

cible in nature. In the solid waste, untanned collagenous matter is present to a considerable degree, and when these are subject to putrefaction, rancid odors become noticeable. The solid wastes from a tannery are aestheticlaly most undesirable and in some cases, if disposed in open dumps, would become harborage for vermins.

*Potentially hazardous waste* is defined as waste which poses a substantial present or potential hazard to human health or living organisms because such waste is lethal, nondegradable, or persistent in nature. The constituents are radioactive, infectious, deplosive, flammable irritants (or) strong sensitizers, corrosive and toxic. Scientific studies of the environmental fate of tannery *solid waste* have not been conducted; and, in many instances, the chemical structure of solid waste in these plants is not well recognized. The total process and potentially hazardous solid-wastes quantities are given in Table 15-37.

INDUSTRIAL SOLID WASTES

Table 15-37. Total Process and Potentially Hazardous Solid Waste Generated by All Types of Tanneries in 1974 (*Metric Tons/Year, Wet and Dry Basis*).

| Year | Production (Thousands of Equivalent Hides) | Total Process Solid Wastes | | Total Potential Hazardous Waste | |
|---|---|---|---|---|---|
| | | Wet | Dry | Dry | Wet |
| 1974 | 35,700 | 2,03,000 | 65,000 | 1,51,000 | 45,200 |
| 1977 | Projected | 2,20,000 | 6,91,000 | 1,73,000 | 51,000 |
| 1983 | | 2,81,000 | 87,700 | 2,14,000 | 68,200 |

| | Constituents of Potentially Hazardous Solid Waste (Metric Tons Per Year) | | | |
|---|---|---|---|---|
| Year | Chromium (III) | Zinc | Lead | Copper |
| 1979 | 909 | 0.46 | 10.6 | 16.9 |
| 1977 | 1,000 | 0.59 | 11.9 | 19.6 |
| 1983 (projected) | 1,300 | 0.94 | 14.7 | 28.0 |

Source: SCS Engineers (Stearns, Conrad, Schmidt Assessment of Ind. Hazardous Wastes Leather Tanning and Finishing Industry, Long Beach, Cal. 1976, pp. 149-150.

### Disposal

The methods adopted for disposal of tannery wastes vary greatly from area to area and country to country depending on many factors. In general, where tanning industries have been long established, there usually has been a simultaneous growth of ancillary industries which utilize tannery wastes. There are a large number of possibilities by which tannery solid wastes could be utilized to produce products of commercial value. As such, re-use of the waste is by far the best method of disposing of the tannery solid waste. The fleshings and hide trimmings are used by rendering plants for the manufacture of animal food, glue, and different grades of gelatin, the latter having a great demand in the pharmaceutical industry for encapsulation. Methods of producing glue and gelatin are well known. The hair is washed, dried, and baled, and subsequently sold for making brushes, carpets, rugs, and so on. Tanned hide trimmings and buffings are used in the manufacture of fertilizer, chrome glue, hog feed supplement, and leather boards. Incineration of leather wastes would provide energy for steam generation. The fuel is clean since, unlike coal, it is sulfur-free. Moreover, a recoverable end product of incineration ash is chromium oxide which, with suitable upgrading in quality, would produce a valuable pigment. There have also been studies conducted which indicate that leather waste heated to 400°C would yield a hard granulated char which could be an economical substitute for activated carbon. Fiberized or ground leather waste has also been found to be a suitable insulation material for buildings and is resistant to ignition. At present the solids are disposed of from the six major processing areas whose origins and concentrations are shown in Table 15-36.

Costs of solid waste treatment and disposal are reported to vary between $37 and $46 per metric ton of solids. In the United States the possibilities of solids re-use appears to be limited to (1) rendering for animal feed, glue, and different grades of gelatin; and (2) heating to 400°C to yield granulated char for use as a form of activated carbon. Some current solids disposal practices from tannery operations in the United States are shown in Table 15-38. According to Muthuswamy (in a private communication, 1981), tannery solid wastes in India are disposed of as given in the Table 15-39. Some sludges and other residues may be utilized for soil conditioner and fertilizer.

## PULP AND PAPER MILLS

Solid wastes from this industry are mainly waste sludges from liquid waste treatment plants. Most of these sludges are able to be re-used back in manufacturing lesser quality or different types of paper products. Some sludges which are composed of very short fiber material and contaminating coagulant chemicals are seldom re-used. Instead they are usually lagooned, landfilled, or subjected to land cultivation.

The solid waste in a paper mill plant originates from the cleaning and preparing of raw materials for pulp. For instance, rags and paper waste that are being re-used must be cleaned before they can be processed. Also when logs first enter the paper process, they are debarked, and large amounts of silt are knocked off the log. There is also residue from incineration of the bark.

Table 15-38. Tannery Solid Waste Disposal.

| S. No. | Mode of Proposal | Chrome Tannery Solid Waste | Ship Skin Solid Waste | Split Tannery Waste | Leather Finishing | Beam House or Tan House Waste | Retan or Finishers |
|---|---|---|---|---|---|---|---|
| 1. | Lagoons and Trenches | 15% | 40% | — | — | 40% | — |
| 2. | Dumps (Open Dumping) | 16% | — | — | — | — | — |
| 3. | Municipal Dumps | — | — | 70% | 40% | — | 30% |
| 4. | Certified Hazardous Waste Disposal | 4% | — | — | — | — | — |
| 5. | Private Landfills | 15% | 30% | — | 25% | — | 20% |
| 6. | Municipal Landfills | — | 30% | 30% | 35% | — | 35% |
| 7. | Off Site Landfills | — | — | — | — | 60% | — |
| 8. | Landfills (On Site) | 50% | — | — | — | — | 15% |

Source: SCS Engineers (Stearns, Conrad, Schmidt) Assessment of Ind. Hazardous Wastes Leather Tanning and Finishing Industry, Long Beach, Cal. 1976, pp. 149–150.

Table 15-39. Disposal of Solid Wastes from Some Tanneries in India.

| Solid Waste | Application |
| --- | --- |
| Salt dust | Heaped outside tannery |
| Hair | Selling to brush, druggist and carpet manufacturers |
| Green fleshings | Dried in tannery and sold for glue and gelatin and fertilizer manufacturers |
| Lime fleshings | Dried in tannery and sold for glue and gelatin and fertilizer manufacturers |
| Untanned shavings | Same as fleshings |
| Waste tan bark | As fuel in tannery |
| Vegetable tan sludge | Sold as fertilizer |
| Tanned shavings and trimmings | Detanned and used as fertilizer; sold for leather board manufacture |
| Tanned splits | Shoe uppers and leather articles |
| Tanned dust | A filler in organic coatings |
| Effluent sludge | Allowed to dry on ground or in sand beds and used as fertilizer |

The solid wastes can be split into organic and inorganic solid waste. The organic material is made up of excess bark and any paper waste that may occur. The inorganic compounds may be divided into three parts: (1) metals—bottlecaps, paperclips, and so on; (2) glass; and (3) plastics. Most of the inorganic material is brought out during the cleaning of rags and old paper waste. Silt and dirt are also deposited in large quantities from the same source.

The three methods used for disposal are incineration, bark-fired boilers, and landfill. Bark-fired boilers get rid of the bark only. Incineration and landfilling are the basic types of disposal. The ashes from the incinerator are hydraulically removed to an ash lagoon. If the plant is using asbestos or any similar item, then landfills must be used. These landfills cannot have any vertical or horizontal migration of groundwater.

300  INDUSTRIAL SOLID WASTES

Most companies use landfill more than incineration. Landfilling is much cheaper and can be done locally. In fact, many companies operate their own landfill on the same ground area as the papermill. Other companies hire local companies to remove the waste to a municipal landfill.

## PHOSPHATE FERTILIZERS

The use of this fertilizer can only continue to increase as a means of providing the world with vital, life-sustaining food.

The phosphate fertilizer industry comprises eight separate processes: (1) phosphate rock grinding; (2) sulfuric acid dissolving; (3) wet process phosphoric acid; (4) phosphoric acid concentration; (5) phosphoric acid clarification; (6) normal superphosphate; (7) triple superphosphate; and (8) ammonium phosphates.

When calcium phosphatic rock is treated with sulfuric acid, the waste sludge produced, gypsum sludge, is composed mainly of calcium sulfate dihydrate and some flouride, uranium, phosphorous, vanadium, and other common rock elements. The sludge has not been widely re-used. Recently, however, recrystallization of the gypsum produces a cementous material which can be re-used in making various building materials. It is also possible to produce sulfuric acid, sulfur, and cement.

Waste byproduct gypsum is normally stored in huge, diked, pond-like areas. Some dikes in Florida extend 100 feet vertically. The earthen-gypsum dikes usually leak a certain amount of contaminated water. Seepage control utilized by the industry is shown in Figure 15-39.

Even with seepage problems under control, the huge mountainous quantities of stored gypsum create unsightly conditions on increasingly valuable land areas. More consideration is being given to the re-use of gypsum into building products. Problems of radioactivity and binding characteristics of the gypsum must be overcome, however, before this solution becomes a reality.

## OILY SLUDGES

Oily sludges and concentrated residues are considered in this text as solid wastes. They consist of waste automotive, industrial, and avia-

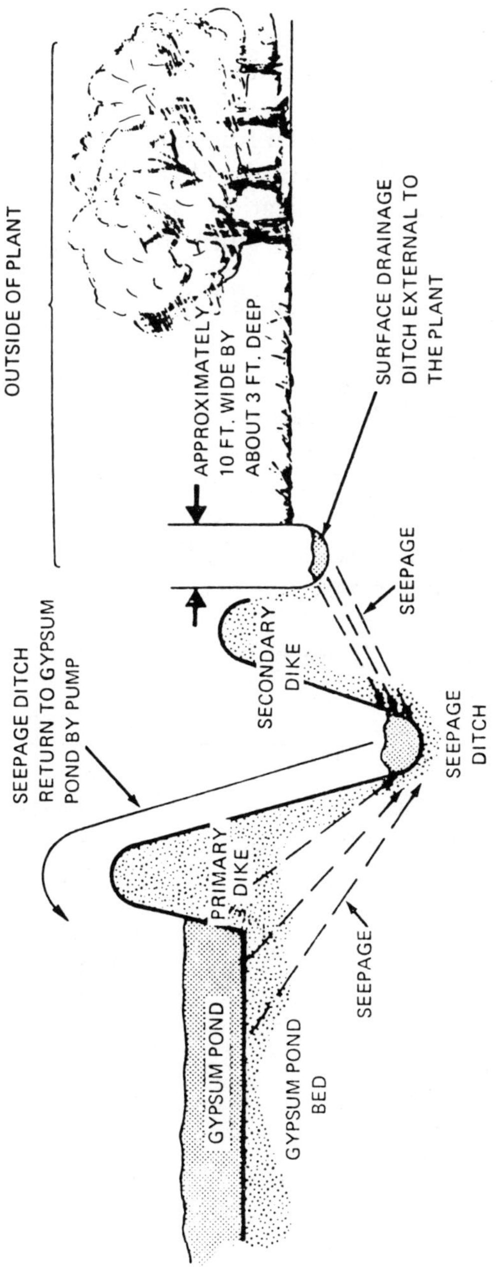

INDUSTRIAL WASTES 301

Figure 15-39. Gypsum Pond Water Seepage Control.

tion oils, and other waste oils. Contaminants present in automotive waste oils consist of volatile products and materials soluble and insoluble in oil. These are shown in Table 15-40. The second category of waste oils is largely waste metalworking lubricants, spent turbine and transformer oils, mixtures of aviation jet fuel and lubricants, gear box oils from industrial rotary machinery, heat transfer fluids, and railroad lubricants. The physical and chemical characteristics of these depend on the specific application and composition of the lubricant. They may include lead, PCBs, sulfur bearing additives, antioxidants, among others.

Oily wastes have been widely disposed of by land cultivation (Dotson et al. 1970; Kincannon 1972; and Lewis 1976). The most extensive of these (Kincannon 1972) was carried out at a Texas oil refinery. Successful degradation rates of 700 barrels of oily sludge per acre per month were obtained at 27°C temperature. Dotson and others (1970) also concluded from three refinery oily-sludge waste studies that soil micro-organisms can degrade a large quantity of petroleum hydrocarbons economically under a wide range of soil and environmental conditions. Further, the soil physical and chemical

Table 15-40. Typical Waste Automotive Oil Composition.

| Variable | Value |
|---|---|
| Gravity, °API | 24.6 |
| Viscosity at 100 °F | 53.3 Centistokes |
| Viscosity at 210 °F | 9.18 Centistokes |
| Flash point | 215 °F (C.O.C. Flash) |
| Water (by distillation) | 4.4 Volume % |
| BS & W | 0.6 Volume % |
| Sulfur | 0.34 Weight % |
| Ash, sulfated | 1.81 Weight % |
| Lead | 1.11 Weight % |
| Calcium | 0.17 Weight % |
| Zinc | 0.08 Weight % |
| Phosphorous | 0.09 Weight % |
| Barium | 568 ppm[a] |
| Iron | 356 ppm[a] |
| Vanadium | 5 ppm[a] |

a. ppm = parts per million.
Source: EPA, PB-257-69 April 1974.

properties are enhanced. Decomposition is enhanced by use of lime and fertilizer, artificial drainage, and tillage.

Recycling waste oils is obviously the most conservative and economical solution to the disposal problem. However, collection and purification are necessary procedures which require study and effort. Collection is said ("Waste Oil Study—Report to Congress" 1974) to be the weakest link in the entire waste-oil recycling and re-use chain. However, there is an improved effort among collectors today as shown in an advertisement which appeared in the June 7, 1981, *Miami Herald* newspaper. Table 15-41 gives an estimate of oil quantities produced from various sources and methods of disposal of each. The average processor has a 28,000-gallon-per-day capacity and operates at 50 percent of that capacity ("Waste Oil Study" 1974). Reprocessing is made more difficult because of additives in the oil, environmental laws and limits on the refining, and disappearance of certain tax advantages. Most of the refined oil is resold for motor or industrial lubricants.

### Oily Solids

Waste oils can also be disposed of into the land since they have been found to degrade slowly and adhere tenaciously to the soil absorbing them. Burning waste oils as fuels has also been gaining in popularity as the "energy crunch" continues.

## RUBBER PLANTS

Pettigrew and Roninger (1971) surveyed solid wastes of the fabricated rubber products industry and related them to the weights of products shipped. These quantities are given in Table 15-42.

## MINING INDUSTRIES

There are many types of mining ranging from strip mining, open-pit mining, quarrying, dredging, and underground mining, all of which produce solid waste. The basic mining procedure is, for example, in strip mining, to remove the overburden and unearth the mineral to

304    INDUSTRIAL SOLID WASTES

Figure 15-40.   Managing Chemical Wastes.

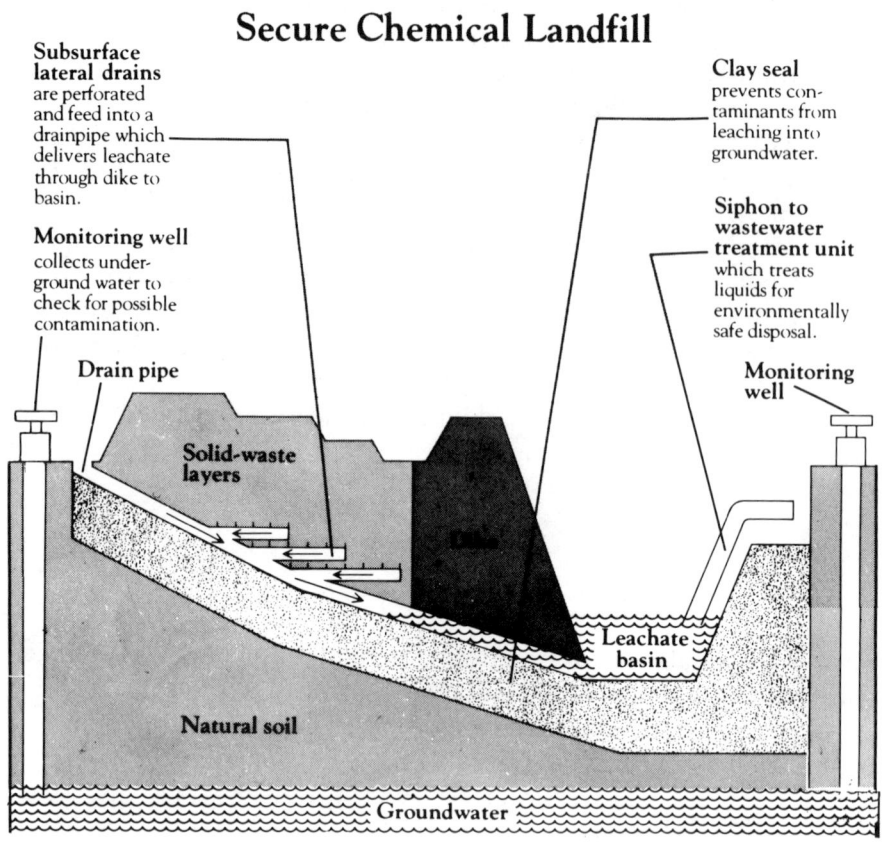

Source: Chemical Manufacturers Association, Box 363, Beltsville, Mo. 20705.

### 1. Eliminating wasteful processes
We're redesigning manufacturing processes and improving efficiency. We're adding on-line treatment systems to neutralize, reduce in volume or change the nature of waste byproducts. We're also using recovery techniques that let us recycle wastes back into the production process. One company, for example, is salvaging phenol, used to manufacture plastics, pharmaceuticals and other useful products.

### 2. Building secure landfills
Secure landfills have a barrier that keeps wastes from seeping out into groundwater and keeps groundwater from migrating through the landfill. Other features may include facilities for recycling liquids or a wastewater treatment unit to clean up liquids for sate disposal. Landfills—if *properly* designed, operated and monitored—are one of the best ways to dispose of many kinds of solid wastes.

Figure 15-40. continued

**3. Continuing industry commitment**
We were finding ways to manage solid wastes long before the nation recognized the need for better waste-disposal methods. In fact, we already had much of the required waste-disposal technology and remedial strategies in place—or being developed—when Congress passed the Resource Conservation and Recovery Act of 1976, which sets forth strict waste-disposal guidelines.

**4. Sharing knowledge and new technology**
As we develop new waste-disposal techniques, we share our knowledge with industry, government and the public. In 1979, we began conducting a series of regional seminars that presented current techniques for solid-waste disposal. Individual companies may use videotapes, visual aids or other techniques to train personnel in waste-disposal methods.

**5. Encouraging solid-waste exchanges**
*Sometimes one chemical company's wastes can become another company's raw material. Fluoride wastes from a phosphoric acid plant, for example, can be used by a company producing aluminum.* So the chemical industry has encouraged the development of waste-exchange organizations, which develop and distribute lists of available wastes.

be mined. In a case like this, the overburden, which is called "spoils," can often create a nuisance. The mineral being mined is carted off to the processing plant where more solid waste is generated by cleaning processes.

About 65 percent of all mineral mining solid waste is overburden, 32 percent is tailings, and 3 percent is other mine wastes. The mining of metallic and nonmetallic ores and mineral fuels is the largest single source of solid waste in the United States. Production of selected ores is given in Table 15-43 and the production of associated solid wastes is given in Table 15-44.

The character of the solid wastes is dependent upon the location of the mine and what is being mined. The majority of the solid waste is earth material such as soils, rocks, and inert material. Care is taken in cases where the spoils must be placed apart from the mines, resulting in hazardous chemicals being washed out by sudden exposure to water. For example, if mining is to occur in salty soils, federal regulations require mining facilities to control environmental degradation due to spoils leachate. Coal-cleaning processes can release high sulfur-concentrated wastes that even though diluted in water must be removed. In ore mining operations one must especially beware of dangerous metal oxides that can be released.

In all types of mining, spoils are characteristic of the method used to mine the ore and of the placement of solid wastes. Usually at the

Table 15-41. Generation, Destination, and Disposal of Waste Oils, 1972 (*millions of gallons*).

| Category | Generated Waste Oil | To Reprocessors | To Re-refiners | Road Oils Asphalts | Fuel | Other |
|---|---|---|---|---|---|---|
| Automotive | 616 | 202 | 105 | 142 | 19 | 148 |
| Industrial and aviation | 394 | 130 | 16 | 25 | 118 | 112 |
| Other industrial | 87 | 28 | 3 | 6 | 25 | 25 |
| U.S. government | 18 | 4 | 3 | 4 | 4 | 3 |
| Total | 1,115 | 364 | 127 | 177 | 159 | 288 |
| Reprocessed oils | | -364 | | 36 | 310 | 18 |
| Re-refinery waste | | | -41 | | 10 | 31 |
| Ultimate Disposal | | 0 | 86 | 213 | 479 | 337 |

Source: EPA, PB-257-69 April 1974.

Table 15-42. Solid Waste Generated by Tire and Tire Products Manufacturing.

| Type of Solid Waste | Lb. Waste/Million Lb. Product Shipped |
|---|---|
| Paper, cardboard, and wood | 13,400 |
| Rubber compound | 11,900 |
| Textile material | 5,900 |
| Metal | 9,700 |
| Other | 14,000 |
| Total wastes[a] | 54,900 |

a. The total wastes amount to about 5.5 percent of the total weight of product made. One million pounds of tire industry product is equivalent to 40,000 passenger tire casings.

Source: Pettigrew and Roninger (1971: 12).

Table 15-43. Projected Annual Ore Production by Selected Industries.

| Industry | Crude Ore to Marketable Product Ratio | Crude Ore Production ($10^6$ Tons) | | |
|---|---|---|---|---|
| | | 1975 | 1985 | 2000 |
| Copper | 193.5 : 1 | 270 | 480 | 730 |
| Iron ore | 2.8 : 1 | 240 | 260 | 320 |
| Uranium | 630.9 : 1 | 7 | 23 | 38 |
| Phosphate rock | 3.8 : 1 | 190 | 300 | 320 |
| Coal[a] | NA | 650 | 1000 | 1660 |
| Other | NA | 1890 | 3100 | 5260 |
| Total | NA | 3250 | 5170 | 8340 |

a. Total for all mineral fuels—anthracite, lignite, and bituminous coal.
Note: NA = Not applicable.
Source: PEDCo Environment Inc. (1979).

end of the mining operation, the spoils are dumped or backfilled into the "pit." Placing the spoils aside and reclaiming land is a project on which the forestry division of the government is working with industry (mining). In quarrying operations all the material is used, mostly for making cement and aggregates. Dredging is generally used for re-

**Table 15-44.** Projected Annual Generation of Mining Solid Wastes by Selected Industries.

| Industry | Waste to Ore Ratio | 1975 | | 1985 | | 2000 | |
|---|---|---|---|---|---|---|---|
| | | Quantity (10⁶ Tons) | Percent of Total | Percent of 1975 Value | Percent of Total | Percent of 1975 Value | Percent of Total |
| Copper | 3.56 | 960 | 41 | 190 | 43 | 270 | 45 |
| Iron ore | 1.73 | 410 | 18 | 110 | 11 | 130 | 9 |
| Uranium | 23.65 | 160 | 7 | 330 | 13 | 550 | 15 |
| Phosphate Rock | 1.9 | 350 | 15 | 160 | 14 | 170 | 10 |
| Coal[a] | 0.17 | 110 | 5 | 160 | 4 | 260 | 5 |
| Other[b] | NA | 330 | 14 | 190 | 15 | 270 | 15 |
| Total | NA | 2,300 | 100 | 180 | 100 | 250 | 100 |

a. Total for tailings from bituminous and lignite coal only.
b. Includes only metal and nonmetal ores.
Source: PEDCo Environmental Inc. (1979).

covering aggregate material. Most surface mining lands are landfilled or backfilled with solid waste and inert materials. Underground wastes are backfilled into the excavated cave by lowering the excavated spoils and inert materials from the processing plant back into the mine. In underground mines the method of sloping or slope-mining is used where a section at a time is mined and that same section is backfilled with wastes and then another section is mined and repeated.

As a general rule, considerable solid waste products come from the processing plant (which contains solids that are in solution). A good portion of this is simply poured to the spoils allowing the water to percolate through, usually to a nearby stream. If treated, the solid wastes are thrown together with the spoils and disposed with them. Treatment is sometimes accomplished by heat or by newer methods such as belt-pressing.

The U.S. government is working to reclaim all mines that are terminating operations in order to make them environmentally suitable.

As shown in Table 15-45, 90 percent of overburden and waste rock is disposed of in piles on or near the mine site with the remaining 10 percent backfilled into the abandoned mine, used for construction, or for holding pond covers. Many holding ponds have clay liners to prevent leakage and may be covered with overburden when filled. The disposal of uranium mill tailings (Uranium Mill Tailings Radiation Control Act of 1978, PL 95-604) requires that these be disposed of in a manner safe for human health and the environment.

### Oil-Shale Solid Wastes

Mine overburden and spent oil shale are the two forms of solid wastes from oil-shale production. Surface overburden (100 to 800 feet thick over the oil shale) produced is about one ton per ton of raw shale. It is normally backfilled into the mine. Currently, about three-fifths of spent oil shale is planned to be returned to the mine site with the rest going to surface holding ponds. If the oil is extracted underground, no spent oil shale will be produced.

Solid-waste refuse from anthracite mining was computed for Pennsylvania refuse banks and given in Table 15-46.

Table 15-45. Disposal Methods for Mining Wastes.

| Type of Solid Waste | Disposal Method |
|---|---|
| Overburden and waste rock | Stockpiles adjacent to surface and underground mines and on the outside slopes of open pit mines (90 percent) |
| | Backfilling of previously excavated areas adjacent to the active overburden removal at surface mines. Backfilling of underground mines with waste rock (0-10 percent) |
| | Utilization as construction material (0-10 percent) |
| Tailings from the mills of both underground and surface mines | Tailings pond (99+ percent) |
| | Backfilling underground mines either by sluicing or truck hauling |
| | Utilization as construction material |
| Miscellaneous wastes | Combination with overburden, waste rock, or tailings |
| | Separate disposal in a sanitary landfill on-/ or off-site |
| | Lake or marine disposal |

Source: PEDCo Environmental Inc. (1979: xxxix-xci).

### Slate Mining

Slate mining is still an active industrial production. Slate is extracted mostly by means of open quarries. For the production of roofing slates Watson (1980) describes the quarry material as first reduced to handleable blocks, then split manually according to their cleavage direction with chisels to form slates of appropriate thickness. Watson gives the chemical composition of slate in Table 15-47. Sometimes granules and powders are produced by crushing, grinding, and screening techniques. To produce expanded slate, high-temperature rotary kilns are used after this size-reduction step.

Watson reports a very high proportion of slate solid waste—in one case, 12.5 tons of block slate required to produce 1 ton of roughly

Table 15-46. Data on Pennsylvania Anthracide Refuse Banks.

| Field | Total | Number of Banks Volume Class Less Than 10,000 yd$^3$ | 10,000 yd$^3$ or More | Area Occupied,[a] Acres | Volume,[a] 1,000 yd$^3$ |
|---|---|---|---|---|---|
| Northern | 372 | 18 | 354 | 3,549.2 | 365,460 |
| Eastern middle | 171 | 33 | 138 | 1,445.9 | 75,160 |
| Western middle | 172 | 9 | 163 | 3,541.2 | 283,380 |
| Southern | 148 | 3 | 145 | 3,502.0 | 186,080 |
| Total | 863 | 63 | 800 | 12,038.3 | 910,080 |

a. Banks containing less than 10,000 yd$^3$ not included.
Source: MacCartney and Whaite (1969).

Table 15-47. Gross Chemical Composition of Most Slates.

| Chemical Composition | Percent by Weight |
|---|---|
| $SiO_2$ | 45-65 |
| $Al_2O_3$ | 11-25 |
| $FeO$ | 0.5-7.0 |
| $Na_2O$ | 1-4 |
| $K_2O$ | 1-6 |
| $MgO$ | 2-7 |
| $TiO_2$ | 1-2 |

Source: Watson (1980).

finished slate. In addition, as much as fifty tons of waste may result from quarrying this amount of block slate. In fact, he quotes a recent U.K. survey in which twenty parts of solid waste were produced for one part of finished slate product. Waste reduction can be accomplished by selecting better quarry rock (absence of planes of weakness and usually found in deeper strata). The disposal of quarry waste becomes even more difficult as space for storage becomes scarce. Major community objections are of an aesthetic nature. Utilization of the solid waste in making other products is a major method of reducing the storage piles. A major product is that of expanded lightweight material for an aggregate in concrete, and also slate or lime brick, resin-bonded molded products, mineral wool, and accoustical tiling. Where there are shortages of conventional road-building materials, the slate waste is highly desirable. Slate dust powder also has been identified as a causative agent in silicosis-related diseases.

## CHEMICAL INDUSTRIES

During the past five years, four major occurrences in the field of hazardous wastes of the chemical industry have opened the door to public and governmental scrutiny of the industry's management practices. These are: (1) the Love Canal tragedy in Niagara, New York; (2) the "rediscovery" of the "Valley of the Drums" in Kentucky; (3) the radon seepage into homes built of phosphate mining refuse in northern Florida; (4) the recent announcement that PCBs are now so concentrated in human milk that the sixfold concentration factor by the baby poses a serious health hazard to the child.

Of course, some of the chemical industry's solid wastes are relatively innocuous. One example is the pulp from the sugar beet industry, where this type of waste causes a problem in its immediate area as well as downstream if it is put into a stream.

The chemical-process industry may be considered as two large industries. First, the *inorganic chemical industry* is classified under the standard industrial code (SIC) as number 281 and consists of alkalies and chlorine (SIC 2812), industrial gases (SIC 2813), inorganic pigments (SIC 2816), and other inorganic chemicals not classified elsewhere (SIC 2819). Many plants produce more than one chemical and often may produce chemicals in various SIC classifications simultaneously. Over 200 chemicals are produced within the SIC 281 classification.

The *organic chemical industry*, SIC 282 through SIC 289, includes: plastic materials and synthetic resins, synthetic rubber, synthetic and other man-made fibers (SIC 282); drugs and pharmaceuticals (SIC 283); soap, detergents and cleaning preparations, and perfumes, cosmetics, and other toilet preparations (SIC 284); paints, varnishes, lacquers, enamels, and allied products (SIC 2851); industrial organic chemicals (SIC 286); nitrogenous fertilizers (SIC 2873); pesticides and agricultural chemicals (SIC 2879); adhesives and sealants (SIC 2891); miscellaneous chemical products not classified elsewhere including explosives and carbon black (SIC 2892, 2895, 2899); and printing ink (SIC 2893).

The nature of the solid waste may be classified as nonhazardous or hazardous. If a solid waste is a hazardous material, much greater care and expense is required in the handling and disposal of the waste. All the data on the following pages will attempt to separate the hazardous solid waste from the nonhazardous. Much of this information was presented by the TRW Report to EPA (Thomas, Ramo, Wallridge 1979) and the Versar Report (1979).

## The Industrial Inorganics

### SIC 2812 Alkalies and Chlorine

*Waste Composition:* Nonhazardous wastes include brine mud (magnesium hydroxide and calcium carbonate from brine purification), and wastewater sludges containing carbonates and chlorides. Hazardous wastes are the mercury and asbestos-bearing wastes from chlorine production in "mercury cell" and "diaphram cell" processes respectively. The amount of hazardous waste generated is 20 kilograms per metric ton. The total amount of solid waste generated is estimated at 9.1 million kilograms per month.

### SIC 2813 Industrial Gases

*Waste Composition:* No hazardous waste is generated from plants producing acetylene, nitrogen, oxygen, noble gases, and nitrous oxide. Plants producing carbon monoxide, carbon dioxide, hydrogen, and so on, may generate hazardous material in the form of spent catalyst containing chromium, copper, and other metals. The waste catalyst is produced at about 0.5 kilograms per month of product. The total quantity of waste generated is 275,000 kilograms per month.

### SIC 2816 Inorganic Pigments

*Waste Composition:* Sludges exist which are corrosive and may contain chromium, cadmium, lead, and other metals. The total amount of solid waste is 38 million kilograms per month.

*SIC 2819 Miscellaneous Inorganic Not Classified Elsewhere*

*Waste Composition:* Hazardous solid wastes consist of water and wastewater treatment sludges, sludges from the purification of raw materials, rejected chemicals, spills and cleanup waste, and dust collected from air-pollution-control devices. These wastes may be highly acidic or basic, and may contain toxic metals, fluorides, cyanides and sulfides. Disposal methods are landfill 19 percent of total generated; recycle, 18 percent; other, 2 percent; and hauled away by contract to unknown disposal, 61 percent. The total amount of solid waste produced is 280 million kilograms per month.

The majority of the inorganic chemical industry disposes of solid waste by either landfill or contract hauler. *The contract-hauler disposal method is often unknown to the producer of the waste.*

## The Organics

*SIC 283 Drugs and Pharmaceutals*

*Waste Composition:* The primary types of solid hazardous waste produced include halogenated and nonhalogenated solvents, organic chemical residues from still bottoms, sludges and tars, heavy metals, test animals, returned pharmaceuticals, low-level radioactive waste, and inert, such as filters which are contaminated with hazardous wastes. The total amount of hazardous waste produced is 296 million kilograms per month. This figure assumes all solid waste is contaminated (Arthur D. Little 1975; TRW Environmental Engineering Division 1979).

*Disposal Practices:* Fifty-two percent dispose of test animals, radioactive materials, solvents, and rejected pharmaceuticals by contract hauling. More than 80 percent of these firms did not know the final destination of the waste. About 17 percent incinerate test animals and solvents; 13 percent use landfills. Other firms use sewers to dispose of acids, bacterial, and viral cultures. On-site lagoons are used for solvents.

*SIC 284 Soap, Detergents, Cleaning Preparations, Perfumes, Cosmetics, and Other Toilet Preparations*

*Waste Composition:* Firms that report 99 percent of raw materials go into the product, and the primary component is the empty containers which may have residues. The SIC group is heterogeneous, and a wide variety of hazardous waste may be present. These hazardous wastes include flammable, toxic or corrosive (organic and inorganic) compounds. The amount of solid waste generated is estimated at 4.3 million kilograms per month (TRW Environmental Engineering Division 1979).

*Disposal Practices:* About 50 percent use landfills; 20 percent do not know what happens to their solid waste; others use lagooning and sewers. Drums and other containers are often recycled.

*SIC 2851 Paints, Varnishes, Lacquers, Enamels, and Allied Products*

*Waste Composition:* Hazardous wastes consist of sludges from process tanks, toxic metals, and toxic and ignitable solvents. These waste originate from raw materials, packaging, cleaning of blending tanks, thinning tanks, and other process equipment, dust from air-pollution-control devices, and waste finished products. The volume of waste produced is estimated a 1.9 million kilograms per month.

*Disposal Practices:* About 40 percent of the solid waste is landfilled; 35 percent is buried; 24 percent is incinerated, and 1 percent is lagooned.

*SIC 286 Industrial Organic Chemicals.* More than 700 chemicals are included in this SIC, and many of the plants change products to meet market conditions. Therefore, the waste generated by a plant may vary with time in quantity and composition.

*Waste Composition:* Solid wastes are generated by still bottoms, wastewater treatment sludges, and various toxic and ignitable organic compounds. The volume of waste generated is estimated at $5.8 \times 10^8$ kilograms per month.

*Disposal Practices:* About 58 percent of solid waste is landfilled; 20 percent is incinerated; and 22 percent is disposed of in some other fashion (TRW, n.d.).

## SIC 2879 Pesticides and Agricultural Chemicals not classified elsewhere.

*Waste Composition:* Waste from pesticide formulation originates from spills, "off spec" batches, equipment cleanup, mixing and grinding operations. This industry produces about 235,400 kilograms per month of solid waste.

*Disposal Practices:* About 80 percent of the solid waste is landfilled; 13 percent is incinerated; 7 percent is recycled, and a small amount is lagooned.

## SIC 2891 Miscellaneous Chemical Products, Adhesives, and Sealants

*Composition of Solid Waste:* Sludges from wastewater treatment processes and cleanup operations, end-of-run operations, "off spec" batches, and solvents.

*Disposal Practices:* 35 percent of plants use contract hauler to dispose of solid waste; 56 percent use evaporation followed by landspreading; 9 percent use deep-well injection. Alkaline waste are disposed of via sewers.

## SIC 2892, 2895, 2899 Miscellaneous Chemical Products not listed elsewhere including Explosives and Carbon Black

*Waste Composition:* Solvent wastes include xylene, acetone, ketones, methylene chloride, benzene, toluene, methanol, and trichloroethylene. Other organic waste include oil-type paste ink, coal tar, binder, and waxes. Metal hydroxides and compounds containing selenium and zinc are present. Inorganic waste include oxidizer, sodium lignosulfonate, incinerator ash silica sands powder pigments, and ink residues. A total of $272 \times 10^6$ kilograms per month is produced.

*Disposal Practices:* 38 percent of the plants generating solvents recycle or incinerate the solvents; 25 percent dispose of the solvents with landfills; 12 percent utilize landspreading for solvent disposal; about 50 percent of plants were not aware of final disposal of solid waste and employed a contract hauler. Lagooning is used for sodium lignosulfonate, sodium hydroxide and miscellaneous chemicals.

### SIC 2893 Miscellaneous Chemical Products, Printing Ink

*Waste Composition:* About 75 percent of the waste is solvent; the remainder is ink residues and miscellaneous resins, hardeners, and adhesives. A total of $2.2 \times 10^6$ kilograms per month of solid waste is produced.

*Disposal Practices:* 10 percent landfill; 10 percent land burial; and 80 percent is unknown.

Problems with solid wastes of the chemical industry seem mainly centered around the sludges which are generated in water purification and must be treated as solid waste. The sugar beet industry now dries pulp and sells it for sixty dollars a ton as animal feed. Lime mud is reburned or used as a sweetner for acid soils or in cement manufacture. In the sugar cane industry, a 96 percent reduction in sludge has been achieved by the use of multiple-hearth kilns operating at 816°C.

### Chemical Composition

Besides the two major groups of chemicals—inorganics and organics—in the past few decades three additional major groups—plastics, fibers, and elastomers—have been designated. Within each of these major groups are subgroups classified by further chemical similarities. They are listed in Table 15-48 along with some of the major products (as classified by production volume) in each. This classification scheme is widely used by *Chemical and Engineering News* in their articles on CPI performance.

The most significant specific organic chemicals are given as of 1975 in Table 15-49 as measured by sales indices.

Table 15-48. Classification of Chemicals by Composition.

1. Inorganics
   a. Acids
      Hydrochloric
      Nitric
      Phosphoric
      Sulfuric
   b. Chlorine and alkalies
      Chlorine gas
      Sodium carbonate
      Sodium hydroxide
   c. Industrial gases
      Carbon dioxide
      Hydrogen
      Nitrogen
      Oxygen
   d. Fertilizer chemicals
      Ammonia
      Ammonium nitrate
      Ammonium sulfate
   e. Other
      Aluminum sulfate
      Phosphorus
      Sodium phosphate
      Sodium silicate
      Sodium sulfate
      Titanium dioxide
2. Organics
   Acetone
   Acrylonitrile
   Aniline
   Benzene
   1.3-Butadiene
   Carbon tetrachloride
   Cyclohexane
   Ethanol, synthetic
   Ethanolamines
   Ethylene
   Ethylene oxide
   Formaldehyde
   Maleic anhydride
   Methanol, synthetic
   Perchloroethylene
   Phenol, synthetic
   Phthalic anhydride
   Propylene
   Propylene oxide
   Styrene
   Toluene
   Urea
   Vinyl acetate
   Vinyl chloride
   Xylenes
   $o$-Xylene
   $p$-Xylene
3. Plastics
   a. Thermosetting resins
      Epoxy
      Melamine
      Phenolic
      Polyester
      Urea
   b. Thermoplastic resins
      Polyamide
      Polyethylene
      Polypropylene
      Polystyrene
      Polyvinyl alcohol
      Polyvinyl chloride
4. Elastomers
   Natural rubber
   Styrene-butadiene
   Polybutadiene
   Butyl rubber
   Polyisoprene
   Neoprene
   Silicone
   Nitrile
   Ethylene-propylene

Table 15-48. continued

5. Fibers
   a. Cellulosic fibers
      Acetate
      Rayon

   b. Noncellulosic fibers
      Acrylic
      Nylon
      Olefins
      Polyester
      Textile glass

Source: *Chemical Engineering*, August 5, 1975, p. 98.

The carbon black industry practicing total recycling has reduced landfilling to zero. In the sulfuric acid industry, landfills receive the slag of combusted sulfur.

The phosphate industry is currently using holding ponds in the size range 0.1-0.4 acres per ton of phosphorous pentoxide per day. These ponds are very acidic (pH of 1.5), and gaseous flouride emissions are high. The system remains, however, for lack of a better one.

Plastic-waste problems, with case studies by the Stanford Research Institute indicate that only one out of seven large plants practice incineration because of the inexpensiveness of landfilling. Some recycling of injection mold scrap is practiced.

The federal government indicates that toxic chemicals, pesticides, acids, caustics, explosives, and biological wastes amount to 10 millions per year (see Table 15-50). Incineration, chemical treatment, and special landfilling are the three basic options for control.

Because of stricter air and water pollution standards, the land-destined industrial waste is increasing dramatically.

Twelve million pounds per year of polychlorinated biphenyls (PCBs) are landfilled. With hydrochloric acid (HCL) control, incineration for two to three seconds at 2000-2400°F destroys the molecule completely. The problems today are coming from PCB that was landfilled years ago. One example is from 1957-1971 National Carbon Research used 45 million pounds in carbonless copy paper.

There is an estimated 290 million pounds in dumps and landfills and 150 million pounds "free." The type of landfilling now practiced utilizes a site with a forty-foot-thick clay base with cells on top constructed to hold polyethylene-lined drums. Five feet of clay covers the cells, and each cell is equipped with a sump. Groundwater

Table 15-49. The Most Significant Organic Chemicals.

| Chemical | Sales Volume (Millions) | Chemical | Sales Volume (Millions) |
|---|---|---|---|
| Tetra (methyl-ethyl) lead | $292 | Bisphenol A | $34.4 |
| Ethylene oxide | 271 | Methylene chloride (dichloromethane) | 32.2 |
| Styrene | 260 | Chlorobenzene | 29.1 |
| Ethylene dichloride | 224 | Carbon disulfide | 28.8 |
| Ethylene glycol | 213 | Glycerol tri (polyoxypropylene) ether | 28.2 |
| Dimethyl terephthalate | 203 | Sodium carboxymethylcellulose | 27.9 |
| Vinyl chloride | 202 | Etheylene glycol | 25.6 |
| Ethylbenzene | 193 | Methyl chloride | 25.4 |
| Urea | 187 | $o$-Xylene | 24.0 |
| Adipic acid | 173 | Diethylene glycol | 23.9 |
| Tetraethyllead | 172 | Methyl isobutyl ketone | 23.9 |
| Acetaldehyde | 145 | Acrylic acid | 22.1 |
| Acetic anhydride | 143 | Nonyl phenol, ethoxylated | 21.2 |
| Methanol, synthetic | 132 | Sorbitol | 21.0 |
| Phenol | 123 | Pentaerythritol | 19.6 |
| Ethanol, synthetic | 118 | Etheylene glycol monoethyl ether | 17.7 |
| Acetic acid, synthetic | 116 | Tridecylbenzene sulfonic acid, sodium salt | 16.6 |
| Isopropanol | 115 | Ethylene glycol monoethyl ether | 16.0 |
| Acrylonitrile | 114 | $n$-Butyl acrylate | 15.2 |
| Propylene oxide | 106 | Cresylic acid, refined | 14.8 |
| Toluene 2.4- and 2.6-diisocyanate (80 : 20 mixture) | 97.6 | Diisodecyl phthalate | 14.8 |
| Dichlorodifluoromethane | 97.6 | Chloroform | 14.4 |
| $p$-Xylene | 95.4 | Decyl alcohols | 14.2 |
| Formaldehyde | 88.5 | | |

| | | | |
|---|---|---|---|
| Cumene | 79.3 | Ethyl acetate | 12.9 |
| Cyclohexanone | 78.6 | Ethylenediamine | 12.4 |
| Vinyl acetate | 72.3 | α- and β-pinenes | 11.6 |
| Tallow acids, sodium salt | 70.7 | Triethylene glycol | 11.6 |
| Phthalic anyhydride | 66.0 | Polypropylene glycol | 11.6 |
| Acetone | 64.6 | Hexamethylenetetramine | 11.5 |
| Alkyl benzenes | 55.3 | Triethanolamine | 10.9 |
| Cyclohexane | 55.2 | Ethylene glycol monomethyl ether | 10.7 |
| Ethylene dibromide | 53.4 | Fumaric acid | 10.6 |
| Dodecyl benzene sulfonic acid, sodium salt | 52.1 | Benzyl chloride | 10.5 |
| | | Diisooctyl phthalate | 10.2 |
| Carbon tetrachloride | 50.6 | Monoethanolamine | 9.6 |
| Perchloroethylene | 49.5 | DDT | 9.5 |
| Coconut oil acids, sodium and potassium salts | 47.9 | Diethanolamine | 9.3 |
| | | Lignin sulfonic acid, calcium salt | 9.2 |
| Trichlorofluoromethane | 46.4 | Isooctyl alcohols | 9.0 |
| Aniline | 43.8 | Nonyl phenol | 8.6 |
| Methyl ethyl ketone | 43.2 | n-Octyl-n decyl phthalate | 8.3 |
| Trichloroethylene | 42.9 | n-Butyl acetate | 8.2 |
| Ethyl chloride | 40.7 | Dimethylamine | 7.8 |
| Di (2-ethylhexy) phthalate | 38.5 | Lignin sulfonic acid, sodium salt | 6.6 |
| Propylene glycol | 38.5 | o-Dichlorobenzene | 6.6 |
| n-Butanol | 37.4 | n-Propanol | 6.6 |
| 2-Ethyl-1-hexanol | 36.6 | p-Dichlorobenzene | 6.3 |
| Glycerol, synthetic | 36.2 | Isobutanol | 4.0 |
| Ethyl acrylate | 35.2 | Dicyclopentadiene | 2.0 |
| Maleic anhydride | 34.4 | | |

Source: *Chemical Engineering*, August 5, 1975, p. 98.

is monitored monthly. Silicate cement and other gelling agents are used to solidify the oily substance.

The Chemical Manufacturers Association proclaims its expenditure of "hundreds of millions of dollars in safer, better waste-disposal methods." It presents five methods for obtaining these goals, especially by landfilling properly and solid waste exchanges. These are described in Figure 15-40, page 304.

The glass industry produces solid waste mainly from the diatomaceous earth filters (up to 20 tons per day). Grinding and polishing wastes are disposed with filter cake in landfills. The detergent industry averages several hundred pounds of solid wastes per day which are landfilled. This comes from the slag of the drying towers which is being recycled more and more.

The paint industry, even at best recycle efficiency, would still produce 17,000-175,000 cubic yards per year to be containerized and landfilled.

Future prospects for the chemical industry wastes in overview seem to be: (1) encapsulation prior to landfilling; (2) special landfills equipped with liners and leachate collection and treatment (continuous monitoring of groundwaters is necessary); (3) ocean and land-based high-temperature incineration; and (4) microwave plasma detoxification.

Table 15-50. Government Estimates of Amounts of Wastes per Year.

| Industry | Amount (Million Tons/Year) | | Percent Solid | Percent Organic |
| --- | --- | --- | --- | --- |
| | Wet | Dry | | |
| Pesticides and explosives | 7.33 | 2.2 | 30 | 99 |
| Paints | 0.42 | 0.11 | 20 | 15 |
| Petroleum refineries | 1.44 | 0.67 | 46 | 99 |
| Pharmaceuticals | 0.2 | 0.62 | 31 | 90 |

## NONFERROUS METALS

### Aluminum

*Origin.* In the process of manufacturing aluminum from bauxite, a residue known as red mud is generated which contributes the major form of solid waste in the manufacture of aluminum. The other sources of solid waste are the sludges resulting from the treatment of effluents emerging from the various manufacturing processes.

*Characteristics.* The red mud produced contains up to 45 percent of various degrees of iron oxides. The sludges produced by the treatment operations contain cryolite, carbon, and calcium fluoride. The quantity of sludge produced varies between 15 to 30 kilograms per metric ton of aluminum manufactured.

*Disposal.* There are three methods of utilizing the red mud. In the first method, when the red mud contains about 45 percent iron oxides, it may be heated with fine grain coal and ground limestone in a rotary kiln to reduce the iron oxide to metallic iron. In the second method of utilization, if the red mud contains low quantities of iron oxides, it could be used successfully for the manufacture of high-quality bricks. The presence of iron oxide in the brick material is bound to increase the compressive strength of the brick to 500 kilograms per square centimeter as against the commonly used bricks which have a compressive strength of 150 kilograms per square meter.

The third method of utilizing the red mud is by reacting it with sulfuric acid whereby a product known as ferrifloc results and is found to be a useful coagulant for the treatment of industrial and municipal wastewaters. The slurry containing cryolite produced by the thickening of the wet scrubbing liquor contains a solids concentration of 200-500 grams per liter. It is filtered to remove the liquid, and the solid cake (about 60 percent solids) is dried in a kiln or multiple-hearth furnace. If the cryolite is pure enough, it is returned to the reduction pots (electrolytic cell); if not, it is landfilled. In some cases the liquid is further treated to reduce the fluoride concentrations present, by precipitation as calcium fluoride, adding calcium chloride or lime. The calcium chloride produced is dried and sold as an industrial chemical of commercial grade purity. In order to reduce the volume of effluents containing cryolite, flouride, and so on, which also increases the solid-waste sludge produced, dry scrubbing methods are being used more and more for scrubbing the gas produced at the pot.

The total of aluminum solid wastes generated in three processes in both the United States and Europe is presented in Table 15-51.

### Copper

*Origin.* The smelting of secondary copper is preceded by a number of pre-smelting operations, some of which generate solid wastes. In the stripping process, when insulation and lead sheathing are removed from electrical conductors such as cables, significant quantities of solid wastes are produced. In the size-reduction process, if insulation material—for instance, from cables—are present, these contribute to the solid waste produced. In the cupola and blast-furnace operations, solid waste in the form of slag is produced.

*Characteristics.* The solid wastes generated in the stripping process contains organic materials such as plastics, paper, and other materials used as protective coverings on the copper cable. In the size-reduction process, similar characteristics of the solid waste are generally observed. The slag produced in the cupola and blast-furnace operations contain limestone, iron, aluminum, copper and silicon, zinc, and so on. The sludge produced in the effluent treatment plants can be as high as 200 pounds per ton of copper produced.

*Disposal.* In general, the disposal of the solid waste from stripping and size-reduction operations has been by landfilling. However, sometimes the solid waste is incinerated. The disposal of wastewater treatment plant sludges has been accomplished mostly by landfilling.

Solid wastes from the copper, lead, zinc, antimony, mercury, titanium, tungston, tin, and ferromanganese production are shown in Table 15-52.

Leonard (1979) quantifies copper and zinc hazardous solid wastes in Table 15-53. He recommends landfilling hazardous and nonhazardous solid materials in separate areas.

Table 15-51. Aluminum Solid Wastes Generated in the United States and Europe.

| Aluminum Total Solid Wastes (kg/metric ton) | United States | Europe |
|---|---|---|
| Prebake | 48 | 20-60 |
| Vertical soderberg | 40 | 20-60 |
| Horizontal soderberg | 49 | 20-60 |

Source: Organization for Exonomic Co-operation and Development (1977).

Table 15-52. Residual Generation Factors for Nonferrous Metal Smelting and Refining.

| Metal Category | Type of Residual | Residual Factor (kg/metric ton) | Hazard Rating Non | Haz. |
|---|---|---|---|---|
| Primary copper smelting and fire refining | Reverberatory slag | 3,000 | X | |
| | Acid plant sludge | 2.7 | | X |
| | Dusts | 17.0 | | X |
| | Miscellaneous slurries | 17.0 | | X |
| Electrolytic refining | Miscellaneous slurries | 2.4 | | X |
| Primary lead | Blast furnace slag | 410.0 | X | |
| | Slag fines | 30.0 | X | |
| | Acid plant sludge | 40.0 | | X |
| | Sinter scrubber | 19.0 | | X |
| Primary aluminum | Shot blast dust | 5.0 | X | |
| | Pot line scrubber sludge | 29.3 | | X |
| | Pot line skmming | 5.5 | | X |
| | Spent pot liners | 53.0 | | X |
| | Cast house dust | 2.5 | | X |
| Primary zinc Electrolytic Pyrometalurgical | Acid plant sludge | 17.0 | | X |
| | Miscellaneous slurries | 9.1 | | X |
| | Retort residue | 1,050.0 | X | |
| | Acid pot sludge | 122.0 | | X |
| | Retort residual (blue powder) | 10.0 | | X |
| | Cadmium plant residue | 1.8 | | X |

(Table 15-52. continued overleaf)

Table 15-52. continued

| Metal Category | Type of Residual | Residual Factor (kg/metric ton) | Hazard Rating Non | Haz. |
|---|---|---|---|---|
| Primary antimony | | | | |
| Pyrometalurgical | Blast furnace slag | 2,800.0 | | X |
| Electrolytic | Anolytic sludge | 210.0 | | X |
| Primary mercury | Kiln or retort residual | 207,000.0 | X | |
| Primary titanium | Chlorinated sludge | 330.0 | | X |
| Primary tungsten | Digester residual | 50.0 | | X |
| Primary tin | Smelting sludge | 915.0 | X | |
| Ferromanganese | Slag | 240.0 | X | |
| | Sludge | 165.0 | | X |
| Silicomanganese | Slag | 1,100.0 | X | |
| | Sludge | 98.5 | | X |
| Ferrosilica | Dust | 338.0 | X | |
| Ferrochromium | Slag | 1,750.0 | X | |
| | Dust | 150.0 | | X |
| Ferronickel | Slag | 31,000.0 | X | |
| | Skull plant | 5,300.0 | | X |
| | Dust | 84.0 | X | |
| | Sludge | 576.0 | | X |

Source: Leonard (1979: 248A).

Table 15-53. Hazardous Waste Generated by the Nonferrous Metals Industries.

Copper—172 kg/MT Fire forming
Copper—2.4 kg/MT Electrolytic refining

Lead—

Zinc—26.1 kg/MT Smelting and electrolytic refining
Zinc—122.0 kg/MT Pyrometallurgical smelting and refining

Source: Leonard (1979: 248A).

## Summary

Solvents are recycled whenever possible to reduce the cost of a new solvent and to avoid disposal costs. Industry is starting to keep hazardous waste from contaminating nonhazardous solid waste and is approaching solutions to the solid-waste problem. The hazardous toxic-waste problem has created opportunities for new developments such as incinerator ships, toxic-waste disposal companies, and process modifications. The "Super Fund" and cradle-to-grave tracking of hazardous material may well contain the hazardous-waste situation within manageable bounds. The backlog of polluting dumps, however, present very costly cleanup problems.

## REFERENCES

Agee, James L., and A. Cywin. 1974. *Development Document for Effluent Limitation Guidelines and New Source Performance Standards for Seafood Processing—Point Source Category*, EPA 44011-74-020-a. Washington, D.C.: U.S. Environmental Protection Agency.

American Cyanamid Co. 1980. "Annual Report." *News Highlights* (July 17).

Arthur D. Little Co. 1975. "Hazardous Waste Generation—Treatment and Disposal in the Pharmaceutical Industry," EPA Contract No. 68-01-2681.

"Assessment of Industrial Hazardous Waste Practices, Textile Industry." 1979. U.S. Department of Commerce.

Baum, B., and C. H. Parker. 1974. *Solid Wastes Disposed*, Vol. 2: *Recycle, Reuse, and Pyrolysis.* Ann Arbor, MI: Ann Arbor Science Publications, Inc.

Brackett, C. E. 1973. "Production and Utilization of Ash in the United States." Proceedings of the Third International Ash Utilization Symposium, Pittsburgh, p. 13.

Danielson, J. 1973. *Air Pollution Engineering Manual.* Washington, D.C.: U.S. Environmental Protection Agency. Cited in *Study of Electrostatic Precipitators Installed in Oil-Fired Boilers*, by Southern Research Institute Study for ERRI, 1978.

Dotson, G. K., et al. 1970. "Land Spreading, A Conserving and Nonpulluting Method of Disposing of Oily Wastes." Fifth Annual International Water Pollution Research Conference, San Francisco, July 26–August 1.

Edison Electric Institute. 1976. *Statistical Year Book of the Electric Utility Industry for 1966.*

Environmental Pollution Laboratory. 1970. Jacksonville, FL.

Environmental Protection Agency. 1971. *Industrial Waste Study of the Meat-Packing Industry.* Washington, D.C.: EPA.

———. 1972. *Salvage Markets for Materials in Solid Waste.* Washington, D.C.: ERA.

———. Office of Solid Waste Management Programs. 1976. "Assessment of Industrial Hazardous Waste Practices for Textile Industry." EPA-PB 258-953, Washington, D.C.: EPA.

———. 1978. *Land Cultivation of Industrial Wastes and Municipal Solid Wastes, State-of-the-Art Study,* Vol. 11, EPA-600/2-78-1406. Washington, D.C.: EPA.

———. 1979. *Economic Impact Analysis of Hazardous Waste Management Regulations on Selected Generating Industries,* EPA (SW/182c). Washington, D.C.: EPA.

Geisman, J. R. 1974. *Techniques for Disposal of Wastes from Fruit and Vegetable Processing.* Wooster: Ohio Agricultural Research and Development Center.

Golueke, C. G. 1970. *Comprehensive Studies of Solid Waste Management.* University of California Sanitary Engineering Research Laboratory.

Golueke, C. G., and P. H. McGauhey. 1970. *Comprehensive Studies of Solid Wastes Management.* Berkeley: University of California, Sanitary Engineering Research Laboratory.

Gouin, F. R., and J. B. Shanks. 1981. "Composted Gelatin Wastes Aids Crops." *Biocycle* 22, no. 4 (July-August): 41.

Grove, C. S., and Antoni, C. M. 1970. *Studies of Modifications of Solid Industrial Waste,* National Technical Information Service Publication No. PB 222-419.

Grove, C. S., Jr.; D. Marlow; M. Saini; J. Mandel; C. M. Antoni; and N. L. Nemerow. 1969. *Survey Review—Solid Wastes: Scope, Quality, and Quantity.* Washington, D.C.: U.S. Department of Health, Education and Welfare, Bureau of Solid Waste Management.

Gutt, W.; P. S. Nixon; M. A. Smith; W. H. Harrison; and A. D. Russel. 1974. *A Survey of the Locations, Disposal and Prospective Uses of the Major Industrial By-Products and Waste Materials,* CP 19/74. Washington, D.C.: Building Research Establishment.

Holland, W. F., et al. 1975. *The Environmental Effects of Trace Elements in The Pond Disposal of Ash and FGD Sludge,* EPRI RP 202. Palo Alto, Calif.: Electric Power Research Institute.

Kaiser, R., and Carotti. 1971. *Incineration of Plastics.* Department of Chemical E.&S.

Kincannon, C. B. 1972. *Oily Waste Disposal by Soil Conservation Process,* EPA RZ-72-100. Washington, D.C.: Environmental Protection Agency.

Lewis, R. S. 1976. "Sludge Farming of Refinery Wastes—Exxon's Bayway Refinery." Proceedings of the National Conference on Disposal of Residues on Land, St. Louis, Missouri, September 13-15.

MacCartney, J.C., and R.H. Whaite. 1969. *Pennsylvania Anthracite Refuse*. Washington, D.C.: U.S. Department of the Interior, Bureau of Mines.

McGinnis, R.A., ed. 1971. *Beetsugar Technology*, 2nd ed. Fort Collins, CO: Beetsugar Development Foundation.

Mehta, A. 1981. "Routes to Metals Recovery from Metal Finishing Sludge." In *Proceedings of the Third Conference on Advanced Pollution Control for the Metal Finishing Industry*, EPA-600/2-81-028, p. 76. Washington, D.C.: Environmental Protection Agency.

Middlebrooks, E.J. 1979. *Industrial Pollution Control*, I, *Agro Industries*. New York: Wiley Interscience.

National Ash Association. 1978. *Ash at Work*.

National Slag Association. n.d. *Processed Blast Furnace Slag*, Bulletin NSA 171-3. Alexandria, VA: NSA.

O'Gorman, J.V., and P.L. Walker, Jr. 1972. *Mineral Matter and Trace Elements in U,S. Coals*. Washington, D.C.: U.S. Department of the Interior.

Office of Science and Technology. 1969. *Solid Waste Management*. Washington, D.C.: U.S. Government Printing Office.

PEDCO Environment Inc. 1979. "Study of Adverse Effects of Solid Waste from All Mining Activities on the Environment." Draft prepared for the U.S. Environmental Protection Agency, Office of Solid Waste.

Pettigrew, R.J., and E.H. Roninger. 1971. *Rubber Reuse and Solid Waste Management*, Pt. 1. Washington, D.C.: U.S. Environmental Protection Agency.

Plank, C.O.; D.C. Martens; and D.L. Hallock. 1975. "Effect of Soil Application of Fly Ash on Chemical Composition and Yield of Corn and on Chemical Composition of Displaced Soil Solutions." *Plant Soil* 42 465-476.

Rousseaux, J.M., and A.B. Craig, Jr. 1981. "Stabilization of Heavy Metal Wastes by the Soliroc Process." In *Proceedings of the Third Conference on Advanced Pollution Control for the Metal Finishing Industry*, EPA-600/2-81-028. p. 70. Washington, D.C.: Environmental Protection Agency.

SCS Engineers. 1976. *Assessment of Industrial Hazardous Waste Practices—Leather Tanning and Finishing Industry*, NT1S PB261018.

*SIC Manual*. 1072. Washington, D.C.: Executive Office of the President, Office and Management and Budget.

*Solid Waste/Disease Relationships*. 1967. Washington, D.C.: U.S. Department of Health, Education and Welfare, Public Health Service.

*Status of Solid Waste Management in California*. 1968. Sacramento: State of California Department of Public Health.

*Technical Economic Study of Solid Waste Disposal Needs and Practices Industrial Inventory*, Vol. II. 1969. Washington, D.C.: U.S. Department of Health, Education and Welfare, Public Health Service.

"The Iron and Steel Industry and the Environment." *Industry and Environment* 3, no. 2 (May-June).

Theis, T. L. 1975. "The Potential Trace Metal Contamination of Water Resources Through the Disposal of Fly Ash." Proceedings of the Second National Conference on Complete Water Reuse, Chicago, p. 219.

Thomas, Ramo, Wallridge. 1979. "Assessment of Solid Waste Management Problems and Practices in the Inorganic Chemical Industry," TRW Report Contract No. 68-02-2613.

Thomas, Ramo, Wallridge Environmental Engineering Division. 1979. "Technical Environmental Impacts of Various Approaches for Regulating Small Volume Hazardous Waste Generators," Contract No. 68-02-2613.

U.S. Department of Agriculture. 1957. *Processing Poultry Byproducts in the Poultry-Processing Industries*, Marketing Research Report No. 181. Wahington, D.C.

Versar, Inc. 1979. EPA Contract No. 68-03-2604.

"Waste Oil Study—Report to Congress." 1974. U.S. Department of Commerce and Environmental Protection Agency, PB-257693.

Watson, K. L. 1980. *Engineering and Environmental Aspects.* London: Applied Science Publishers.

Weston, R. F. 1970. *A Statewide Comprehensive Solid Waste Management Study.* Albany, N.Y.: New York State Department of Health.

Willson, G. B., et al. 1980. *Manual for Composting Sewage Sludge by the Beltsville Aerated Pile Method*, EPA 600/8-80-022. Cincinnati: Office Research and Development, Environmental Protection Agency.

# 16 ENERGY-PROCESS SYSTEMS FROM SOLID WASTES

In Chapters 5, 6, 8, and 9 we discussed the theories and general practices for utilizing incineration, pyrolysis, anaerobic digestion, and composting for treatment methods. We point out here that these same methods may be incorporated as parts of systems to recover products for re-use in our industrial cycle. All through this text I have advocated that optimum treatments of solid wastes are those which incorporate recovery and re-use as an integral part of the process. In fact all treatments for ultimate disposal should be discouraged when valuable materials are not recovered. The systems are presented here—largely aided by schematic drawings—were originally presented in 1977 by the U.S. Environmental Protection Agency's Office of Solid Waste Management programs. The publication is entitled *Resource Recovery Plant Implementation, Guide for Municipal Officials—Part 2, Technologies*, and was prepared by Steven J. Levy and H. Grigor Rigo.

As far as recovering energy from burning municipal solid wastes, Payne (1976) concluded that the practice is becoming increasingly widespread. He predicted that there could be over fifty such plants operating in North America utilizing approximately 60,000 tons per day. However, he reports that the entire country's municipal solid waste, if used as an energy source, could only provide less than 1 percent of the energy demand.

In its simplest—and most wasteful—form, a typical incineration diagram is shown in Figure 16-1. Here solid waste is unloaded into the refuse pit and transferred by an overhead loading crane directly into the furnace. The gases of combustion heat the boiler water circulating through the overhead convection section of the furnace before passing through electrostatic precipitators for air-pollution control and to the smoke stack. The unburned ash residue is collected from under the furnace grates for ultimate disposal, usually in a landfill. Stack gases and ash residue represent potential environmental contaminants of some concern to the solid-waste engineer. Steam represents the only recovered product from this process. Payne (1976) finds that each ton of municipal refuse contains about $9 \times 10^6$ BTUs ($9.5 \times 10^6$ kT) of heat energy which can be recovered from the steam at an overall efficiency of between 10 and 60 percent.

Before solid waste can be disposed of more effectively in landfills or incinerated efficiently or re-used for various products, it must be properly pre-treated. One such system is proposed and shown schematically by Levy and Rigo in Figure 16-2. Solid waste is shredded and then separated by an air-classification unit. The heavies (metals, glass, and so on) are removed from the bottom and recovered for sale or are landfilled. The air-classified lights are trommelled to remove the grit (which is combined with the heavy rejects for landfilling). The grit-free lights are shredded or ground again for producing fluffed or dusted raw dry fluff (RDF). Dust from this system is a major environmental concern, especially to the workers. Ferrous and nonferrous metal recovery makes this process economical especially when combined with the sale of the raw dry fluff.

This raw dry fluff can be re-used to produce electrical energy. In the $150 million, Dade County (Florida) Solid Waste Resource Recovery Facility which took three years to build, this process is occurring (*Miami Herald*, March 26, 1981, p. 4C). Two overhead cranes, in this plant, pick up the refuse and dump it onto a conveyor belt. The belt carries the refuse to a tubelike trommel, twelve feet in diameter and eighty-five feet long, which rotates and grinds the garbage as it pounds its sides. Nearly all the glass falls through two and one half-inch holes in the trommel, ready for cleaning and recycling. The rest is mashed up into small pieces, then sent through cyclones that shoot combustible, organic solids like paper out the top and metals, stones and other inert material out the bottom. The inert chunks move

ENERGY-PROCESS SYSTEMS FROM SOLID WASTES 333

Figure 16-1. Typical Waterwall Furnace for Unprocessed Solid Waste.

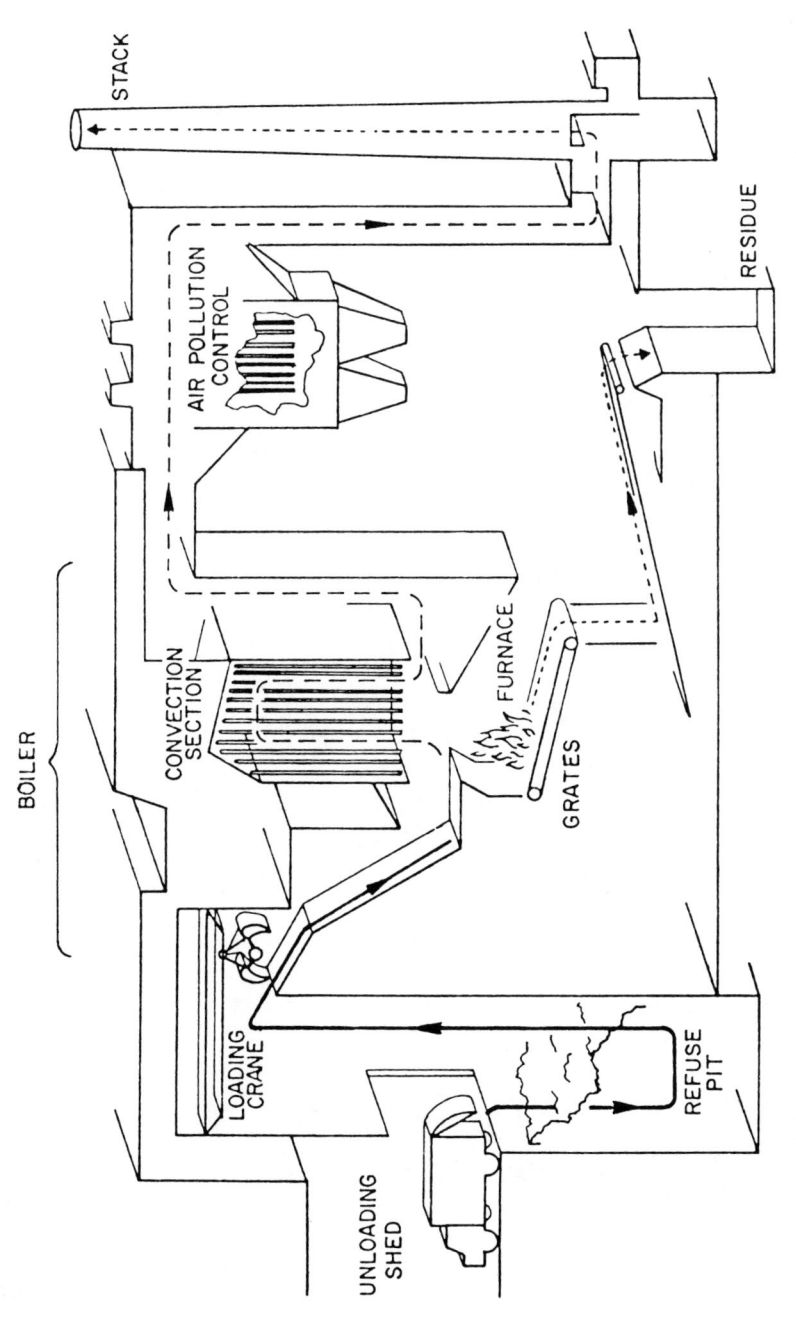

Source: EPA, 1977).

334  INDUSTRIAL SOLID WASTES

**Figure 16-2.** Simplified Flow Diagram Showing How the Dry Processing Approach Is Used To Produce Fluff, Densified, or Dust RDF.

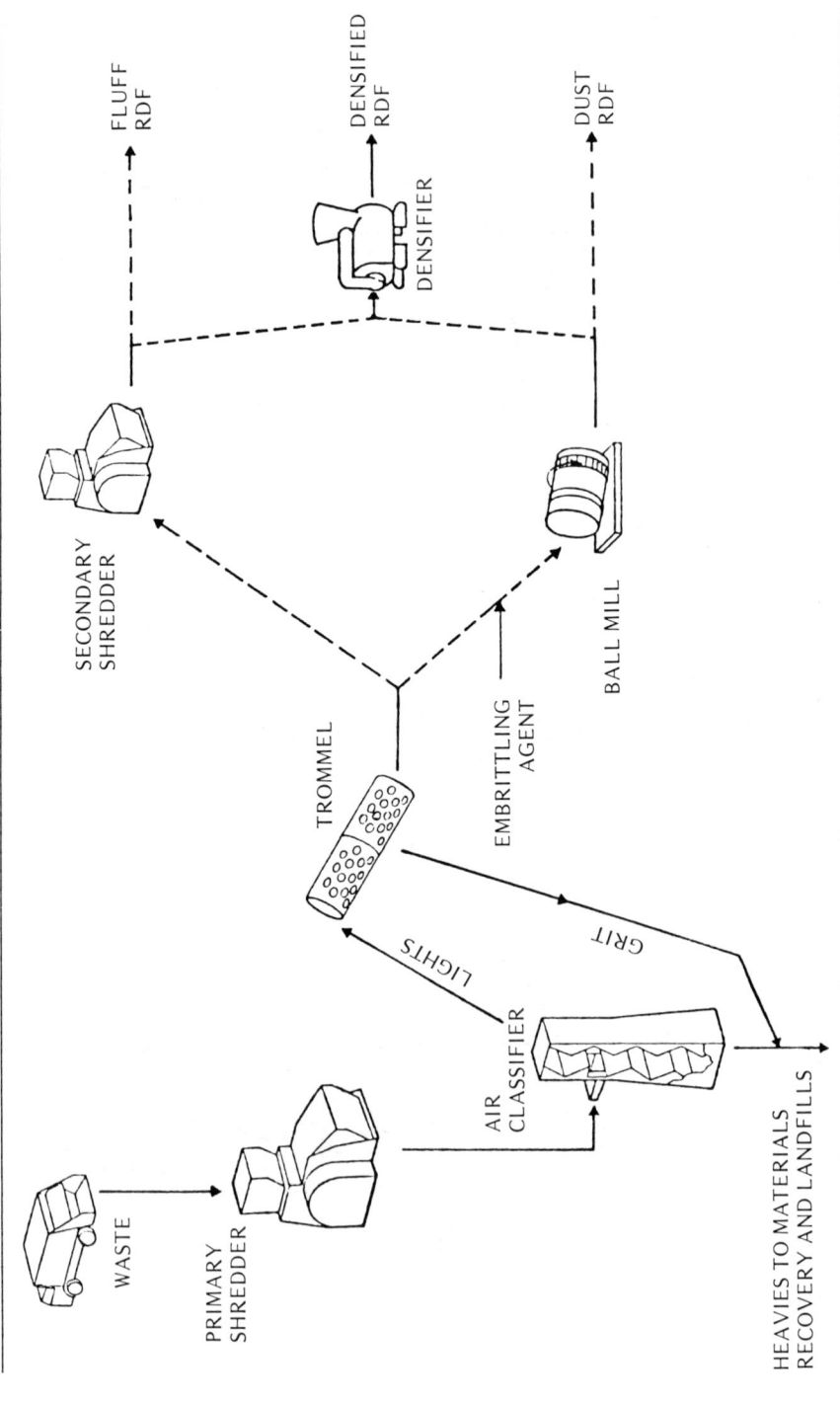

Source: Levy and Rego (1977).

along trays three feet wide and bounce wildly on huge springs. Magnets pull out the ferrous metals. A Tampa firm pays $50 a ton for that. Mechanical sorters pick out the prized aluminum. Another company has contracted for this process at about $600 a ton. A third firm will buy the glass.

The combustible portion is shredded into a fluffy gray mix, with tiny shards of wood, bits of plastic bags, and pieces of cardboard. This mixture is called RDF-refuse-derived fuel.

The RDF serves as fuel for the boilers which encase miles of pipes carrying 75,000 gallons of water per minute. The water is also recycled. The water, when converted to steam, spins the blades of turbines which produce electricity to run the plant, and, at a capacity of 3,000 tons per day, will supply current for more than 40,000 homes. Thirty feet away from the plant's output lines is Florida Power and Light's high-voltage system. As soon as legal problems are solved, F.P. & L. will purchase this electricity from the plant.

While the previous process was a dry one, another system is gaining in acceptability for burning solid waste while recovering metals and power. It is an hydrapulping wet process as depicted by Levy and Rigo in Figure 16-3. Solid wastes are loaded into a hydrapulper into which water is also added. Ferrous and nonferrous metals are removed by a combination of traveling magnets and liquid cyclones. The residual solid waste is fed to a surge chest, thickened by a combination of barrel press and cone press before being fed to the boiler. The pressed water is recycled to the hydropulper. Steam from the boiler drives a turbine which turns a generator while the low-pressure steam is returned to the process. The system theoretically produces a minimum of environmental impact since the water is completely recycled. However, boiler ash must still be disposed of, and microbiological degradation must be controlled in the recirculating press water.

Combustion of solid wastes yield gases which can be used directly to produce steam for energy manufacturing. Such a system has been derived by Monsanto Langard and is shown in Figure 16-4. Solid wastes are shredded and after storage are fed to a kiln for burning with the aid of auxiliary fuel oil. The burned solids are passed through magnets for ferrous metal recovery while the glassy aggregate and char remaining must be disposed of. Kiln gases are burned again to ensure complete combustion and the heat liberated is used to heat boiler water to produce steam for sale. Water is fed into the

**Figure 16-3.** Wet Process Energy Recovery System.

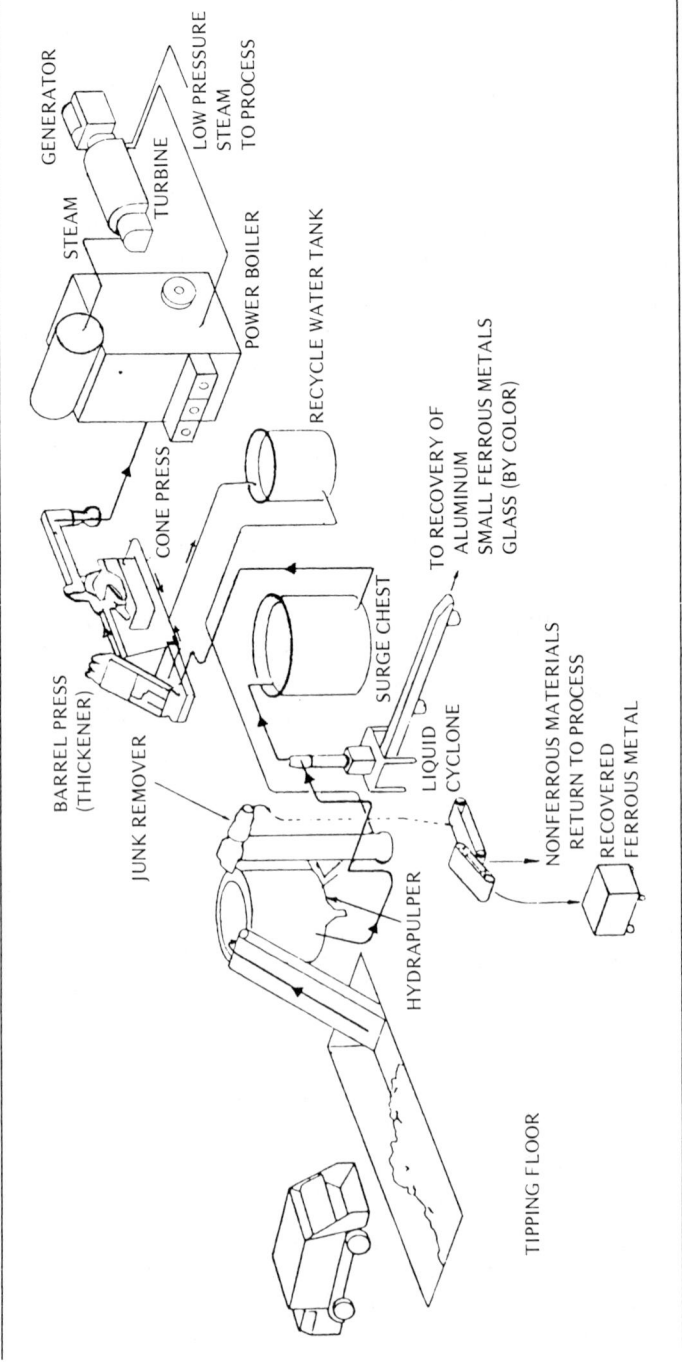

Source: Levy and Rigo (1977).

ENERGY-PROCESS SYSTEMS FROM SOLID WASTES    337

Figure 16-4. The Monsanto Landgard System.

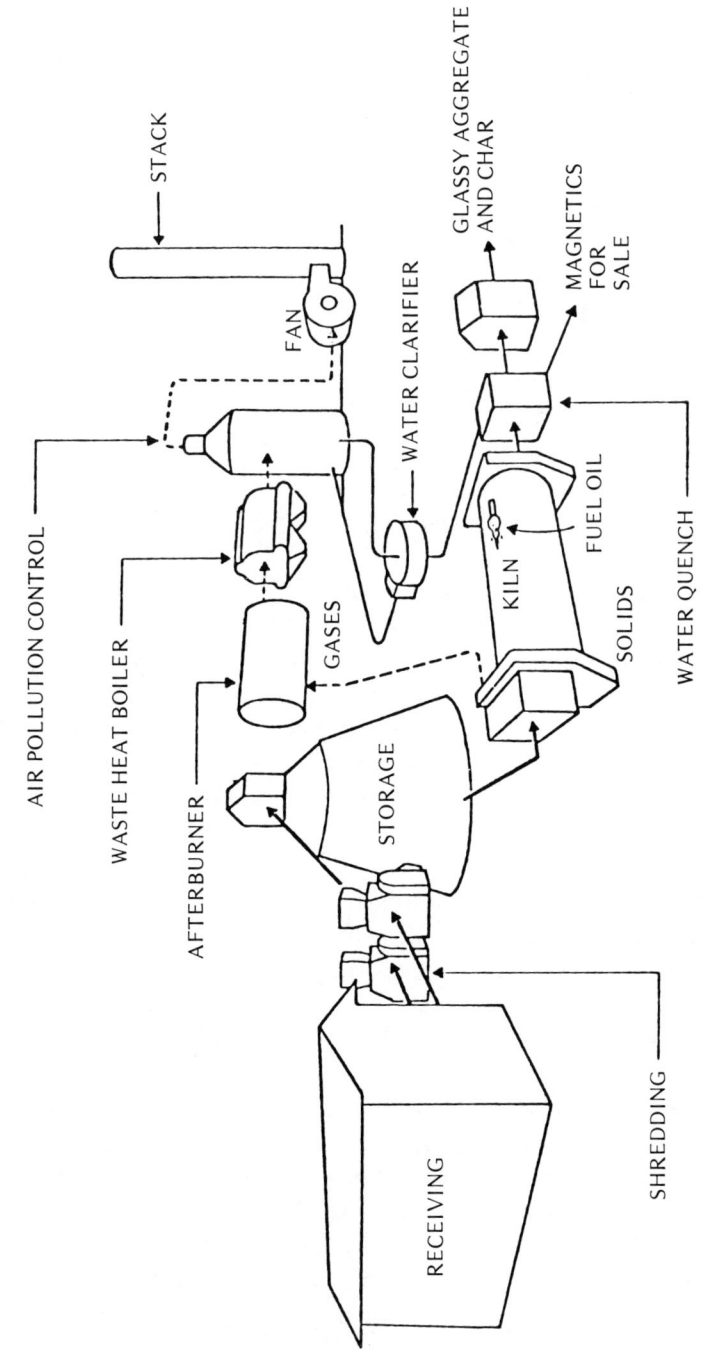

Note: This system produces a low BTU gas which is burned for steam production.
Source: Levy and Rigo (1977).

system to quench the kiln-burned solids and is clarified to remove entrapped solids. The clarifier overflow water is used to scrub the waste boiler gases to prevent air pollution before existing at the smokestack. Steam and ferrous metals are recovered while char and stack gases still enter the environment.

It is not always necessary to burn (incinerate) solid waste completely. The Torrax System utilizes pyrolysis to produce steam for direct re-use by industry (Figure 16-5). Solid waste is fed directly into a degasifier through which compressed and heated air is forced counter-current to the downflowing refuse. Inert residue exists at the bottom while heated refuse and gases are piped directly to the pyrolyzer which combusts (at somewhat lower temperatures than used in incinerators) the entire organic solid wastes. The hot gases are led to the boiler where water is converted into steam for sale to industry. Residual gases are cleansed by electrostatic precipitation prior to discharge. Pyrolyzed slag, gasifier inerts, and electrostatic precipitated solids all remain to be disposed of ultimately while waste boiler gases represent some threat to atmospheric quality.

Another pyrolysis system, using pure oxygen rather than air, is available from Union Carbide Company, a major manufacturer and supplier of pure oxygen. The Union Carbide Purox System is illustrated in Figure 16-6. Shredded, magnetically separated refuse is fed into the pyrolysis furnace up through which pure oxygen is forced. Solid residue is withdrawn at the bottom and quenched. The furnace off gas is scrubbed by recycled water, purified further by electrostatic precipitation, and finally cooled before being sold as product gas to consumers. Off-gas solids recovered from the scrubber and precipitator are recovered by physical separators and returned to the furnace. The only major solid waste from this system is the furnace residue, while liquid wastes also result from quench water overflows and excess scrubber water.

Still another system using pyrolysis makes more use of recovered materials and produces oil rather than gas for its energy product. It is called the Occidental Process and shown in Figure 16-7. After shredding and air classification, the heavies are collected from the bottom. Ferrous metals are separated magnetically from the heavies while the glass is separated from aluminum by trommeling. The glass is further concentrated by froth flotation which floats the glass of all colors, while the heavier residuals are collected at the bottom and landfilled.

ENERGY-PROCESS SYSTEMS FROM SOLID WASTES 339

Figure 16-5. Torrax Slagging Pyrolysis System.

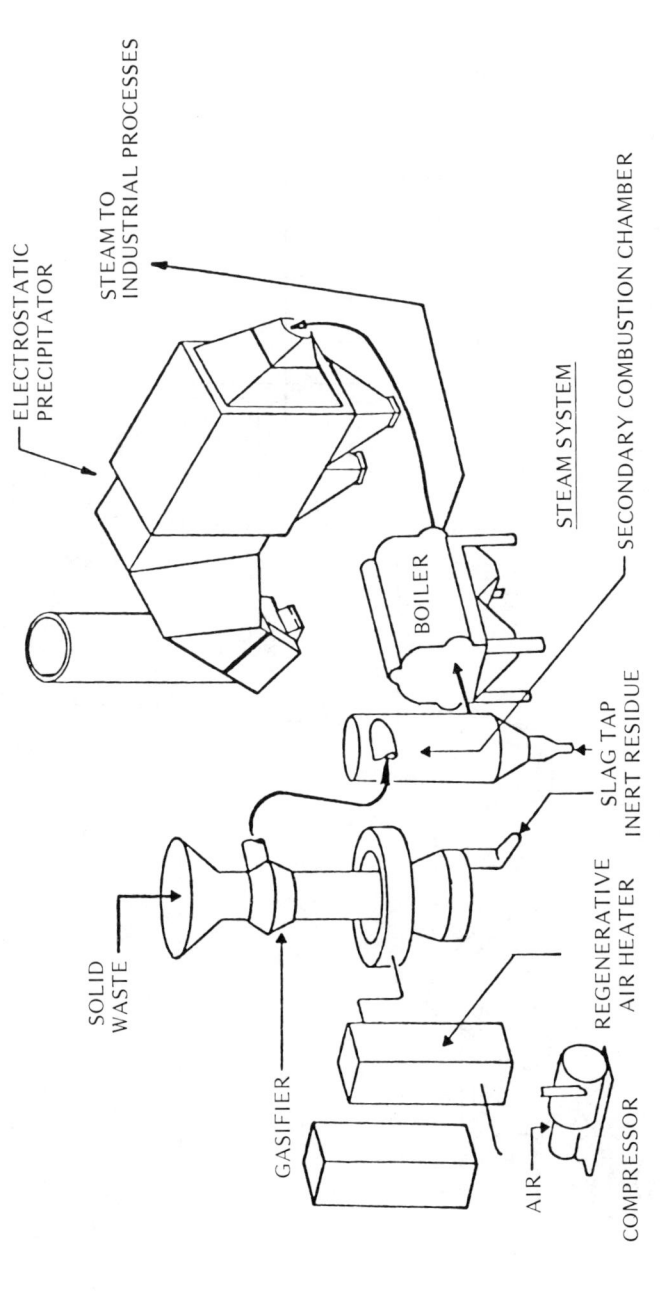

Source: Levy and Rigo (1977).

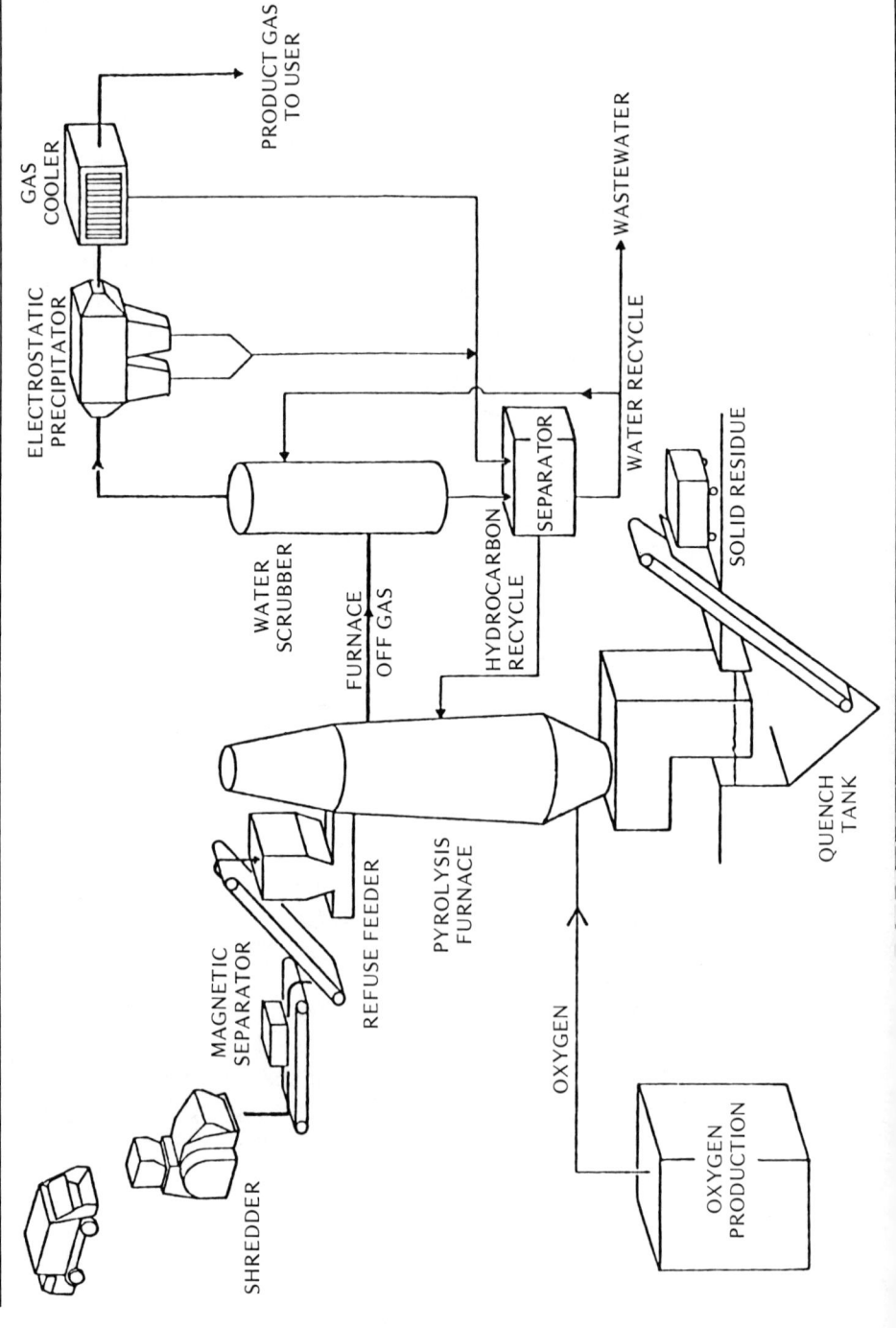

**Figure 16-6.** Union Carbide Purox System.

ENERGY-PROCESS SYSTEMS FROM SOLID WASTES 341

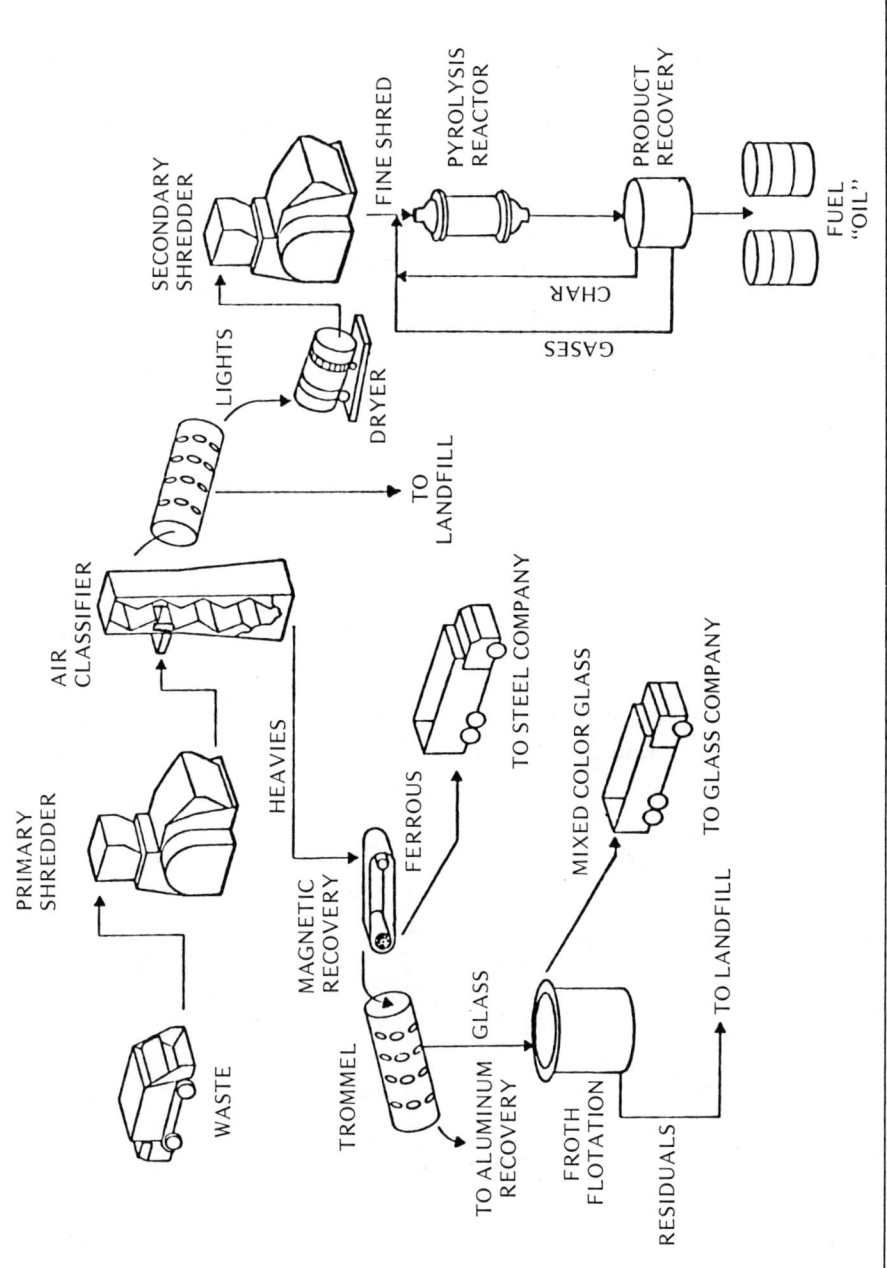

Figure 10-7. Production of "Oil" from Solid Wastes Using the Occidental Process.

Source: Levy and Rigo (1977).

The air-classified lights are also further trommeled lights with the heavy rejects also being landfilled. The trommeled lights are dried, shredded again, and fed to the pyrolysis reactor. The char and gases are recirculated into the reactor, while the fuel oil is recovered and sold along with ferrous and aluminum metals and glass. Only the landfilled rejects represent a strain on the land environment. No air or water contaminants result.

When municipal solid waste is relatively high in organic matter and can readily be landfilled, energy can be obtained by directly collecting and utilizing the gas generated after microbial decomposition. Such a system is presented schematically by Levy and Rigo (1977) in Figure 16-8. A perforated gas-collection tube is placed nearly to the bottom of the landfill. The gas collected in this tube rises to the surface either to drive a reciprocating engine coupled to an electric

**Figure 16-8.** Production of Electricity from Landfill Gas.

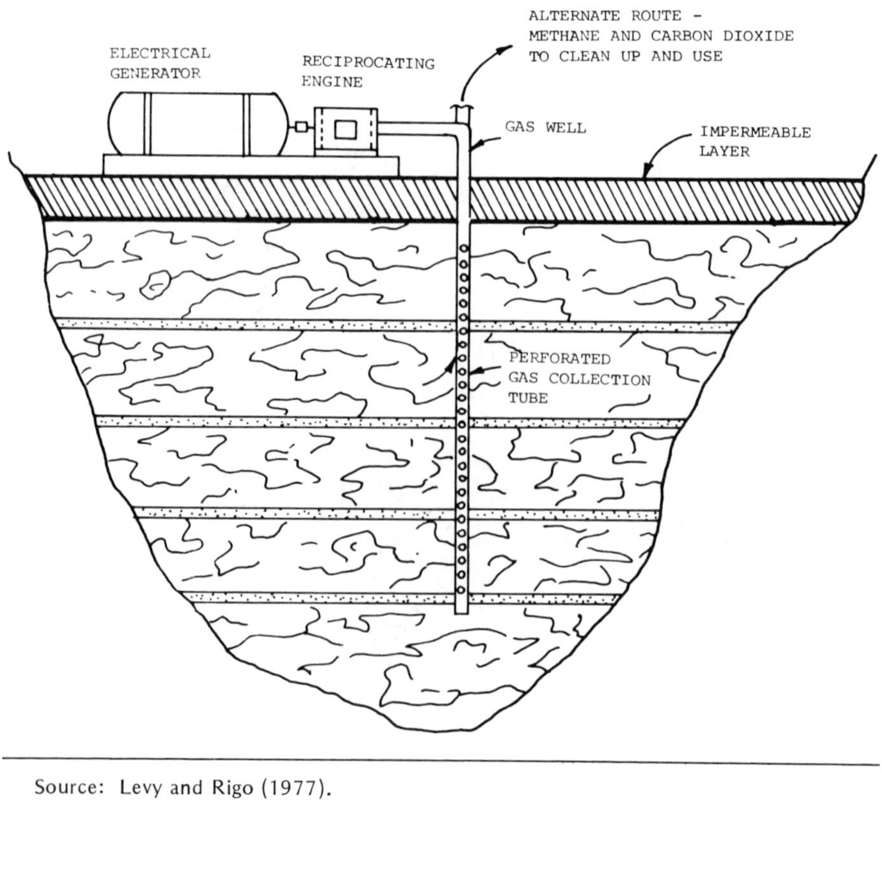

Source: Levy and Rigo (1977).

generator or to gas purification and subsequent use of the methane. This system is relatively inexpensive and efficient, but is low in yield and is land-consuming.

Anaerobic digestion of solid wastes can be used above ground and under controlled feed, temperature, and retention conditions. The highly methanated gas can be sold directly or used to produce steam in a boiler for direct sale or conversion to electricity through a turbine-generator hookup. Such a system is being studied by the University of Miami team on a 100-ton-per-day pilot plant in Pompano Beach, Florida. Schematic drawings of the system and mass balances are shown in Figures 16-9 and 16-10. Shredded municipal refuse is air-classified, the heavy rejects being landfilled. The air-classified lights are shredded again and fed to the continually mixed anaerobic digester. Digester slurry is vacuum-filtered, the sludge being landfilled regularly while the filtrate is recirculated within the dilution and digestion system. At present the digester gas is monitored and flared to the atmosphere, but in the future it will be utilized at a full-scale installation.

In another system to produce electricity the gas generated by a special combustion chamber is used to power a gas turbine which, in turn, turns the generator producing electrical energy. The Gas Turbine Generating System as sketched by Levy and Rigo (1977) is shown in Figure 16-11. Fluff RDF is produced from solid waste by a combination of shredding and air classification. This is fed to fluid pressurized bed combuster into which air is also fed from the turbine compressor.

The combustion gases are cycloned and gravel-bed filtered before feeding directly into the turbine compressor. The cyclone rejects are sand and bag house-filtered. These rejects, along with the shredded and air-classified heavies, must be landfilled or ultimately disposed of in another manner.

Wet fiber can either be recovered from municipal solid waste for resale or burned in a fluidized bed reactor. Levy and Rigo (1977) describe the rather complicated Wet Process Fiber Recovery System in Figure 16-12. Solid waste is fed from the tipping floor into the hydrapulper from where the slurry is sent to the liquid cyclone. The heavies are separated by a belt-magnet separator into magnetics (ferrous) and mainly nonferrous (aluminum and glass). The lighter fraction is sent to the fiberclaim system for cleaning, screening, densification, pressing, and final dilution for wet-fiber production and

344  INDUSTRIAL SOLID WASTES

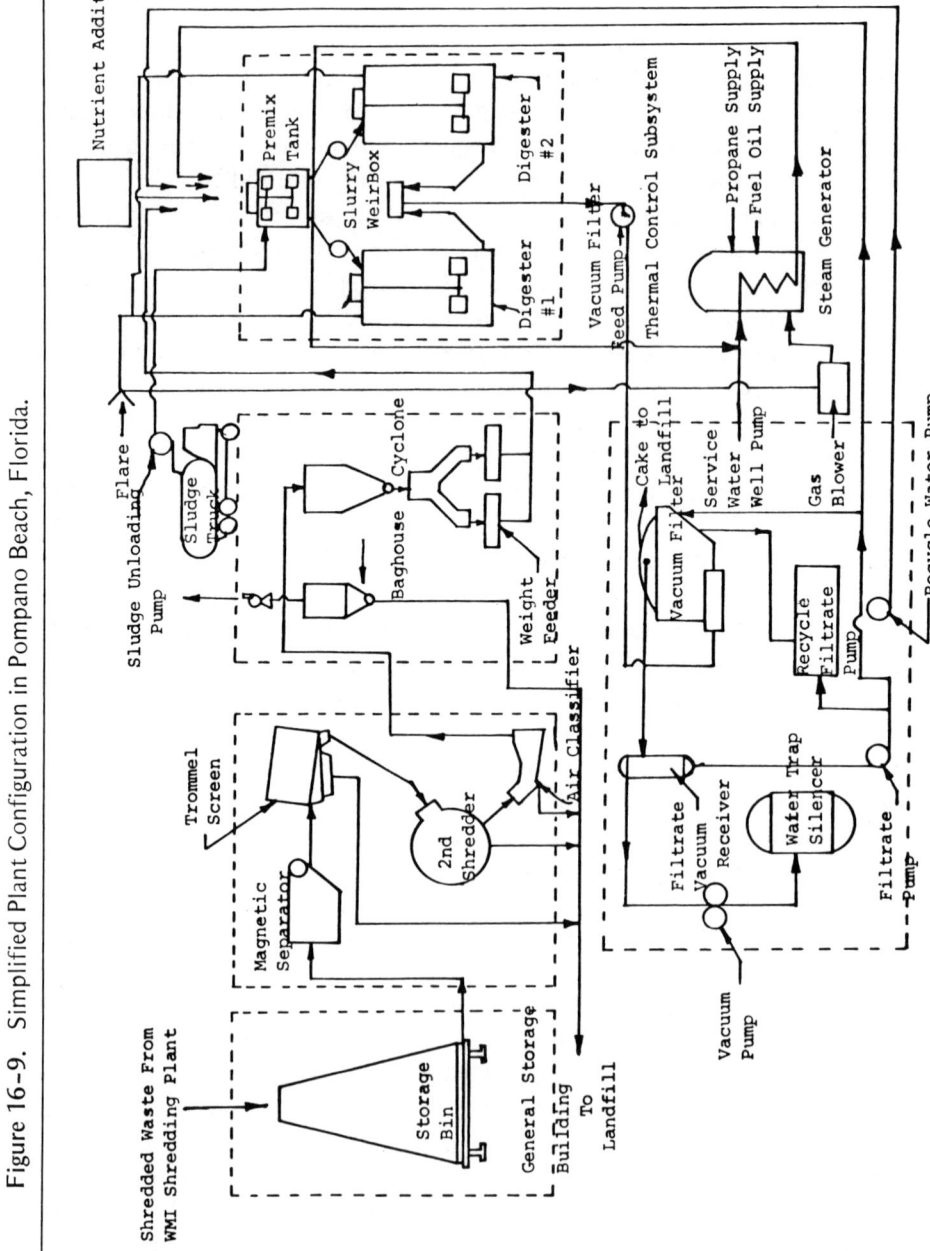

Figure 16-9. Simplified Plant Configuration in Pompano Beach, Florida.

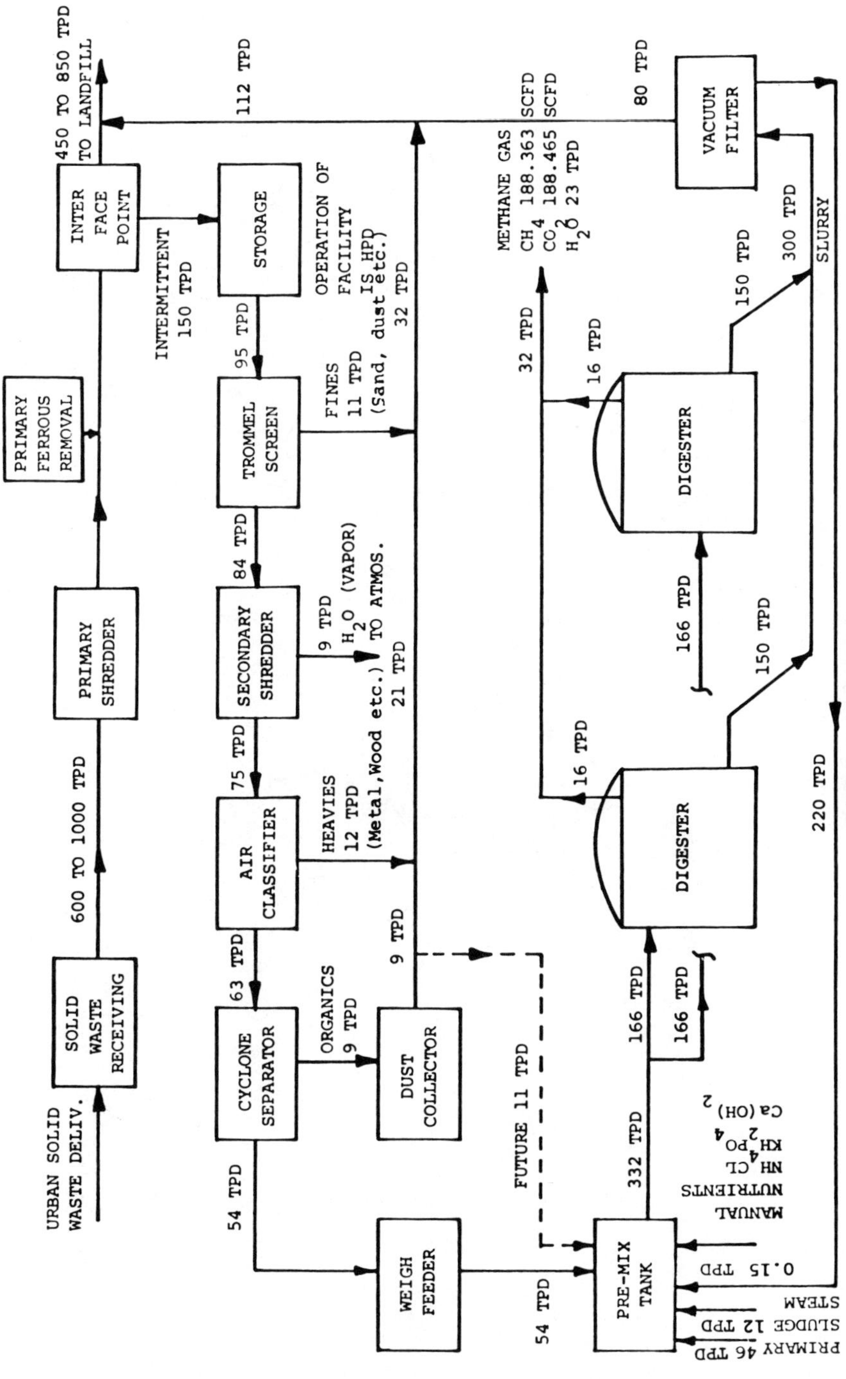

Figure 16-10. RefCOM Process.

346  INDUSTRIAL SOLID WASTES

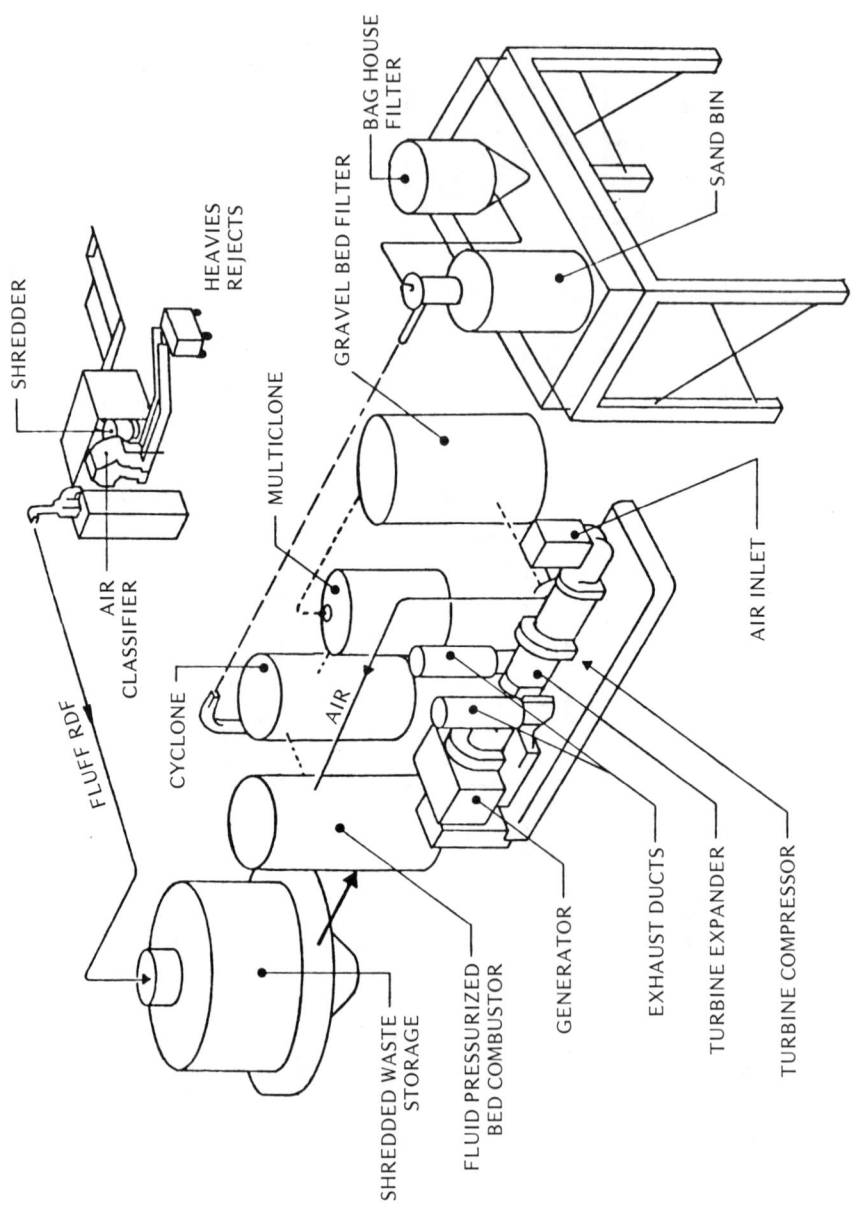

Figure 16-11. Gas Turbine Generating System Using Refuse as a Fuel (CPU-400).

**Figure 16-12.** Wet Process Fiber Recovery System (*after Levy and Rigo, 1977*).

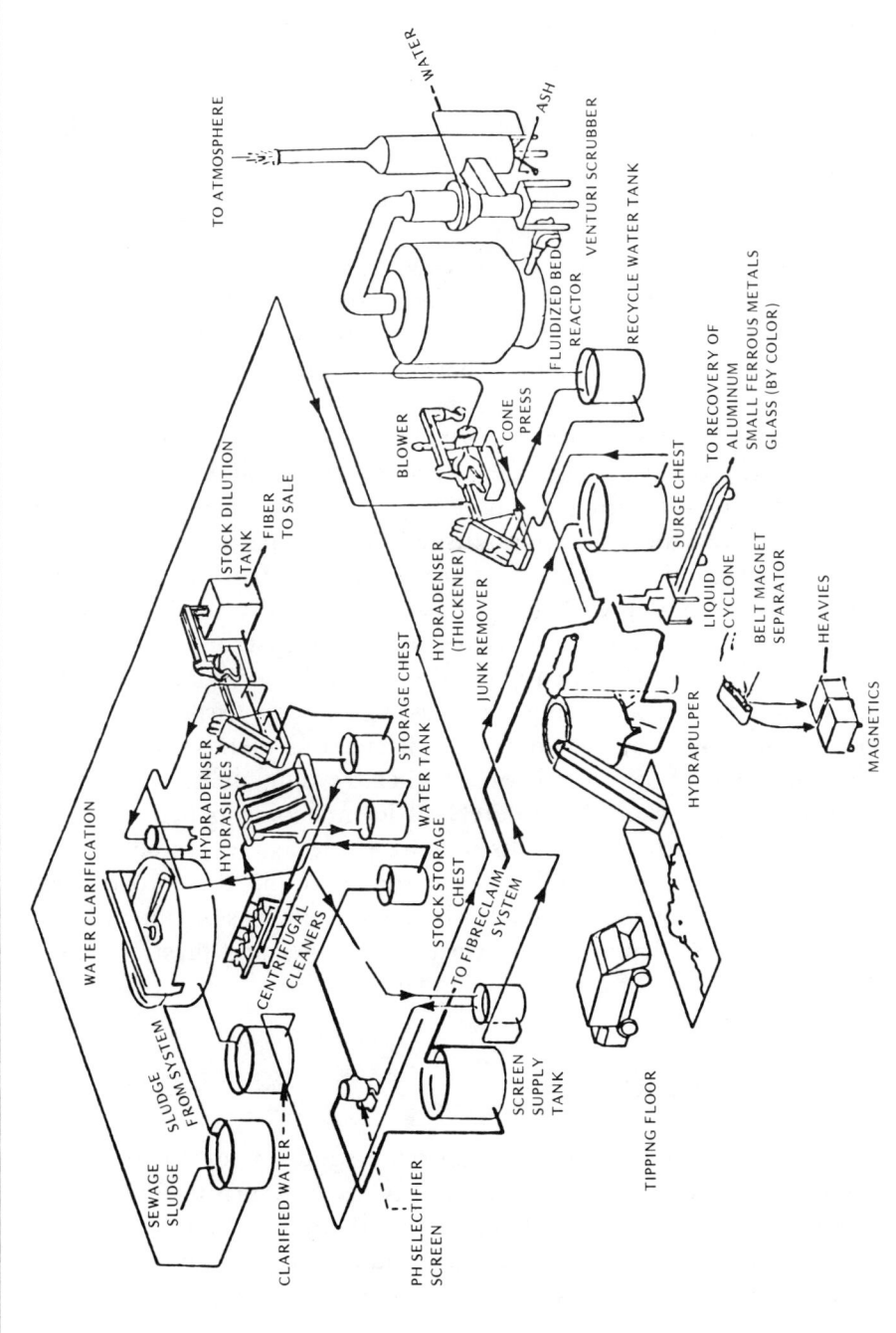

Source: Levy and Rigo (1977).

sale. The wastewater sludge and municipal sewage sludge are returned to the thickener and cone press for feed to the fluidized bed reactor. Ash is disposed of—usually on a landfill—while the reactor gases are vented (after Venturi scrubbing) to the atmosphere.

When a system is operated primarily as a recovery system, it can be referred to as a Dry Process Paper and Materials Recovery operation as shown by Levy and Rigo (1977) in Figure 16-13. The stored refuse is conveyed to a primary drum separator. The heavies are reprocessed in a secondary drum separator, and again the heavies are passed on electromagnets to recover ferrous metals. The lights from both primary and secondary separators are flailed before magnetized for additional ferrous metal recovery. Nonferrous materials are fed into both air classifiers. The lights from the one air classifier are recovered for fiber use; from the others, for plastic recovery. The heavies from the other air classifier are recovered as aluminum, and from the third to a rotating drum separator before finally producing a waste solid residue for disposal, usually by burial.

Since magnetic separation of ferrous metals from paper and plastics is an integral pre-treatment for most solid-waste recovery and treatment systems, a schematic detail of the system is presented by Levy and Rigo (1977) in Figure 16-14. Shredded solid waste enters from the left on a moving belt passing under a suspended rotating belt magnet. The magnet picks up the ferrous metals and allows the demagnetized waste to exit on another travelling belt. Since the separation system is not completely efficient, some paper and plastic are removed with the ferrous metals. These are separated by an air knife retaining the metallic pieces and directing them to travelling belt solely for magnetics, while the paper and plastic films exit at a third belt. The demagnetized waste and paper and plastic films can be treated further or disposed of in landfills.

## SUMMARY

The future of pollution potential from solid wastes remains dependent upon the willingness and ability of enterprising industries in recovering and re-using valuable constituents of the various wastes. This involves spending some capital and operating monies in order to recover these byproducts and to avoid environmental contamination.

Figure 16-13. Dry Process Paper and Materials Recovery (*after Levy and Rigo, 1977*).

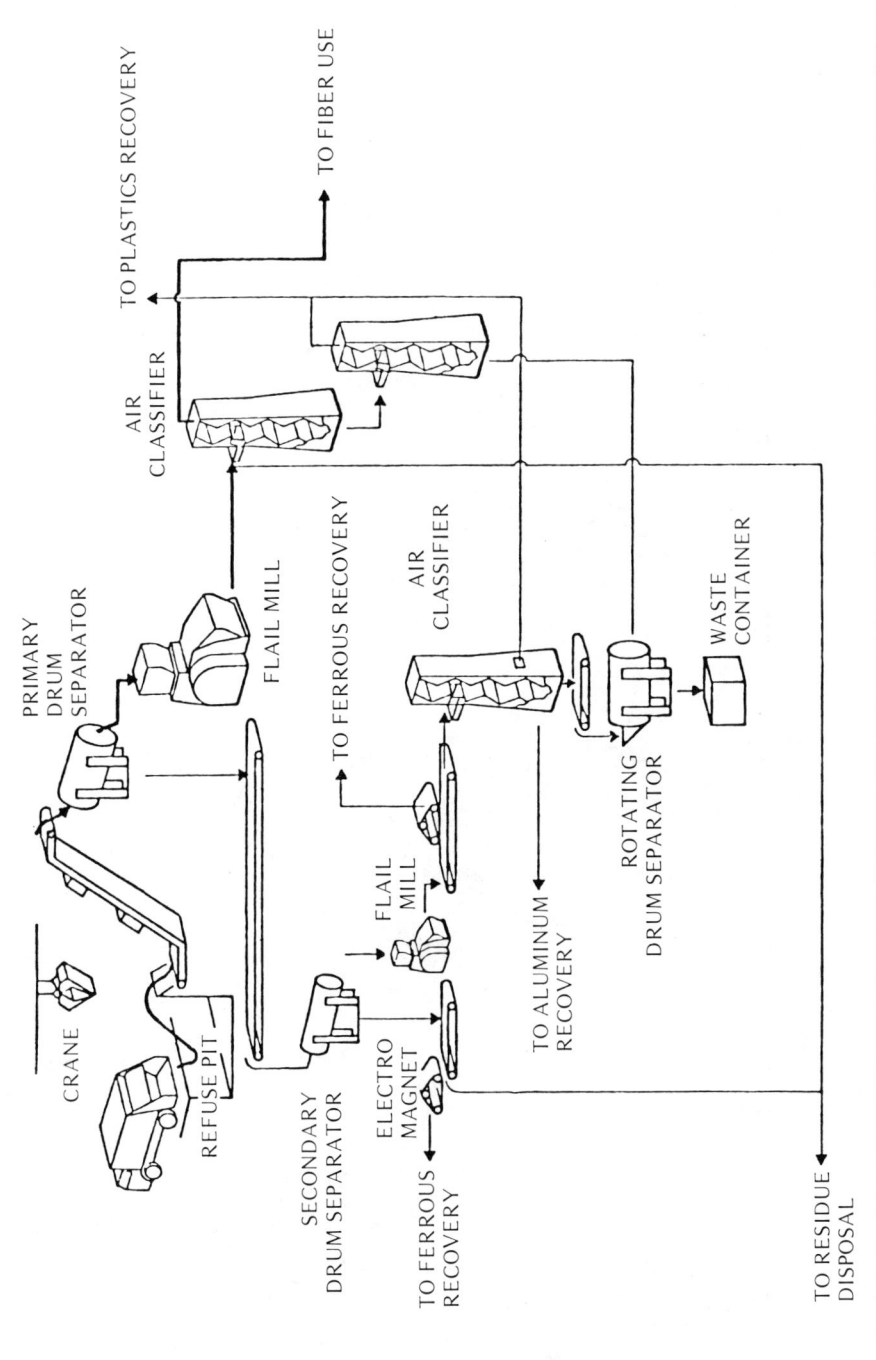

Source: Levy and Rigo (1977).

350   INDUSTRIAL SOLID WASTES

Figure 16-14.  Magnetic Separator Configuration.

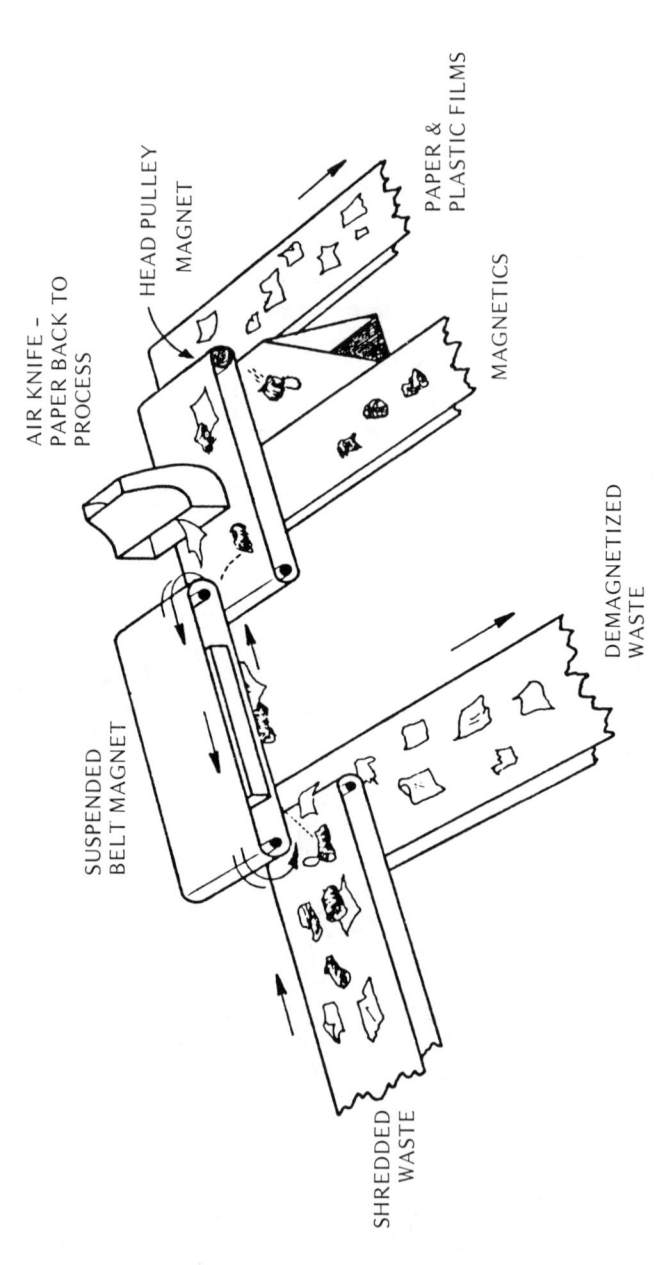

Source: Levy and Rigo (1977).

The rate of development of re-use systems will be influenced by the state of the economy and the prices for materials recovered and the scarcity and price of energy.

## REFERENCES

Levy, Steven J., and H. G. Rigo. 1977. *Resource Recovery Plant Implementation, Guide for Municipal Officials*—Part 2, *Technologies*, SW-157.2 Washington, D.C.: Environmental Protection Agency.

Payne, J. 1976. "Energy Recovery from Refuse—State of the Art." *Journal of Environmental Engineering* Div. A.S.C.E. Vol. 102, EE2, 281, April 1976.

# INDEX

**A**

Air pollution control, 84
Algae, 46
Anaerobic digestion of refuse, 344
Asbestos dust, 137

**B**

Bacteria, 44
Bagasse, 216
Baled refuse, 68
Bark, 299
Bottom ash, 30
Bulgur Process, 226
Burges, 47
By Products, 253

**C**

Carbon reactions, 47, 48
Chemical conversion processes, 39
Collection fees, 22
Collection of wastes, 13
Collection operations, 18
Collection routes, 19
Compaction in landfills, 65
Compactors, 100, 168
Composting, 10, 25, 26, 53, 55, 76, 223
Container capacities, 14

Continuous mechanical composting, 59
Corrosivity of wastes, 93, 94
Cost of incineration, 78, 167, 170
Cotton crops, 146, 148
Crop waste, 143, 147
Crushers, 11
Cryogenics, 30

**D**

Dano Biostabilizer Process, 59
Detoxification process, 103
Diseases of human wastes, 145
Dry Process Paper and Materials Recovery, 348, 349
Dust-steel mill, 200

**E**

Economics of landfilling, 68, 70, 76
Energy recovery, 131, 331
Environmental Protection Agency, 135, 240
Extraction Procedure for Toxicity, 93, 94, 116, 119

**F**

Fairfield-Hardy Process, 58
Feathers. 253

353

Fire and explosions, 95, 98
Fleshings, 294
Flue Gas Desulfurization Sludge, 285, 290
Fly ash, 30, 82, 85, 287
Fly borne disease, 140
Frazier-Ericson Process, 58
Fungi, 46

## G

Garbage grinders, 7, 24
Gas hazards from landfills, 72, 98
Gas Turbine Generating System, 343, 346
Gypsum sludge, 300, 301

## H

Hair, 294, 296
Hammermills, 10
Hauled Container Systems, 13
Hazardous wastes, 93, 103, 113, 114, 126, 127, 132, 135, 296, 314
Hazardous Waste Bill of 1980, 132
Health hazards, 135, 136, 139, 140, 141, 145
Heat value of refuse components, 82, 332
Heil-Gonard System, 9
Hoist truck, 15
Hospital Solid Waste Handbook, 158, 160, 161, 165
Hospital wastes, 157
Hydrasposal, 10

## I

Ignitability of wastes, 93, 94
Illegal dumping, 20
Incineration, 8, 12, 39, 76, 77, 78, 80, 82, 83, 84, 100, 112, 167, 168, 170, 331, 333
Industrial Wastes, 175
  Agricultural, 143, 144, 147, 148, 149, 215
  Aircraft, 182, 192
  Aluminum, 5, 24, 27, 28, 29, 33, 323, 325, 335
  Animal, 143, 146, 147, 149, 150, 152, 154, 155, 202, 230, 240, 241, 242, 243, 245, 246, 247, 248-250, 251
  Asbestos, 137
  Ashes, 285, 287, 288, 289, 290,

Ashes, 285, 287, 288, 289, 290, 292, 293
Automobiles, 31, 176, 192
Auto batteries, 41, 113
Beet sugar, 219, 222, 317
Benzopyrene, 138
Beryllium, 138
Beverages, 177, 208
Breweries, 179, 208, 210, 212, 213, 214
Cannery, 176, 178, 203, 205
Carbon refinery, 220, 319
Cement, 283
Chemicals, 113, 176, 181, 188, 304, 312, 318
Citrus, 207
Coal, 113, 139, 285, 287, 289, 311
Coke Byproducts, 186
Construction and Demolition, 175, 177, 282, 284
Copper scrap, 28, 29, 324
Cotton ginning, 193
Dairies, 178, 206
Dirt, ashes, brick, 38
Electrical, 113, 176, 184, 188
Electroplating, 113, 184, 195
Energy, 184
Explosives, 113, 185
Ferrous, 5, 27, 33, 38
Fertilizers, 186
Fibers, 139
Fish and fish products, 204, 206
Furniture, 176, 188, 192
Garbage, 74, 175
Glass, 5, 24, 30, 33, 38, 74, 131, 176, 322, 335
Grain mills, 223-229, 231, 232, 233
Hazardous, 93, 101, 114
Hides, 28, 294
Hospitals, 157-159, 162, 163, 164, 170
Inorganics, 313
Insecticies, 187
Laundries, 180
Leather, 5, 38, 75, 113, 176, 183, 188, 289, 293, 294, 295, 299
Lumber and Wood, 180, 188
Manganese, 138
Meat processing, 176, 194, 202
Mercury, 138
Metals, 5, 74, 113, 138, 140, 176, 188, 335

Industrial Wastes (*continued*)
  Mining, 113, 131, 139, 174, 303, 305, 307, 308, 309–311
  Newspapers, 28
  Nickel, 28
  Non ferrous metals, 33, 38, 323, 325
  Office packing, 175, 177
  Oil drilling, 182
  Oil shale, 309
  Oily wastes, 300, 302, 303, 306
  Organic, 103, 312, 314, 315, 320
  Paint, 113, 193, 315, 322
  Paper, 5, 24, 29, 33, 38, 75, 113
  PBB, 138
  PCB, 138, 319
  Pesticides, 103, 316
  Petroleum, 103, 113, 176, 182, 322
  Pharmaceuticals, 103, 312, 314, 315, 320, 322
  Phosphates, 185, 300, 319
  Plastics, 5, 31, 33, 38, 75, 90, 113, 131, 160, 176, 253, 255, 256, 317
  Poultry, 202, 244, 250, 252, 254
  Power plants, 285
  Printing, 113, 176, 317
  Pulp and Paper, 176, 188, 297
  Radio and TV, 192
  Rubber, 5, 28, 30, 33, 38, 74, 113, 176, 185, 303
  Sawmills, 176
  Slaughter houses, 178
  Sludges, 131, 256, 258, 281, 294, 300
  Soaps and Detergents, 315
  Soft drinks, 208
  Solvents, 316, 327
  Steel mills, 197
  Steel scrap, 27–29
  Stone and Clay, 176, 188
  Sugar cane, 215, 217, 218
  Supermarkets, 193
  Textiles, 5, 33, 74, 113, 176, 179, 188, 257, 259, 262–282
  Tin, 28, 38, 176
  Tires, 131, 307
  Toxic wastes, 103, 104, 114
  Transportation equipment, 176, 188
  Waste treatment, 182
  Wood, 5, 38, 74, 103, 176
  Wine and Brandy, 227, 230, 234, 235, 236, 238, 239
  Yard wastes, 5, 38
  Zinc, 28
Insurance for industry, 119

## L

Land cultivation, 73
Landfill gas, 343
Landfilling, 10, 51, 61, 63, 68, 69, 71, 72, 74, 76, 101, 128, 223, 343
Laws-U.S.A. for toxicity, 96, 97, 103
Leachate control from landfills, 71, 98
Lees, 238
Legal aspects, 121, 128, 132
Levy and Rigo, 343, 351
Livestock wastes, 149, 237, 239, 248

## M

Magnetic separation, 348, 350
Market prices for commodities, 28
Materials recovery systems, 34, 35, 36, 40
Metal market prices, 26
Microbiology of solid wastes, 43
Middlebrooks, E. J., 143, 156, 241
Midland Ross Pyrolysis System, 92
Milling, Grinding, Shredding, 9, 10, 67
Monsanto Landgard System, 337
Mosquito-borne diseases, 140

## N

National Bureau of Standards, 129
National Pollutant Discharge Environment Standard, 291
Naturizer Process, 58
Nitrogen reactions, 47, 50

## O

Occidental Process, 338, 341
Occupational Mycoses, 139
Odors animal wastes, 151
Offal, 253
Office of Solid Wastes, 126
Oily wastes, 103

## P

Pathogens, 141
Pomace, 228, 231
Protozoa, 46
Public Health of landfills, 69

Pulping of refuse, 8
Pulverator, 10
Pyrolysis, 25, 39, 87, 88, 89, 90, 91

**R**

Rasp mill, 11
Reactivity of wastes, 94
Recycling, 32, 129, 194, 256, 303, 335
Recycling energy saving, 29
Refcom Process, 345
Refuse chutes, 99
Refuse compaction, 7, 8
Refuse Densified Fluff, 334, 335
Resource Conservation and Recovery Act of 1976, 93, 103, 119, 123
Resource Conservation Committee, 132
Resource Recovery, 23, 25, 41, 101, 131
Resource Recovery Act of 1970, 121
Riker Process, 58
Rodent-borne diseases, 140
Ruther MSA System, 60

**S**

Scarab Process, 58
Scrap, 200
Scrubbers, 166
Secretary of Commerce, 129
Ship incinerator, 112
Shredders, 101
Slag, 198, 199
Slate mining, 310–312
Solid wastes composition in Spain, 4, 5
  in U.S.A., 24
Solid wastes, collection and disposal costs, 3
Solid wastes light and heavy fractions, 38
Solid wastes quantities, 2, 176, 188
Soliroc Process, 196
Spills, 120

Spontaneous combustion, 95
Stationary Container Systems, 15
Steffen Process, 220
Stillage, 235, 238
Structural integrity procedure for toxicity, 118
Sulfur reactions, 51, 52

**T**

Tchobanoglous, G., 4, 18
TIKI burners, 73
Tilt frame container, 15
Torrax System, 338, 339
Toxic Substances Control Act, 103, 104
Transfer operations, 16
Trash Trailer, 15

**U**

Union Carbide Purox System, 338, 340
Uranium mill tailings, 309

**V**

Vector control in landfills, 70
Viruses, 46
Volume and size reduction, 7
Vulcanus cargo ship, 112

**W**

Wastebaskets, 99
Wastes from industrial processes, 174, 175
Wastes from raw materials, 173
Wastes from used materials, 174
Water pollution from landfills, 71
Water quality limits, 104
Wet Process Energy Recovery System, 336, 343, 347
Wet pulping, 169
Wet scrubbers, 166
Windrow composting, 57